ENGINEERED
WORK MEASUREMENT

DEDICATION:

DR. HAROLD BRIGHT MAYNARD

MR. GUSTAVE J. STEGEMERTEN

MR. JOHN L. SCHWAB

FOURTH EDITION

ENGINEERED WORK MEASUREMENT

The principles, techniques, and data of Methods-Time Measurement, background and foundations of work measurement and Methods-Time Measurement, plus other related material

DELMAR W. KARGER
Dean Emeritus and Ford Foundation Professor of Management Emeritus
Rensselaer Polytechnic Institute
Troy, New York

FRANKLIN H. BAYHA
Private Practice
Ann Arbor, Michigan

INDUSTRIAL PRESS INC.

Library of Congress Cataloging-in-Publication Data

Karger, Delmar W.
 Engineered work measurement.

 Bibliography: p.
 Includes index.
 1. Work measurement. I. Bayha, Franklin H.
II. Title.
T60.2.K36 1986   658.5'4   86-10555
ISBN 0-8311-1170-4

INDUSTRIAL PRESS
200 Madison Avenue
New York, New York 10016

First Printing
FOURTH EDITION

ENGINEERED WORK MEASUREMENT

Contents

Foreword

The status of work study has undergone a considerable change over the years. Originally, Frederick W. Taylor used it as a means of replacing guess-work and estimate by facts as he developed his ideas on the scientific man-agement of industry. The work measurement aspects of his work attracted the greatest amount of public attention. The social, ethical, and moral rightness of work measurement became a controversial issue of the day, culminating in a full scale Congressional investigation of scientific manage-ment in 1912.

The contemporary work of Frank B. and Lillian M. Gilbreth in the field of motion study brought new and original ideas on the role of method in productivity. Because it was new, it also provoked controversy although less intense than that engendered by work measurement. The Gilbreths contributed a deeper understanding of work, methods, and fatigue than had hitherto existed and above all a greater understanding of the impor-tance of the human element in all work study.

During the pioneer era, brilliant work study installations were made by Taylor, the Gilbreths, and their associates, installations which aroused great interest on the part of management for what these procedures could contribute to the ever-present task of reducing costs and increasing produc-tivity. Although there were two schools of thought as to whether the main emphasis should be placed on time study or on motion study, the proce-dures began to receive increasing acceptance on the part of the more pro-gressive leaders of business and industry.

Unfortunately for the orderly progress of work study, World War I came along and with it an unprecedented demand for increased production. The record which had been compiled by the application of time and motion study in bringing about increased production, although not yet extensive, was sufficient to cause managements to turn towards work study in impor-tant numbers. So sudden was the increased demand that there were not enough qualified time and motion study practitioners to meet it. The demand continued, however, and presented to unqualified practitioners the opportunity to capitalize on it. Self-styled efficiency experts sprang up overnight. They had no difficulty in selling their services to management, for management was in the mood to buy. What they did, however, was all but disastrous. By applying unskillfully a powerful tool which they understood but poorly, if at all, they presently aroused the antagonism of both management and labor. Their monumental failures with time and motion study and wage incentives reduced the regard in which these procedures were held to a low point. Managers, whose hopes for these procedures had been destroyed, were not inclined to try again. Workers lost confidence in the benefits the procedures were said to bring to them and wanted no more of them. It was a dark moment in the history of work study and one from which a long period of recovery was necessary.

But it is impossible to keep a sound idea submerged for any length of time, no matter how badly it may have been mishandled. During the interval between Word War I and World War II, an increasing number of sound work study installations were made by well-qualified practitioners. These demonstrated beyond question what a properly engineered installation could do for a company. It was no coincidence that in industry after industry, the leading companies were those who used work study procedures the most extensively.

During this period, the work study procedures themselves underwent considerable development. Emphasis swung more and more from work measurement to methods study although both were recognized as being necessary to each other. Research was undertaken to develop procedures capable of doing an ever-better job of methods improvement and work measurement. General understanding of the procedures also grew. Partly this was due to the application experience of a growing number of people in industry. Partly it was due to the colleges and universities, who by adding work study courses to their curricula, turned out an increasingly large number of graduates with at least an appreciation-level knowledge of the field. A gradual increase in books and articles dealing with work study subjects also made its contribution to increase in understanding.

As a result of all this, when World War II came along and brought another great surge in the demand for increased productivity, industry was better prepared. Managers, many of them former time and motion study men, knew better what constituted sound work study practices. Many trained practitioners were available. Even more important, perhaps, was the fact that adequate training courses were available so that the work study force could be and was expanded in a fairly short time. Work study was applied extensively and in the majority of cases soundly. The result was a tremendous increase in productivity which contributed importantly to the outcome of the war.

The war, of course, brought many changes. It stimulated research in the work study field just as it did in so many other fields. New procedures were developed and improved application practices for the older procedures came into being.

In addition, the attitude of labor towards work study became much more receptive. That work study resulted in increased production was beyond question. The need for increased production was equally evident, especially to the men and women who had sons and daughters in the armed forces. Training courses at the appreciation level were widely given which took much of the mystery out of work study and removed fears of the procedures based on lack of understanding. The War Production Board, too, made an important contribution. Under the wise guidance of men with an extensive background in the work study field, an understanding of the principles on which methods improvement, work measurement, and wage incentives should be based was widely spread. The Board insisted upon cooperation between management and unions whenever new installations were made and saw to it that both groups understood and accepted the work study practices which would best meet the needs of their particular situation.

It should not be inferred, however, that all things were satisfactory in the work study field during the war period. Many distortions of sound practice were countenanced in order to meet temporary conditions. Certain managements, for example, feeling that they had somehow to get around the restrictions imposed by the wartime freezing of wages in order to attract or hold the needed workers used distortions of their wage incentive systems as a means of providing higher take-home pay. Labor, too, in some instances, used its strength and the confusions of the times to force distortions that were to its immediate but not longrun advantage. The resulting misapplications tended to make certain installations appear unsound and probably slowed down the general progress of work study to a certain extent.

But these distortions were temporary disturbances which while perhaps serious in their effect on an individual company could not permanently stay the advance of work study. In the post war period, work study spread not only in the United States of America, but especially in the war-ravaged countries of the Western world. There the need for increased productivity was beyond question. Production in unprecedented amount was necessary to replace the destructions of war if the people were to survive. So work study know-how was sent abroad under the technical assistance program sponsored by the United States government. It gave great impetus to the installation of work study methods, particularly in Western Europe.

And presently the work study practices which had been so freely exported returned to the United States, often reinforced and improved by sound research and engineering work done abroad. The very term work study that is used so freely in this book was first brought into general use in Great Britain by Mr. Russell Currie, C.B.E., of Imperial Chemical Industries Limited, to express to his management the basic purpose of all of the techniques and procedures which had been developed for the purpose of increasing productivity and reducing costs.

The techniques and procedures, in the meantime, multiplied further during the post war period. The literature also multiplied as many experienced practitioners and teachers of work study methods contributed their ideas and thinking to the general advancement of the art. Each book and article which is produced adds to the general understanding of work study and what it can do and thus is to be welcomed by those who believe that increased productivity and increased standards of material well-being are inseparable.

So it is with the pages that follow. There will be found a reflection of the impact that work study developments have had on two men who have lived through several decades of work study application. As practical men in daily contact with the work study application problems faced by the dynamic industry of which they were a part, they have had the opportunity to test and use many of the procedures which are a part of work study. Their description of these procedures, colored as it is by firsthand practical experience with them, provides an unusually useful guide to their successful application.

Because of its practical flavor, the book will encourage practicality in approach to all who study it carefully. Of necessity, it describes in some detail the procedures which it advocates. It does so, however, from the standpoint of those who have used them to solve the practical problems with which they have been confronted. The book reports no new research done by the authors themselves and advocates no untested ideas which they may hold. Instead, patiently and painstakingly, it tells the reader step by step what he must know and do to handle a work study assignment acceptably under a wide variety of conditions.

The authors support the procedures they advocate in no uncertain terms, although at the same time they recognize that other procedures, properly used, may be equally acceptable. They endeavor to integrate the older time and motion study practices with some of the new developments, particularly in the field of predetermined motion times and thus arrive at a combined procedure which is more complete than either of its parts.

In total, they depict in practical terms the current evolution of work study. The book thus presents a valuable summarization of the influences of the past on the practices of the present to the serious student of work study and its use by management.

H. B. MAYNARD

Pittsburgh, Pa.
September 1, 1957

Preface

Since the publication of the third edition of *Engineered Work Measurement* in 1977, a number of developments have occurred that had to be taken into account when this revision was planned.

One development directly affecting what could be added to this text was the publication of *Advanced Work Measurement* in 1982. It was authored by one of the present authors (D.W.K) and Dr. Walton M. Hancock, Professor of Industrial Engineering at the University of Michigan.

Advanced Work Measurement was originally conceived by D. W. Karger and F. H. Bayha. However, Mr. Bayha was unable to participate in the details of its writing. As a result, Dr. Hancock joined D. W. Karger in the writing of *Advanced Work Measurement*.

Advanced Work Measurement covers the entire spectrum of the MTM family of systems that were in existence at the time of its writing, except for MTM-1. Moreover, it covers many less well known, but very important, areas of work measurement knowledge. In fact, it was designed as a companion text to *Engineered Work Measurement* to be used by industrial engineers around the world and as a college-level text.

Next, in the decade prior to 1977, a number of "first course" work measurement texts (time and motion study oriented) appeared and, since then, they have been improved substantially. These are relatively large books containing a wide array of work measurement knowledge, and if this information were added to the MTM-1 coverage that has always been the "heart" of *Engineered Work Measurement*, the size of the resultant text would necessitate it being published as a two-volume text.

Finally, taking the preceding into account, including the availability of suitable texts covering time and motion study as well as much important related material, it seemed appropriate to substantially revise the contents of the fourth edition of *Engineered Work Measurement*. Since Mr. Bayha was unable to assist in its revision, the revision was the work of D. W. Karger.

xi

The fourth edition of *Engineered Work Measurement* retains the heart of the third edition—the MTM-1 material that has been refined and tested for more than three decades. It essentially answers all the questions about MTM-1 that are found in the MTM Association for Standards and Research MTM-1 examinations. The remaining material is the minimum work measurement background material that all who apply MTM-1 should understand.

The MTM-1 material extracted from the third edition was jointly developed and written with F. H. Bayha between 1955 and 1977 and consists of approximately 350 pages. Therefore, his name rightly appears as an author.

The fourth edition recognizes, with regard to the college and university market, the current use of a beginning text covering time and motion study plus other now well-accepted material as proper for a first course. For this market, *Engineered Work Measurement* is now an obvious text for a second course by itself or used in conjunction with *Advanced Word Measurement*. In the latter case, it could even be a two-semester course or a one-semester course of 4–5 semester credit hours.

Secondly, *Engineered Work Measurement* can be used in company and industrial engineering consulting organizations as a training course text or as a student reference when MTM-1 is taught using an MTM Association approved training manual. (Such manuals are available through the MTM Association for course background and training if a college or university desires to use them.) There are many such courses, since MTM-1 is rarely taught at the college level. While MTM-1 courses do provide Applicator Training Manuals, much additional information, benefit, and *scope* can be obtained through the use of this text.

Some application personnel trained in one of the higher-level systems often do not have MTM-1 training, and they *do* need some MTM-1 knowledge in certain application situations. Since *Engineered Work Measurement* was designed to answer all ordinary, and most extraordinary, application situations, it is a logical text to use to solve the previously described problem.

Finally, *Engineered Work Measurement* and its companion volume *Advanced Work Measurement* will be valuable reference sources for researchers exploring human capabilities and for use as references in almost any library housing technical books.

In this edition, as in the other editions, the authors owe a debt of thanks to the USA/Canada MTM Association for Standards and Research, its Training and Qualifications Committees, and its Director, James P. O'Brien. Special thanks for her understanding and support are also due Mrs. Ruth L. Karger.

DELMAR W. KARGER

DeFuniak Springs, FL
February 3, 1986

PART I

Introduction, Background, and Fundamentals of MTM

CHAPTER 1

Origins and Background of Work Measurement

ORIGINS OF WORK MEASUREMENT

The search for a better way to do things has progressed since the birth of mankind. The inventor of the wheel, undoubtedly, was motivated by a desire to eliminate some of the hard work involved in dragging things about on a sled or travois. Better methods, which are integral to proper and correct work measurement procedures, were thus early allied with the benefit and welfare of human individuals. Indeed man's self interest demands more efficient, less fatiguing procedures by which to do work.

From its earliest beginnings, industry always has been concerned in some degree with methods, particularly manufacturing methods. Today, the methods or procedures associated with every activity of every kind of organization are of concern. The measurement of performance in relation to methods has expanded from manufacturing into the clerical and indirect-activity areas. The next expansion probably will be into the overhead and staff work areas.

The interest in methods always is greatest during periods of low profit and severe competition. When such urgency for reducing costs by finding better methods and establishing sound work measurement subsides, there is a tendency to be satisfied with existing conditions. This management error tends to peak when high profits are the general experience. This attitude is aided and abetted by the short-range emphasis (timewise) by U.S. industry. In part this is caused by the way top management is paid— premium bonuses are paid on the basis of yearly profits, and these in the short run are maximized by emphasizing the short-range view of what should be done.

To an important extent, if worker productivity fails to increase in approximate proportion to wages, then productivity becomes one factor causing restrictions in the whole economy. Capital, management, and labor together constitute any productive enterprise; and each has a profound stake in an ongoing effort to find and institute better methods.

In the 19th century and the early part of the 20th, Taylor, Gantt, Emerson, the Gilbreths, and others originated the scientific management concept. They introduced and developed many new principles of management together with techniques designed to systematize and standardize the planning, operation, and control of industry. Essential to implementing many of these progressive steps was the establishment of an adequate basis for work measurement; since, without measurement, one cannot exert effective corrective control—a fundamental function of management, universally recognized. Control is simply comparing some performance or condition (machine, human, environmental, etc.) against a standard and exerting corrective action when deviation from the standard occurs.

The work measurement engineer's output is used to establish as a standard the method and time required to perform a work task. No qualified analyst would consider standardizing an incorrect method that does not optimize the usage of resources. In fact, a time standard is defective unless it is based directly on recognized specific methods. Hence, the techniques of analyzing and optimizing methods are emphasized along with the techniques of time determination in this text. Also, while this book deals most directly with manual work in the factory and office, the measurement of work which includes mental elements is becoming increasingly vital; thus an approach to analyzing the performance of this kind of work also will be covered.

The word "standard" often is used with a variable meaning, so that a clarification of its context in work measurement is desirable. When the work is used in ordinary engineering parlance, it signifies codified solutions to design problems or operating characteristics, usually those which are most repetitive. In work measurement, standards are method and time solutions to work design problems, often of a unique or non-repetitive nature. Standards generally are established to record the resolution of such problems; but they also have an active aspect of facilitating the operating measurement of work and providing a basis for applying corrective control action when deviation is detected. Standardization as an activity is concerned with the conception, formulation, distribution, revision, and usage promotion of standards.* Not only are work measurement standards and techniques used as described, but also they are a necessary and integral part of producibility engineering* and value engineering* activities.

Frederick W. Taylor, generally known as the father of scientific management, was responsible for the first definitive approach to work measure-

---

*A meaningful discussion of standardization, producibility engineering, and value engineering can be found in *Managing Engineering and Research,* 2nd and 3rd eds., by Delmar W. Karger and Robert G. Murdick (New York: Industrial Press Inc. 1969 and 1980).

ment. He first measured his worker's performances and established production levels or operation times from past performance records. He found such measurements and the production goals established on ordinary historical performance records to be unreliable because they were based on the same poor performance that he had noted originally. His next step in the development of work measurement followed the example of M. Coulomb, who, around 1760, used a stopwatch to establish the time needed to perform a given operation. While Taylor was not the first to so time operation elements, he was the first to develop a complete technique for standardizing performance times by using a stopwatch for times tied with a specific method of operation. Taylor's full description of Time Study was part of his discussion on "The Present State of the Art of Industrial Management" for a Subcommittee on Administration of the ASME[1] as follows:

The analytical work of time study is as follows:

(a) Divide the work of a man performing any job into simple elementary movements.

(b) Pick out all useless movements and discard them.

(c) Study, one after another, just how each of several skilled workmen makes each elementary movement, and with the aid of a stop watch select the quickest and best method of making each elementary movement known in the trade.

(d) Describe, record, and index each elementary movement, with its proper times, so that it can be quickly found.

(e) Study and record the percentage which must be added to the actual working time of a good workman to cover unavoidable delays, interruptions, and minor accidents, etc.

(f) Study and record the percentage which must be added to cover the newness of a good workman to a job, the first few times that he does it. (This percentage is quite large on jobs made up of a large number of different elements composing a long sequence infrequently repeated. This factor grows smaller, however, as the work consists of a smaller number of different elements in a sequence that is more frequently repeated.)

(g) Study and record the percentage of time that must be allowed for rest, and the intervals at which the rest must be taken, in order to offset physical fatigue.

The constructive work of time study is as follows:

(h) Add together into various groups such combinations of elementary movements as are frequently used in the same sequence in the trade, and record and index these groups so that they can be readily found.

(i) From these several records, it is comparatively easy to select the proper series of motions which should be used by a workman in making any particular article, and by summing the times of these movements and adding proper percentage allowances, to find the proper time for doing almost any class of work.

(j) The analysis of a piece of work into its elements almost always reveals the fact that many of the conditions surrounding and accompanying the work

[1]*Transactions,* ASME, vol. 34, 1912, pp. 1199, 1200.

are defective; for instance, that improper tools are used, that the machines used in connection with it need perfecting, that the sanitary conditions are bad, etc. And knowledge so obtained leads frequently to constructive work of a high order, to the standardization of tools and conditions, to the invention of superior methods and machines.

From the 1912 date of the Taylor summary, one can see that Time Study is not a new technique. His summary is as valid today as when it was originally penned.

Soon after Taylor began his work, Mr. Frank B. Gilbreth left the construction field in which he had been a successful building contractor to devote his time to the study of methods as they are related to scientific management. The story of Gilbreth's discovery, on his first day as a bricklayer's apprentice, of the many different methods used by bricklayers for the simple task of laying a brick is known to everyone familiar with the field of industrial engineering. This simple discovery, plus the encouragement of his wife. Dr. Lillian M. Gilbreth, finally provided the incentive to give up his profitable business so he could devote his time to promoting scientific management and conducting research and application work in the field of Motion Study. The Gilbreths made many detailed laboratory studies of motions and methods before they eventually developed the Micromotion Study procedure that forms a basis for much of the material included in this text.

Both Taylor and the Gilbreths won many followers who saw fundamental differences in the new procedures developed by these outstanding people. At length, two groups of practitioners developed who considered themselves irrevocably opposed to each other. This lasted roughly from 1910 to 1930. One group was known as the Time Study group, the other as the Motion Study group. Neither grasped the fundamental truth that motions take time and both ideas are uniquely interdependent. Eventually, the groups became better acquainted and found that they really differed little in their fundamental approaches. This ultimately resulted in combining the best features of both procedures into what is now known as "Methods Engineering." Before entering into a general discussion of Methods Engineering, a more detailed look at the acknowledged shortcomings of Time Study is advisable, since the same topics are involved in Methods Engineering as well.

For the United States Commission on Industrial Relations[2], Mr. Robert F. Hoxie, in 1914, made a thorough study of Scientific Management, which was then synonymous with practicing or applying Time Study. He listed 17 factors in Time Study practice of the era which could be varied subject to human will. These factors are as true today as they were in 1914:

(1) The general attitude, ideals, and purposes of the management and the consequent general instruction given to the time study man;

[2]Robert Franklin Hoxie, *Scientific Management and Labor.* (D. Appleton & Company, 1921) pp. 46, 47.

(2) The character, intelligence, training, and ideals of the time study man;

(3) The degree to which the job to be timed and all its appurtenances have been studied and standardized looking to uniform conditions in its performance for all the workers;

(4) The amount of change thus made from old methods and conditions of performance, e.g., the order of performance, the motions eliminated, and the degree of habituation of the workers to the old and the new situation when the task is set;

(5) The mode of selection of the workers to be timed and their speed and skill relative to the other members of the group;

(6) The relative number of workers timed and number of readings considered sufficient to secure the result desired;

(7) The atmospheric conditions, time of day, time of year, the mental and physical condition of the workers when timed, and the judgment exercised in reducing these matters to the "normal";

(8) The character and amount of special instruction and special training given the selected workers before timing them;

(9) The instructions given to them by the time study man as to care and speed, etc., to be maintained during the timing process;

(10) The attitude of the time study man toward the workers being timed and the secret motives and aims of the workers themselves;

(11) The judgment of the time study man as to the pace maintained under timing relative to the "proper," "normal," or maximum speed which should be demanded;

(12) The checks on the actual results used by the time study man in this connection;

(13) The method and mechanism used for observing and recording times and the degree of accuracy with which actual results are caught and put down;

(14) The judgment exercised by the time study man in respect to the retention or elimination of possibly inaccurate or "abnormally" high or low readings;

(15) The method used in summing up the elementary readings to get the "necessary" elementary time;

(16) The method employed in determining how much should be added to the "necessary time" as a human allowance; and

(17) The method of determining the machine allowance.

Time Study practice today is still open to active criticism on each of these factors. That such criticism is crucial in nature is affirmed by serious strikes, files of grievances, and other positive evidence of Time Study failure due to one or more of the foregoing factors.

In order to combat the shortcoming listed above, industrial engineers have developed a reasonably uniform concept of normal pace, Time Study personnel have been trained to exercise proper judgment, statistics have been used to validate the accuracy of time studies, etc. Much has been learned about improving Time Study practice and this has helped to build faith in Time Study as a recognized and accepted yardstick of measurement. The fact still remains, however, that Time Study is a very rough measure which in no way compares in accuracy to a vernier caliper or a micrometer.

As production methods have been refined and the work volume per worker has increased, the standards developed through work measurement techniques had to be more precise. The preciseness of measurement and the tolerance inherent in normal Time Study procedure were once acceptable, but they are rapidly becoming unacceptable in many applications. For example, today it is not unusual to find one or more production workers filling key operational positions in a partially automated production system. Obviously, management must know precisely what to expect of the worker under such a circumstance.

Taylor recognized the relationship existing between methods and time. It was one of his chief reasons for subdividing an operation into such Time Study elements as "pick up piece," "place piece in jig," etc. In a very real sense, the early disagreements would have been avoided by closer study of his works.

Frank and Lillian Gilbreth refined Taylor's concept of work elements by further subdivision into "therblig" (their name spelled backwards) elements, which to them were basic manual work segments describing the human sensory-motor activities or other basic elements of an operation. Therbligs form a generally recognized basic language for methods description usable in expressing motion-time data. The original 17 therbligs were:

Search	Release Load	Use
Select	Position	Unavoidable Delay
Grasp	Pre-Position	Avoidable Delay
Transport Empty	Inspect	Plan
Transport Loaded	Assemble	Rest for Over-
Hold	Dissemble	coming Fatigue

These therbligs basically emphasized a better understanding of the *composition* of normal industrial operations from the operator's standpoint. They were not developed with great concern for their absolute performance times, but rather primarily to provide a methods improvement tool. They did, however, lead in a modified form to a classification of elements adequate for what is known today as *predetermined time standards,* a subject treated later in this text. The Gilbreths also contributed toward method improvements by establishing the currently well-conceived and generally accepted Principles of Motion Economy, plus the previously mentioned Micromotion Study technique.

While these are the origins of the scientific management concept and of work measurement, it should never be forgotten that this is a dynamic world. It is a world of increasingly rapid change and development owing to ever expanding knowledge, the result of increasing research and development efforts in every field of human activity, including work measurement. Lately, another factor has become important, the effect of computers on virtually every human activity.

HUMAN FACTORS

Human factors have a great and dominant effect on the business world. They loom as large or larger than the purely technical aspects of a given problem or situation—especially for organizational problems. Guidance and measurement of human performance are thus necessary and important. For these reasons, study of the specific factors underlying group and individual behavior of workers is a very practical and useful pursuit, especially to practitioners of work measurement.

Not only are many "human equation" factors involved in determining human performance, but *these factors must be understood by the person(s) directing and using work measurement.* There are many texts dealing exclusively with this subject and no attempt will be made here to discuss substantively this area of concern.

THE AIM OF WORK MEASUREMENT

The aim of work measurement is to aid everyone—managers, workers, and consumers: (1) it accomplishes this by greatly assisting in methods improvement, which, in turn, reduces the labor content or cost to produce; (2) it provides a standard of performance that greatly facilitates rational management of an organization's labor operations; (3) it (the standard of performance) makes possible the rewarding of better than standard performance through associated incentive systems, which, in turn, still further reduce associated labor cost (workers typically achieve 120–125% of standard performance and therefore earn a 20–25% bonus) by reduced overhead charges (for space, energy, capital equipment, unnecessary hiring costs, less fringe benefits, etc.); and (4) all of the preceding items lower product cost so that the price to the consumer is reduced and at the same time the employer is more competitive—this protects the employee's job.

Work measurement offers one of the most reliable ways to achieve the benefits of increased production at lower cost for the advantage of everyone. While this is a good generalization, the fact is that the factors and systems involved are very complex and will eventually affect our individual economic lives very directly. This aspect of the problem is treated at the end of this chapter.

Before proceeding further there is a need to ensure the reader's grasp of the term "management" as well as its strong right arm "industrial engineering"; these are explored further.[3]

[3]Much of the following has been paraphrased from Prof. William J. Jaffe's introduction to ANSI (American National Standards Institute) Standard Z94 and Z94.1-12, "Industrial Engineering Terminology." The Z94.1-12 designation is the 1982 version, which contains all the ANSI definitions in this text. These standards will be frequently referenced by ANSI Z94.1-12 IE Term. and/or designated in a footnote by "Footnote 3, Chap. 1."

To aid Justice Louis D. Brandeis in the celebrated 1910 Eastern Rate Case Hearings before the U.S. Supreme Court, Messrs. Gantt, Gilbreth, Emerson, and others espoused the term "Scientific Management," which Taylor in 1903 had used in a more limited sense. They meant to support the concept that *management* can be systematically studied and practiced deliberately rather than remaining a capricious, rule-of-thumb, seat-of-the-pants meandering from crisis to crisis. However, the full term "Scientific Management" has had limited use during the last two decades; instead, simply *management* (both as a noun and an adjective) has been most widely used. A few of the many definitions that have appeared are:

"Management is a broad term and covers almost all the factors in the operation of an enterprise."[4]
"Management is the science of applied human effort . . . "[5]
"Management is involved in the avoidance of waste in any human effort."[6]
"To manage is to forecast and plan, to organize, to command, to coordinate and control."[7]

The definition given in ANSI Z94.1-12[8] in the volume on *Organization Planning and Theory* says:

"Management: 1. The process of utilizing material and human resources to accomplish designated objectives. It involves the activities of planning, organizing, directing, coordinating, and controlling.
2. That group of people who perform the functions described in 1, above."

In recent years management has come to be viewed in terms of *systems,* which can be defined as:

. . . an organized or complex whole; an assemblage or combination of things or parts forming a complex or unitary whole.[9]

One can therefore say that:

Management is a system that utilizes men, money, materials, and machines to produce goods and/or services.

Moreover, management can be viewed as a series of subsystems such as control systems, decision systems, communications systems, etc. It is therefore not surprising—since industrial engineering always has been associat-

[4]Richard H. Lansburgh. *Industrial Management,* 2nd ed. (New York: John Wiley & Sons, Inc., 1928) p. 4.
[5]Alexander H. Church, "The Science and Practice of Management." (New York: The Engineering Magazine Company, 1918) p. 282.
[6]Oliver Sheldon, "The Philosophy of Management." (London: Sir I. Pitman & Sons, Ltd., 1923) pp. 101 ff.
[7]Leon P. Alford, "Scientific Industrial Management," Paper No. 341.
[8]See Footnote 3, Chapter 1.
[9]Richard A. Johnson, Fremont E. Kast, and James E. Rosenzweig. *The Theory and Management of Systems.* (New York: McGraw-Hill Book Company, 1963) p. 4.

ed with the concepts of engineering, engineering design, and management—that the Institute of Industrial Engineers officially defined its profession in 1955, revised 1975, as follows:

> Industrial engineering is concerned with the design, improvement, and installation of integrated systems of people, materials, equipment, and energy. It draws upon specialized knowledge and skills in the mathematical, physical, and social sciences, together with the principles and methods of engineering analysis and design, to specify, predict, and evaluate the results to be obtained from such systems.[10]

Scientific Method

The use of the Scientific Method—the application in a systematic manner of expert knowledge and skill—is central to work study and/or work measurement. Inherent in the scientific method is the recognition that true knowledge about anything must be based on facts. However, since work measurement is concerned with the actions of people, it must also be recognized that they are governed by mixtures of facts and opinions. That part of an action or reaction which is based on fact is more likely to be an objective kind of thing, whereas that part resting on opinion tends to be subjective in nature and hence influenced by personal feelings. Thus, personal effects which seem likely to result from compliance with the acts and order of others have a definite influence on the actions of people.

Good work study techniques deal in facts, not in personalities. Facts about the human being involved in work situations permit the use of objective skills, while avoiding the prejudicial effects accompanying efforts to form an opinion or feeling about the same human being. In this respect, the scientific approach seems and tends to be impersonal.

Remember, however, that while objective skills are properly employed to get the right answers. implementing those answers requires the use and understanding of many subjective techniques based on enlightened opinion. The role of objectivity thus becomes one of providing the enlightenment needed to promote justice in subjective action. People have a natural tendency to become so enamoured of quantitative measurement techniques that they fail to recognize the necessity for supplementing such techniques by subjective approaches, especially when evaluating such a difficult-to-measure item as human performance. This failing is particularly acute when quantitative measurement is being applied to a departmental unit as a whole. The complete elimination of all subjective elements in decision-making never will be possible, however, since our total knowledge of human action always will be limited to some extent.

Another attribute required of scientific procedure, since it basically proceeds on fact rather than opinion, is that a means of determining facts must

[10]American Institute of Industrial Engineers; 25 Technology Park/Atlanta; Norcross, Georgia 30071.

exist. It must also be established that the facts discovered are meaningful and useful. The scientist essentially determines the facts by measurement. The meaning of the measurement is established by validation of the procedures used to determine these facts. Work studies, therefore, to be objective, must involve both adequate measurement and validation in order that the resulting facts can permit necessary subjective application without fear of gross error.

Measurement itself includes two major problems. First, is the matter amenable to measurement? Secondly, do the measurements taken accord with actual performance in the changing fabric of existing life? A major contribution of Taylor and the Gilbreths was a reproducible demonstration that the performance of human beings can be measured. To some extent they also achieved validation of the reported results of their studies. Since their time, however, the means of measurement have become increasingly finer, more sensitive, and revalidation has been found necessary. The improved ways of measurement in modern time study, motion study, and predetermined time systems plus some of the necessary revalidation will be explained in this volume.

Finally, the study of work measurement, being concerned with human activity, is closely allied with such other human sciences as economics, biology, bio-mechanics, sociology, physiology, psychology, and philosophy. Indeed, the work analyst must rely heavily on his knowledge in the related fields to isolate his facts as well as to validate them.

WORK STUDY STEPS

The optimum procedure to follow in performing a work measurement study is well defined by the experiences of analysts, and any first course in the work measurement text will adequately describe the steps in the procedure. However, many of the issues or things that must be taken into account for proper time-study-oriented work measurement were essentially presented earlier in this chapter. More specifically, the following are preliminary steps:

1. Receiving the request for the study.
2. Obtaining the cooperation of the departmental foreman and the immediate supervisor of the work to be studied.
3. Select the operator to be studied; usually he or she should be a fully trained operator performing the work with average skill and effort—this latter is more important in a time study.
4. Determine whether the work is ready for study, i.e., are any changes desired or needed before the work study is implemented.
5. Obtain and record all general information that will completely define the work operation to be studied.

A predetermined time study follows essentially the same procedure as a time study except that motions are identified and classified in a structured manner rather than timing the segments (elements) of the operation.

A proper work study itself involves determining the equipment, tools, workplace, and other physical conditions best suited for the task at hand. The next logical step is to utilize these in the best possible way, which means deciding the exact method or process to be followed by the human operator. A detailed knowledge of human capabilities is vital to this second step. When this approach is practiced by competent analysts, and the resulting work method is installed with appropriate worker training and enlightened supervision, the attainment of the maximum human performance becomes possible.

Establishing work methods involves decisions regarding such scientific topics as, for example, the time for the eye to recognize objects or the possible simultaneous motions of which the body members are capable. In its entirety, the field of work measurement involves knowledge in all the branches of engineering and mechanical skills related to the task being studied. It also intimately involves all fields of knowledge regarding the human operator.

This text, as it attempts to promote the aims of work measurement, essentially concerns itself with two major concepts. First, what are the human motion capabilities and acts? Secondly, what is the time required to perform these functions? The laws or principles governing human motion must include both answers as well as due appreciation for the moral and ethical values by which human activities are guided and limited.

THE VALUE OF ANALYZING WORK

So far, the presence and need for knowledge of the laws and principles governing human motions are evident. If the aims of work measurement are realized, what value do they then have in the social structure? The trend toward more extensive use of work measurement is unmistakable; it is typical of modern observers to wish to know what desirable results will accrue from this trend. Or, to state the question in everyday language, what good is all this, anyway?

One obvious factor is that by setting a rational goal or objective standard for the worker, the degree of attainment provides a yardstick for measurement of performance. It will thereby indicate when there is need for corrective action and the area in which it will be most wisely applied. Conversely, it indicates when good performance, possibly meriting a reward, has occurred. The unequal degree of motivation and ability of workers in an enterprise obviously makes these approaches desirable.

Work measurement tends to remove purely opinionated or prejudicial treatment of a given individual in the organization. Not only will it show up good or bad performance; but, as later outlined in this book, it will help locate the true source of most performance difficulty. In many cases, competent usage of the technique of work measurement will also materially aid in determining the proper solution to a given work-performance problem.

Not only is it desirable to derive standards for gaging performance and thereby make it possible to approach the higher attainment; but experience has also amply demonstrated that rewarding the worker financially, for above-standard performance, can result in an average of 20% to 30% increase in productivity. In the factory this usually is done on a directly proportional basis—that is, 120% output for a given time yields 120% of the basic pay for that period. Therefore, another value of work measurement can be direct financial benefit to the worker. In turn, the company will gain a greater financial return per unit of investment in the worker's job. Ultimately, this can mean lower prices for the consumer at the same level of product quality. Thus, all share in increased productivity.

Human output directly affects production control, materials control, and the functioning of the personnel department. The problems here involve determination of the number, kind, and quality of employees required to meet schedules. Superior results in such management decision can evolve from adequate usage of work measurement. Conversely, the possible level of production or amount of materials required for the available labor force and equipment can be more accurately prescribed when the productivity per worker is known. The personnel functions are directly affected in either event.

Hosts of such specific examples could be cited to illustrate the value of work study. It was with direct labor in factories that Taylor, the Gilbreths, and other work measurement pioneers were concerned, and the long usage of work measurement in such areas has somewhat reduced the potential for further improvement there, at least to the extent attained in the past. Recently, the growth of indirect personnel has tended to focus more attention on work measurement in the office, where past neglect offers greater potential for savings. In the last decade or two, the measurement of clerical and other indirect work in most industries, and even in many governmental offices, has gained greater interest and importance, both because of the relative increase in the number of such workers and the fact that improved ways have been found to measure such work. The greatest further potential for labor reduction now lies in the overhead and staff work areas.

For information on work measurement applied to other work areas, including the professional, staff, and managerial areas, the reader should consult *Advanced Work Measurement*.[11]

[11]Delmar W. Karger and Walton M. Hancock, *Advanced Work Measurement*. (Industrial Press Inc., New York, N.Y., 1982).

It should not be assumed from the preceding discussion that work measurement is the only factor influencing the changing labor ratios or that the only reason to reduce labor costs is to increase profits. The fact is that if a business cannot compete successfully in the world markets, it will eventually be forced to close—to the detriment of the owners, the workers, and the nation. This still holds true, even if the firm does not market outside of its national borders, since the potential foreign competitors will bring their goods here. Just recall what happened to the U.S. radio and TV business, much of our automobile business, a large segment of our semiconductor business, cameras, etc.

Today, automation of both the factory and the office has become of great importance, even in countries that have low labor rates; MTM can help design products for automated production. In the United States with our higher labor rates (as much as 2–10 times that of some foreign competitors) we need to do much more than merely set labor standards in order to compete. We must produce quality products that have quality designed in. In addition, we must utilize effective and optimum management techniques, and do our very best to design products for maximum producibility—only then can we use the very best manufacturing methods.

Currently, our real competition is Japan, South Korea, Taiwan, Hong Kong, and Singapore (at the time of the writing of this text we had very large trade deficits with each of these countries)—but more such places are on the horizon. These competitor countries not only have lower labor rates than those found in the United States, but they also have a unique cooperation between their management, labor, and government that promotes being *very* competitive in the world's markets with quality products.

Probably of more importance than work standards, automation, etc., is that the citizens of these countries are almost universally committed to the "long-term view." This will gain them ultimate victory in the competition unless we beat them in the competitive battle, a battle they are winning with their long-range strategy borrowed at least partially from the writing's of our "best minds." While all this is vitally important, and somewhat related to techniques promoting and producing labor efficiency, it is not germane to the subject area of this text and we must return to our main subject matter.

HUMAN ENERGY AT WORK

The preceding material has oriented the reader regarding various aspects of work measurement that should always remain in the forefront as the field is mastered and practiced. He must, however, focus attention on the human being as a very efficient and marvelous system in order to achieve the goals of work study which are the benefits just discussed.

The product or service is the *result* of work, not the *work*. Work itself is energy directed toward an object or purpose to shape, alter, form, or transform it into a condition in which its utility, operation, or state will command the buying power of a consumer. That energy can result from mechanical, chemical, electrical, or human action. Other fields of engineering deal adequately with all but human energy; this source of power is the exclusive province of work study analysts. From what has been said earlier, it is easy to see that industrial engineers are responsible for designing effective usage (methods) for what is perhaps the most complex power source of all, as well as the most precious—the human being. Not only is the direct power of the human body used to shape a product or other output of a system, but also the mechanical, chemical, electrical, and other forms of energy used are always directed and/or aided by humans.

Business exists as an activity because the owners desire to make a fair profit while rendering service to society, and a concern will survive only by satisfying this desire. Other organizations exist to serve society in many different ways, yet the problems of managing the work of humans are exactly the same as in business. To achieve or maximize profits or results, the industrial engineer must know how to obtain maximum performance from the human resources. Yes—for many practical purposes of work study, the worker must be viewed as an energy machine. *Under no circumstances, however, can or should the social, physical, and other aspects of the human worker be ignored.*

Human labor is expensive, yet it is the agent of production which adds the greatest value to the finished product. Not only is it responsible for its own actions, but must govern and control the materials and machines as well. Higher degrees of skill, knowledge, and/or ability are even more expensive; and workers possessing them are hardest to find and hold on the job. The need for analysis of this cost, therefore, is obvious. No enterprise could profitably ignore this major factor in product or service cost.

In observing and evaluating workers as merely energy machines, human relations problems tend to be simplified rather than aggravated. The work analyst measures the worker only with respect to his actual performance and output, keeping this endeavor apart from any delving into personal affairs or judging other private concerns of the worker in an attempt to achieve a total picture of the worker's entire nature. He must, indeed, generally understand people and know well how to cooperate and guide them into better work habits; but such matters are the *art* of work study, the *science* of work study being restricted to impersonal measurement techniques.

The *art* of work study is equally important as the *science,* and much of the knowledge relating to this will either be covered or mentioned in such a manner that the proper area of study can be readily consulted in additional reference books. However, since the *art* of any subject can be described only partially in writing or discussion, a largely personal element in mastering it can be acquired only by direct experience and practice.

Finally, it is important to note that all people differ in many ways, including their physiological behavior. To allow for this factor, industrial engineers have developed the creditable concept of the "average worker." This concept is especially important when measuring manual work. No one likes to admit he is just average, although he may believe that average men and women do exist. The novice, especially, experiences difficulty in forming the correct concept of an average worker. With perseverance, experience, and unbiased examination of many workers, however, he soon develops the idea and matches it with that held by acknowledged work analysts. there are definite aids to this that will be discussed under performance rating. The real problem is not achieving agreement with an individual analyst, but rather in formulating a clear definition which distinguishes the average worker in a way that is acceptable to work analysts, other employees, and the general public. This is a psychological problem beyond the scope of this book, but the industrial engineering profession has recognized it and is taking steps toward achieving this desirable goal.

CHAPTER 2

Rating an Operator's Performance

VARIABLES AFFECTING HUMAN OUTPUT

In order to understand the worker as an energy machine, the specific factors influencing individual output and their relative importance must be considered. Figure 1 represents the consensus of many competent industrial engineers, wherein the listings are a synthesis of many sources. While it was originated for manual work, it applies almost equally well to mixed mental and manual work, and many entries are common variables in purely mental work. Reference to this figure in the following discussion will clarify the text.

The result of the application of human energy is *production;* and this can be a product, a service, a performance, a calculation, a design, etc. In manual work, the amount produced depends directly on motions and the times these motions consume. Since the time required to perform a given motion under set conditions is demonstrated later to be essentially a constant for the average worker, it follows that the kind of *motions used* will be the largest single factor determining the amount of output that can be anticipated. The *motion speed* is subject to natural limits, and thus has less influence on the net output. It can readily be demonstrated in a number of ways that these statements are true. For example, the style of penmanship employed more directly determines the time required to write a letter than does the swiftness of the hand and fingers alone, composing time being excluded. As a matter of fact, a writer can greatly increase his writing speed by adopting a simpler style of penmanship even though he makes no effort to write faster. *Method* is understood by industrial engineers to connote the motions used when speaking of a manual operation, with tooling eliminated as a consideration. When motion speed is subdivided into its components, the industrial engineer believes they are *effort, skill,* and *conditions.*

Effort is defined as the "will to work," the drive or impetus behind the motions of the worker.

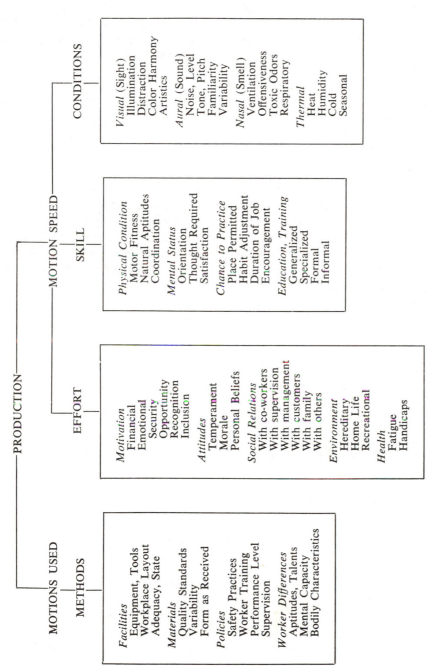

PRODUCTION

MOTIONS USED

METHODS

Facilities
Equipment, Tools
Workplace Layout
Adequacy, State

Materials
Quality Standards
Variability
Form as Received

Policies
Safety Practices
Worker Training
Performance Level
Supervision

Worker Differences
Aptitudes, Talents
Mental Capacity
Bodily Characteristics

MOTION SPEED

EFFORT

Motivation
Financial
Emotional
Security
Opportunity
Recognition
Inclusion

Attitudes
Temperament
Morale
Personal Beliefs

Social Relations
With co-workers
With supervision
With management
With customers
With family
With others

Environment
Hereditary
Home Life
Recreational

Health
Fatigue
Handicaps

SKILL

Physical Condition
Motor Fitness
Natural Aptitudes
Coordination

Mental Status
Orientation
Thought Required
Satisfaction

Chance to Practice
Place Permitted
Habit Adjustment
Duration of Job
Encouragement

Education, Training
Generalized
Specialized
Formal
Informal

CONDITIONS

Visual (Sight)
Illumination
Distraction
Color Harmony
Artistic

Aural (Sound)
Noise, Level
Tone, Pitch
Familiarity
Variability

Nasal (Smell)
Ventilation
Offensiveness
Toxic Odors
Respiratory

Thermal
Heat
Humidity
Cold
Seasonal

Fig. 1.   Variables determining productive output.

Skill is defined as the proficiency with which an operator can follow a given method. Note that this work measurement definition of skill does not match the common meaning of the term, which denotes the ability of a worker to perform difficult tasks easily. In the parlance of work analysts, however, a janitor can be just as skilled as a watchmaker, provided each is following a *prescribed* set of motions with equal facility and ease.

Conditions is defined as those factors in a productive situation which are external to the worker himself, but affect his performance. Evident examples are the environment and sensory influences surrounding him.

RATING

It should be obvious from the preceding material that if someone times a worker's performance, then the probability of the performance time being representative of what the average person working with average skill and effort under average conditions would require is very low. As a result, industrial engineers have developed a concept of normal or average performance as well as a teachable and usable model. This chapter is concerned with rating an operator's performance in such a manner that it can be used to adjust actual performance time to normal or average time.

The adjustment of the timed performance to make it representative of what the average operator could achieve when working with average skill and effort is usually accomplished by applying a percentage factor. ANSI Z94.12[1] defines this procedure and several related terms as follows:

Performance Rating: (1) The process whereby an analyst evaluates observed operator performance in terms of a concept of normal performance. (2) The performance rating factor. *Syn:* leveling, pace rating, effort rating, objective rating.

Performance Rating Factor. The number (usually a percentage) representing the performance rating.

Performance Measurement. The assessment of accomplishments in terms of historical or objective standards or criteria.

INTRODUCTION TO WORK MEASUREMENT RATING OF OPERATOR PERFORMANCE

Work measurement rating would not be discussed except for the fact that some readers may not be familiar with time study procedures (including rating) and because the rating of operator performance was involved in normalizing performance times of MTM-1 and many of the other predetermined time systems—regardless of how the timing was accomplished.

[1]From ANSI Z94.1-12 IE Term. (Footnote 3, Chap. 1).

Moreover, in the case of MTM-1, not only were a variety of timing devices used, but several methods of rating operator performance were also involved. Hence, the reader/student trying to learn MTM-1 and its application certainly needs to understand the various approaches to rating.

In the initial research, most *timing* was accomplished by counting the number of motion picture film frames involved in the performance of a given motion. Additionally, some use of the stop watch was involved, for example, in the analysis of walking time. In later research electronic timing, very-high-speed film analysis, and a computer managed performance timing.

The original research utilized the LMS (standing for Lowry, Maynard, and Stegemerten) system to normalize the times (rate the operators)—a perfectly workable and usable system, but one which is not in vogue today. Later research performance times were normalized using the "pace" or "speed rating" system developed by the Society of Management (SAM). Also used were some statistical techniques referenced and/or described in the various research reports that are available from the U.S./Canada MTM Association.

There is reason to look closely at the various approaches to the rating of operator performance since MTM analysts must resort to timing devices to establish standards for work that is not covered by the MTM system and for process time. For this reason the MTM analyst must know how to normalize such work times (not process times) so as to bring them into congruence with the MTM data.

By now you know that there must be at least two basic approaches or "schools of thought" in regard to leveling or rating the operator's performance. These are the "speed rating" technique as used with the SAM[2] rating films and the LMS system developed by Lowry, Maynard, and Stegemerten at Westinghouse which ascribes numerical weights to skill, effort, conditions, and consistency as found during a study. The SAM procedure attempts to judge only the over-all performance, whereas the LMS system treats the human variables in a work situation separately. Both techniques, of course, rely on gaging the performance variables in relation to a mental concept of average performance held by the rater. Actually, the SAM system now enjoys the widest usage and probably is today's standard approach.

Since the original MTM time values were leveled by the LMS system, it is appropriate for analysts to study this system in detail, even though they may not use it directly. Moreover, a good understanding of this leveling system virtually guarantees a rather complete understanding of the factors affecting the *performance* of an operator and the substance of a rating by either system. Complete data by the originators of the LMS system, two of

[2]Society for Advancement of Management.

whom also participated in inventing MTM, may be found in a textbook.[3] Each variable of the LMS system will be identified and enough of the evaluation procedures required to evaluate or define them will be discussed so that the reader will understand what they are and how they are used. Also, how to produce a rating by combining the effect of the variables is explained. However, full details should be obtained from either the third edition of *Engineered Work Measurement* or the previously referenced text. *Knowledge of the extent to which the variables discussed qualitatively in Chapter 1 affect the productivity of a worker quantitatively help the analyst make a more precise total evaluation.* This basic advantage of the LMS system should not be overlooked, even though the SAM system is simpler and more frequently used today.

SKILL

Skill is one of the LMS variables requiring the rater's judgment as to the degree that the attribute is possessed by the operator being rated. In the restricted sense of proficiency at following a given method, skill cannot be varied at the will of the operator over a short time span. It is the basic aptitude of the individual for the given operation. Practice or training can increase it or it may be temporarily curtailed due to illness, dissipation, worry, excessive fatigue, and similar causes.

A worker is most capable of following directions—methods—when he is in excellent physical condition. His reflexes and other evidences of motor fitness are a gage of the responsiveness of his nervous system, much as the sensitiveness of temperature controls are judged by the tolerance of setting they will control. Natural aptitudes may either be beneficial or detrimental to successful action of muscles in a given task. Coordination, or the lack of it, may make the difference between a master craftsman or a duffer working with the same tools. The work analyst must, therefore, either fit his methods to the available human power or else recommend the correct human capabilities for a desired method.

When instituting a new method, or revising existing methods, the industrial engineer must consider the mental status and equipment of the worker. Whether the workers on the task are capable of exerting the required thought often will decide which method will be most practical; however, the actual "think time" in most factory and clerical productive jobs usually is far less than most persons suspect or realize. This is especially true for many quasi-professional jobs not ordinarily found on a factory floor. Some employees love to think, others shun the appearance of being intellectual. It is also a helpful truth that when job satisfaction or pride in workmanship

[3]Steward M. Lowry, Harold B. Maynard, and Gustave J. Stegemerten. *Time and Motion Study,* 3rd ed. (New York: McGraw-Hill Book Company, 1940).

result from following a prescribed method, the mental problems of method enforcement tend to dissolve. Sensible appeal to this factor is thus a good industrial engineering procedure promoting increased skill.

To follow a given method, much depends on the opportunity for practice given the operator. Skill cannot be consciously varied by the worker over a short span of time or training, but long practice and special training in a work technique can greatly enhance his skill. Industry meets these challenges with training schools, vestibule schools, or utilization of training personnel on the shop floor; the work analyst must fit his efforts into the particular policy on practice of new methods which prevails in his plant situation.

Finally the education and training of workers enter into the problems of skill and its relation to work methods. Workers cannot be expected to perform skillfully on tasks which their background has not fitted them to do. Likewise, a watchmaker might well enthusiastically follow the set method for assembling a complex control device, but balk and procrastinate when expected to grind gates off rough iron castings. The skill of a worker must be complemented by the method to assure compliance.

According to authorities in psychology, there is a wide range of manual dexterity between the best and poorest individuals. Basic manual dexterity as a characteristic of an individual has been studied by many psychologists and standard tests are available for measuring this aspect of the human machine.

Since skill involves manual dexterity most directly, it is logical to expect a like ratio in the worker population. However, the screening of applicants by employment departments, limitation of workers to appropriate job categories, normal attrition of layoffs and discharges which tend to reduce the number of poorer workers, and other modern personnel policies tend to reduce this range in industry. It actually falls in about a 1.5 to 1 ratio.

Skill *of the degree ordinarily met in industry* has been classified in the LMS system into six categories. These are as follows:

1. Poor skill.
2. Fair skill.
3. Average skill.
4. Good skill.
5. Excellent skill.
6. Super skill.

In order to illustrate how further aids are provided in the LMS system for judging this attribute, the guides for poor skill, average skill, excellent skill, and super skill are now presented.

Poor Skill:

1. Cannot coordinate mind and hands
2. Movements appear clumsy and awkward

3. Seems uncertain of proper sequence of operation
4. Untrained on operation
5. Misfit
6. Hesitates
7. Errors occur frequently
8. Shows lack of self-confidence
9. Unable to think for himself

Average Skill:

1. Self-confident
2. Appears a little slow in motion
3. Work shows effects of some forward planning
4. Proficient at the work
5. Follows sequence of operations without appreciable hesitation
6. Coordinates mind and hands reasonably well
7. Appears to be fully trained and therefore knows his job
8. Works with reasonable accuracy
9. Work is satisfactory

Excellent Skill:

1. Self-confident
2. Possesses high natural aptitude for the work performed
3. Thoroughly trained
4. Works accurately with little measuring or checking
5. Works without errors in action or sequence
6. Takes full advantage of equipment
7. Works fast without sacrificing quality
8. Performance is fast and smooth
9. Works rhythmically and with coordination

Super Skill:

1. Naturally suited to the work
2. The operator of excellent skill perfected
3. Appears to be super trained
4. Motions are so quick and smooth that they are hard to follow
5. Work has machine-like appearance and action
6. Elements of operation blend into each other
7. Appears not to think about what he is doing
8. Conspicuously an outstanding worker

EFFORT

The "work readiness," or willingness of an employee to expend his energy in effective work, is a complex of human behavior worthy of close attention by industrial engineers. Lack of this attention has historically been the root of many production floor problems, management difficulties, work stoppages, union friction, and disastrous strikes. Management increasingly is recognizing its fair share in the problem of providing the psychological climate in which its personnel can and will put forth their best efforts. Much of the bargaining material with which unions deal concerns this factor of productive output.

Typically, people work because of financial motivation. When their minimum monetary needs are slightly exceeded, however, they discover the desire for additional satisfactions such as better light, ventilation, safety, relief from excessive fatigue, understanding, recognition, and encouragement by supervisors, etc. The hue and cry for security has been prominent since the early 1930's in many fields of life and is considered even more important today.

People want the chance to realize the opportunities for self-improvement and advancement that arise in the normal course of their tasks—in fact, engineers rate the type of job and the opportunity for professional growth almost as high as salary. Recognition will increase effort, as is attested by such present-day industrial activities as clubs, newspapers, company awards, personal letters from the executives, and the old "pat on the back"—the standby of the foremen of Taylor's days. Workers like to feel included in the group and will strive to produce better when they believe they "belong." The use of motivation by all parties concerned in a productive enterprise has indeed graduated from theoretical to practical emphasis in recent years.

The effort an operator makes also results from his attitudes. His temperament must be consonant with the task and his co-workers before his effort will be more than lackadaisical. High morale in the shop tends to increase effort. A person working in a tolerant economic and religious climate that permits and aids human freedom will tend toward excellent effort; the reverse is true under environments of despotism and suppression. The maintenance and improvement of employee attitudes thus are realistic goals which aid in raising output through extra effort.

The Japanese phenomenon has given us a real-life national example of how the cultivation and promotion of a "good attitude" toward the individual, the company, and the job can provide benefits to all parties. To make this work there must be sincere concern for the welfare of both the workers and company.

For a man to show his best efforts, he must be happy in the social sense. All of the people with whom he comes into contact both at work and away have an influence in this respect. Disappointing or frustrating social relations, whether of the brief or long-lasting type, are destructive of working effort. Carefree and satisfying activities with other people keep the worker attuned to the need and value of doing a fair day's work. Common courtesy between employees often makes the difference between good and poor effort in a factory.

His job is but one part of a worker's total environment. Some people are by nature, or family custom, lazy and indifferent to appeals to increase their effort. The hereditary energy of other types of people is such as to assure that they will hardly ever fail to put everything they have into the work at hand. Most folks are somewhere between, but even hereditary factors permit change by intelligent management seeking to benefit from higher effort levels.

Difficulty at home may likewise militate against good job performance, whereas a happy home life usually is the hallmark of a good worker. Conditions at home are being given more and more consideration by management. Prospective employees, particularly executives, are now often being thoroughly investigated concerning this item. Counsellors are also often provided to help solve these and other general problems of the employees.

Modern industry has recognized the needs of employees in recreational activities by providing such opportunities within company groups, and by generally encouraging private recreation. Sports, entertainment, and cultural pursuits help fit a worker to try to give fair output.

For humanitarian, as well as legal and practical reasons, the importance of health has induced management to provide many medical services and much related advice to their personnel.

Effort—the "work readiness" or drive of a worker to expend his energy—is a variable which the operator can control at will. The effect on effort of various human attributes discussed earlier are not as great as generally supposed. Only a 25 to 30 per cent spread between the slowest and fastest workers was found to be attributable to effort alone by Lowry, Maynard, and Stegemerten. Naturally, slow workers—unless other human factors loom larger—tend to be eliminated whereas excessive effort by even the most conscientious worker cannot be maintained over long periods of time without harm to himself.

Good managers do not desire or encourage the "speed-up" which labor unions decry, rather they achieve most of their production increases by attention to motion-saving methods and mechanization. On the other hand, there is no denying that the opportunity for operators to earn more by expending a greater effort does occur when an incentive system is installed. It is a universal "fact of life," and management often elects to install and operate such a system.

That a worker must be worthy of his hire is also a fact, and management must have the right to rid their organizations of obvious shirkers. All of us who are among the employed must provide the employing organization with services that are worth more than our wages.

When making a work measurement study it is not unusual to find that some workers attempt to "fool" the work analyst by deliberate pacing of their work while under study, but the vast majority of employees are honest and try to be fair in this respect. Indeed, some workers tend to give better effort while being studied than they would normally exert.

As was done for skill factors, six ranges of effort were isolated and defined. The key characteristics outlined permit ready reference and identification of the effort level. They are poor effort, fair effort, and again through to excessive effort. The good effort guide statements are as follows:

Good Effort:

 1. Works with rhythm
 2. Idle time is minor or non-existent
 3. Conscientious about his work
 4. Interested in his work
 5. Works at a good pace that can be maintained throughout the day
 6. Actions indicate faith in time-study man
 7. Readily accepts advice and suggestions
 8. Makes suggestions for improvement of operation
 9. Maintains good order at workplace
10. Uses proper tools
11. Keeps tools in good condition

It is noteworthy that, while skill and effort are evaluated separately in this rating system, the levels of skill and effort tend to complement each other rather than being contrasting. Poorly skilled workers attempting to compensate by excessive effort usually interfere with what little skill they possess and this appears as fumbles, false starts, and frustration. A highly skillful worker, however, appears to work with deceptive ease; every motion counts and produces useful effects on the workpiece. Operators with higher levels of skill can reduce their effort successfully without losing effectiveness. It is usual, however, for poor effort and poor skill to combine and for high skill and high effort to coexist when training has been adequate.

CONDITIONS

The questions of effort and skill, as are readily apparent from the preceding discussion, largely revolve around human capacities and relationships. However, people also work in conditions or surroundings which, while not primarily composed of human factors, do have direct bearing on their productiveness. It devolves upon management to enhance the working environment in order that the speed of motions may be maintained and improved. *The important factor in the following conditions is not what their absolute value might be, but whether one condition or another is normal and the best that can be provided for the work being done.* People will adapt themselves to an amazing range of conditions and react normally as long as they feel that humane and just consideration in adjusting those conditions has been taken by the management.

Conditions basically affect the sense mechanisms of the human machine.

To adequately consider conditions thus implies that the impact of the surroundings on the sensual apparatus is involved. The senses of sight, sound, smell, and thermal comfort are amenable to a degree of control by changes in the surroundings directly, thus being affected by external or mechanical control.

Visual perception is aided by the correct kind and amount of illumination, and much expert data and advice on this subject are available. Reduction of distraction of the eyes can be achieved by shielding flashes, drawing the sight of the worker to the more important areas by effective workroom design, and/or proper use of color and similar strategies. The effect of harmonious and contrasting colors in workrooms has been given much study, and specific rules are obtainable from experts on the subject.

The assault of noise in modern industry on the aural senses can be terrific, as witness the ear troubles often found among industrial workers in steel mills, machine shops, transportation centers, and even many offices. The noise level alone is not entirely to blame for this. Sounds might be loud, but of such tone and pitch as to not displease or distract the worker. On the other hand, much trouble can be caused by even low intensity noises that are dissonant, rasping, whining, or gruff. Public address systems are often misused in this way. Familiar sounds, such as the music played at intervals during the workday, help combat the strain caused by unusual noises which shock the listener. Acoustical treatment of work spaces is also being employed for such purposes so as to reduce the aural fatigue associated with highly variable or high frequency noises. With modern acoustics providing so many relieving methods, management can ill afford to lose production because of sound problems within the work surroundings. The industrial engineer can do much to recognize noise troubles and suggest their elimination.

The problems of nasal irritation are closely allied with health and safety areas of work study. The supply of fresh air directly aids or inhibits production. Toxic odors should be kept below safe physiological levels. Offensive smells, although not necessarily dangerous, are usually undesirable as related to good working conditions. The presence within work areas of agents which cause such respiratory troubles as colds, sinus attacks, dry throats, and allergic reactions should also be discovered and eliminated where possible. Alert management can do much to promote higher production by attending to the nasal problems.

Finally, thermal conditions can either engender productiveness or inefficiency by their effects on the workers. Extremes of heat or cold should either be avoided, or action taken to make them tolerable. The dress of workers may often be dictated by the temperature usual to their tasks. Providing protective clothing where dangers to the body and/or ordinary dress exist and providing temperature control to permit working in light clothes will aid human output. Humidity must be kept appropriate to the temperature, or else fans and other moisture and/or heat combating devices must be employed to assure worker comfort. Modern heating and air conditioning has progressed to the stage where a good thermal environment will pay for itself in the extra production it makes possible. Seasonal and cyclic weather variations, while most vital to outdoor workers, also affect the personnel who work indoors. They enter and leave from seasonal extremes or sometimes must frequently pass in and out of doors in their normal task routines. Good conditions in this respect will help maintain the work pace, and the work analyst can recommend changes in jobs which will help to minimize interference to production from thermal variations.

Industrial conditions today are generally such that the great majority of time studies will be rated as involving good to excellent conditions—and a few might even be rated ideal. A variation of 15 per cent is about the maximum that can be ascribed to the range between poor and ideal conditions. Also remember when judging conditions that the "target" is a moving one, in the better companies conditions generally keep getting better.

Again, six classes of condition have been segregated: poor, fair, average, good, excellent, and ideal. The important aspect for the analyst to note is the relative departure from what are usual conditions prevailing in the workspace concerned.

Notice that effort and skill are attributes of the worker himself, whereas the job conditions are external influences on his output. The conditions rating factors have nothing to do with the method employed, functioning of machinery, propriety of tools used, sharpness of tools, and similar considerations. Such matters are subjects for separate attention by conscientious and cost-conscious managers. They relate specifically to tools and methods, having no relation to the concepts gaged as conditions in the leveling procedure.

CONSISTENCY

As is true of any series of numerical observations, the elemental times taken during a time study exhibit a variability that can be subjected to statistical analyses of various kinds. Such variation has little to do with the operator, so long as he is giving average performance. It is noticeable, however, that attempts by a worker to influence the time taken will result in inconsistent variation of the elemental times read from a watch. Obviously, the analyst must make an effort to reduce the effect of this ranging of times on a standard.

According to the LMS rating system, the range of inconsistency for valid readings will not exceed about 8 per cent from low to high range. Six classes of consistency have been assigned to adjust for acceptable variations: poor, fair, average, good, excellent, and perfect.

The work analyst can best determine the consistency while he is summarizing the study away from the workplace. If no variation exists in the readings for each element, consistency is perfect; such times are very unlikely to be encountered often and the standard should be increased to recognize this fact. Process- or machine-controlled elements are the only ones in which perfect consistency is not unexpected. Poor consistency is apparent when the times range widely from the mean reading in a random manner. Most studies have average consistency, which means that the readings are very close but have some high and low variations from the average observed time.

Occasionally, serious variations in readings are found in study elements. These variations are often caused by non-repetitive or non-productive events which a properly made study should include to permit adjustment of the standard. Another standard procedure is to strike out from consideration in the averaging of element times any readings that are obviously incorrect or influenced strongly by extraneous facts. The remaining readings are then judged as to consistency.

When judging the consistency, the nature of the work elements should be taken into account. Simple operations such as "move lever," "toss part into container," and the like should show little or no variation. More complex elements such as "obtain lockwasher from parts pan" and "fit shaft into bearing" will obviously show greater variation.

The skill of the operator should also be considered. Highly skilled workers normally work more consistently than poor employees. In fact, highly skilled operators, providing parts fit properly, etc., will work with good to perfect consistency unless they deliberately attempt to show a poor performance.

PERFORMANCE RATING DATA

The factors in human behavior that affect productivity were discussed in detail in Chapter 1; to this point in this chapter, more specific comments have been added that shed quantitative light on the subject. What remains, therefore, is to see the numerical facts assigned in the LMS system and to learn how to combine them to provide a useful rating factor. This rating factor applied to the mean value of the observed times will then yield the select time desired for any given element.

During the development of the LMS system, some 175 men made comparison studies at every opportunity available for approximately a year's time. In this manner, the effects of varying degrees of skill, effort, conditions, and consistency were evaluated. When the results of these studies were first published in 1927, the leveling factors had been used successfully in many industrial organizations.

Table I.   Performance Rating Data—LMS System

FACTOR ———→		SKILL	EFFORT	CONDITIONS	CONSISTENCY
Category	Code	Additive Percentage Values			
SUPER	A1 A A2	+.15 +.140 +.13			
EXCESSIVE	A1 A A2		+.13 +.125 +.12		
IDEAL	A			+.06	
PERFECT	A				+.04
EXCELLENT	B1 B B2	+.11 +.095 +.08	+.10 +.090 +.08	+.04	+.03
GOOD	C1 C C2	+.06 +.045 +.03	+.05 +.035 +.02	+.02	+.01
AVERAGE	D	.00	.00	.00	.00
FAIR	E1 E E2	−.05 −.075 −.10	−.04 −.060 −.08	−.03	−.02
POOR	F1 F F2	−.16 −.190 −.22	−.12 −.145 −.17	−.07	−.04

When considering the numerical data, remember that it measures the typical industrial worker and that the range of variation will be more limited than might be found in the population as a whole. Many workers in the fields of agriculture, commerce, transportation, government, and distribution probably work differently; inclusion of these groups might materially increase the range and lower the level of job activity considered as normal in this system. However, the values apparently cover adequately the industrial situation, since the LMS system has been used in widely dispersed geographical and political locations with rather uniform success.

The ratings factors for the LMS system are shown in Table 1. The data is used in the following manner: Each level of performance under one of the four major factors corresponds to the descriptions in the referenced text.[4] These categories are coded by letters and numbers for ease of usage. When the times study man is observing the operator, he notes with the code sym-

[4]*Ibid.*

bols the effort, skill, and conditions for each element and/or the job as a whole. Later, when summarizing the study, he also assigns a consistency rating.

The numerical rating for each of the factors is then added *algebraically* to the nominal 100 per cent to produce a finished rating. Whether this composite value is one that applies to an element or an entire job, the procedure is the same. Note, however, that for skill and effort more than one value is given for each category. Actually the letter and the number combinations are the LMS data as it appears in most published sources and on many time study forms; but the LMS system also contains the intermediate values for those who would rather use the letters only. To employ the double symbols, an analyst must judge whether the high (1) or low (2) range of the category in question is appropriate.

As an example of how the table is used, suppose a time study element was rated according to the descriptive definitions in this chapter as B1A2C and a rating of B was assigned for consistency during the study summary. The combined rating for the element is then B1A2CB for skill, effort, conditions, and consistency in that order. It is conventional to write the symbols in this order. The per cent rating would result as follows:

B1	Skill	+0.11	(Excellent)
A2	Effort	+0.12	(Excessive)
C	Conditions	+0.02	(Good)
B	Consistency	+0.03	(Excellent)
		1.00	
	Total LMS Rating	1.28 or	128%

Further examples are:

Fair Skill	E1	−0.05	Fair Skill	E	−0.075
Good Effort	C2	+0.02	Poor Effort	F	−0.145
Fair Conditions	E	−0.03	Average Conditions	D	0.00
Good Consistency	C	+0.01	Fair Consistency	E	−0.02
		1.00			1.00
Total LMS Rating		0.95	Total LMS Rating		0.76
		or 95%			or 76%

It is notable that the maximum combination of skill and effort in the table shows a value of +0.28 or 128 per cent. Conversely, the minimum combination has a value of −0.39 or 61 per cent. By division, the spread between the best and worst effect the operator can produce is better than 2 to 1. Such differences ordinarily are not likely to be seen in industrial operations. It is important to emphasize, however, that even this spread in productive output can be greatly exceeded by methods changes. It is not uncommon to devise methods by which average operators can triple or quadruple their output at the same performance level; this, obviously, is not possible due to the operator alone.

The time study observer may sometimes note changes in the performance level during the course of the study. Since conditions remain relatively
constant, and the worker cannot vary his skill level consciously during the
study without departing from the method under study, such changes are
normally due to the operator's effort or the use of motions which really constitute a different method. Such variations should be noted in the space for
remarks on the study sheet and keyed to the time observations to permit intelligent interpretation or correction later.

Evaluating skill, effort, and conditions as separate items makes it possible to apply LMS leveling factors to a one- or few-cycle study. Although
timing of only one or a few cycles is not a good time study practice, practical
conditions sometimes necessitate the procedure. For instance, limited
parts availability, lack of time for an intensive study, the desire for indications of the over-all time in a hurry, and like situations may require that
only one cycle or a few cycles can be timed. If the factors could not be evaluated separately, no such leveling would be possible.

On this score, a note about the SAM rating procedure—the more popular
rating system—is in order. As explained earlier, this method of rating
judges over-all operator performance in relation to what is considered average in the given plant under the conditions involved. This makes it feasible
to rate not only the average observed time for each element of the operation, but also each observed time for the element. Of course, this would normally be carrying the SAM procedure to an unwarranted extreme; but it can
be done validly and is sometimes helpful in analyzing a particular operation.

THE CONCEPT OF AVERAGE

Before any rating procedure will permit the equating of one man's time
study values to that of another observer, a proper concept of average must
be established upon which all industrial engineers can agree. Both men
must have the same concept of what constitutes average performance.

One way of achieving this is to pair a novice and a veteran rater to study a
variety of operations. Their ratings, upon comparison back in the office or
privately on the factory floor, will show whether the novice has the proper
concept of average. The novice can then, with the help of such comparisons
and practice, adjust his mental concept to agree with that of the more experienced rater.

A better method is to utilize, by purchase or rental, some of the performance rating films obtainable from the Society for Advancement of Management, universities, private industrial firms, or consulting engineering
firms. The mere fact that a man is experienced is no guarantee that he can
level correctly. Rating films are excellent for training both the novice and

the experienced time study analyst. The scheme of pairing can then be used as a supplement to these rating films if the experienced man has proven that he can consistently level correctly.

The first known (to the authors) industrial company[5] marketing its own rating films (in color or black and white, with sound track) provides ten scenes of different industrial operations being performed at various paces. Complete user instructions accompany these films.

The Society for Advancement of Management presently has available three sets of rating films. The first set contains 6 reels covering manufacturing and service operations plus 1 reel each of clerical and laboratory operations. The second set of 4 rating films is devoted entirely to clerical operations. The third set of 6 rating films is devoted to the general area of production and indirect labor, and it includes element rating. The three sets illustrate rating on 24, 12, and 18 operations, respectively. The SAM rating films[6] are provided with a set of standardized instructions and standard rating values which account for the allowances and per cent of attainment expected in the particular plant where the rating practice will be done. A graphical scheme (see Fig. 2) also permits plotting of ratings made by the film viewers in such a way that it is easy to see whether the observer is rating "tight," "correct," or "loose." "Tight" means that a man rates too low and

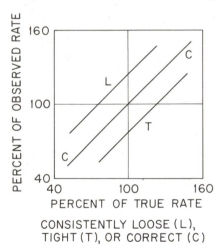

CONSISTENTLY LOOSE (L),
TIGHT (T), OR CORRECT (C)

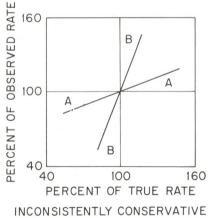

INCONSISTENTLY CONSERVATIVE
(A) OR LIBERAL (B), BUT
NORMAL AT 100% FOR BOTH

Fig. 2.   Rating plots based on SAM rating films.   The five general possibilities are shown.

[5]Vought Aeronautics, Personnel Development, Unit 1-97030, P.O. Box 5907, Dallas, Texas 75222.

[6]These films can be rented or purchased from the Society for Advancement of Management, 135 West 50th Street, New York, N.Y. 10020. They are 16 mm films and are supplied with the "correct" ratings developed by a consensus of many raters.

would thus allow too little time for fairness to the operator. "Loose" means that the rater's generosity to the operator would be reflected in too much time allowed for the operation—this is unfair to the management. For the best interests of both management and worker, it is essential that the performance rater achieves "correct" ratings of a task.

The greatest sin for a rater is to be inconsistent. When he is "consistently tight" or "consistently loose," it is possible to adjust his readings when he knows his degree of error. When he is inconsistent, however, it is difficult, if not impossible, to reconcile his final time data with an objective standard.

Two other organizations (other than manufacturing firms) are known to have rating films that probably are satisfactory and available by rental and/ or purchase. We are also reasonably sure that still other such films exist, some for private use and some available for a fee.[7,8]

One remaining method of achieving an average concept has arisen since the usage of MTM has become widespread. This consists of standardizing an operation and setting a reliable MTM rate on it; this time comprises the time required of a 100 per cent operator before the addition of any time allowances. The stopwatch will then show accurately the true actual time, and by division the true performance level, which will enable a crosscheck of the performance rating made by either new or experienced time study men.

A second MTM approach to leveling is to establish one element of the study so that it includes the motions defined in a standard MTM motion category. The actual time taken by the operator to perform this element can then be compared (divided) by the MTM time (detailed data) to obtain a performance rating of the operator for that element of the study. If the operator's consistency is average to excellent it would then be reasonable to expect the regular time study rating of the operator to compare closely to this element rating.

Finally, the MTM Association has a library of 16 mm practice analysis films[9] which may be adapted to rating practice. The data package for each film loop includes an operational description, the industrial source, a complete MTM motion pattern analysis with explanations, and miscellaneous photographs and workplace layouts. Since the film speed for 100% pace on a daywork basis is specified for each film, the rating trainer can use a variable speed projector to simulate different working paces with a constant work method. This latter point helps to overcome one source of possible error in judging the performance level in the films cited previously, where the method for all working cycles may not be truly constant.

[7]Rath & Strong, Inc., Lexington, MA 02173.

[8]Tampa Manufacturing Institute, Holmes Beach, FL 33510.

[9]*MTM Film Loop Library,* MTM Association for Standards and Research, 16-01 Broadway, Fair Lawn, N.J. 07410. This catalog contains full film descriptions and prices, and is available on request.

These films are also available in BETA and VHS video tape for MTM practice analysis purpose. The *video tapes* are not recommended for *pace rating purposes* because the conversion process from film to tape may have introduced some speed variations—also most video tape equipment do not have built-in variable speed drives.

CHAPTER 3

Predetermined Time Systems

The background necessary to an understanding of work measurement and its practical application was given in Chapter 1. Additional information required to understand one aspect (the work pace on which the system times are based) of the predetermined time system known as MTM was provided in Chapter 2. In this chapter we will resume the history of work measurement.

A period of adjustment followed World War I. The economy began to expand, and boomed for almost the entire 1920 decade. The pressure of the times, together with political and social trends toward optimism and satiation of desires, caused an ever-increasing demand for goods and services of every description. Industry grew rapidly, but not wisely. This, in turn, generated a demand for the application of the promising approaches of work measurement in order that profits might survive in spite of increasing financial pressure, competition, and inflation. Another significant trend was the rising influence and power of labor unions.

Reliable work analysts of both the time study and motion study groups were enlisted to solve such problems with the hope of quick answers and certain results. However, it became increasingly apparent that the answers of work measurement were neither quick nor easy. Good analysis took patient, long hours of qualified effort and, in addition, areas of uncertainty persisted. The truism that *there is no substitute for good management* was discovered. Work study was no panacea, in fact its application without informed management was dangerous.

As often happens in human affairs, charlatan or well-intentioned but ignorant persons stepped into this situation which was ripe for exploitation. "Salesmen" with little real understanding of either management or work measurement grabbed stopwatches, "pat" analysis reports, newly created jargon, and other deceptive paraphernalia and took to the "pickins." The companies served benefited little, and true professional management and/

or industrial engineering was seriously damaged. Thus evolved the era of the "Efficiency Expert" that damaged the good repute of work measurement, alienated workers and their unions, and bred other undesirable effects which still partially persist. Work study's promise and its promotion of managerial competence seemingly failed its first real test, resulting in public discredit and distaste for the whole approach; progress was dealt a hard blow.

What went wrong? Obviously, adequate and reliable work measurement is a professional activity of high caliber. It is no plaything to be exploited by the uninitiated or untrained person working without professional restraints. While the lack of professional approach caused most of the trouble, two other factors also were involved. One was that managers of this period had no prior experience with engineered work standards and associated incentive systems. They were unfamiliar with good standards and what management responsibility toward incentive installations really entails. A second factor, more subtle and difficult to identify, also seemed to be present—a possible fundamental weakness in the measurement technology evidenced by all the trouble its usage apparently caused. This motivated the more astute practitioners to analyze the nature of work measurement itself. Initial tools (in this case time study) serve until natural processes of curiosity, or calamity, engender closer scrutiny that finally reveals faults and sets the scene for future refinement and improvement.

The shortcomings of both time study and motion study as separate entities were elaborated in the first chapter. Now old tools are not necessarily invalid, but are usually proper for the kind of job they were intended to do provided they are wisely and correctly used.

The basic precepts of the pioneers themselves pointed the way to salvation. Dr. Taylor insisted that:

(1)   The right operator be found to permit reliable study.
(2)   The correct tools should be properly used by the operator.
(3)   The operator should use the proper *motions* during performance.
(4)   Only then should a stopwatch be employed to find the task *time*.
(5)   No value could accrue from the study unless it was followed by teaching and enforcing the use of the method on which the time was taken.

To these guiding light, the Gilbreths added several other ideas:

(1)   If the best and least involved *motions* are used for a task, the *time* required will automatically be minimal.
(2)   Performance times are subject to natural variation and the laws of probability; out of the range of times for a given motion, the most useful value is the mean time. The average operator is most numerous.
(3)   No work study or other means of specifying job content is adequately done unless the human factors have been considered and humane answers found. Cooperation is superior to coercion.

(4) Human beings differ markedly in many ways; but, in the statistical sense, there are points of similarity between all people which permit the discovery of laws of behavior that can predict actions with calculated reliability. The physiological mechanism involved in motion performance is such a point of likeness.

The ideas of Taylor, father of time study, and the Gilbreths, parents of motion study, given above did not deserve the division and disunity that occurred in work measurement. Often in human affairs, the middle course proves wisest. Logically, then, attention should properly be given to *both* the *motions* and the elapsed *times* involved in work performance if major benefits are to result from any form of work measurement. This principle is followed in today's methods engineering, and it is the avenue to all predetermined time systems having a sound, scientific basis.

The basic key to all predetermined time systems is the fact that variations in the times required to perform the same motion are essentially small for different workers who have had adequate practice. Mathematical relationships can therefore be established between motions and times subject to predictable statistical limits. The credit for first stating this key is generally attributed to Mr. A. B. Segur, who wrote in *Manufacturing Industry* during 1927 as follows:

> Within practical limits, the time required for all experts to perform true fundamental motions is a constant.

The desirability of predetermining both the most basic motions and the time they require for an average operator was indirectly acknowledged almost from the beginning of work measurement. Many noted analysts achieved a degree of success by amassing large amounts of standard data for typical jobs, assuming that the same data would permit method establishment and setting of task times on similar jobs. The work of Carl Barth, Harrington Emerson, The Industrial Management Society, and Ralph M. Barnes typified this approach. They focused the problem sufficiently to specify that, when basic human motions have been gaged in terms of method effect and mean performance times, a practical set of work study data has been achieved.

This is a good point to consider the official[1] definition of predetermined time systems. The Work Measurement and Methods Standard Z94.1-12 is Z94.12 and it says:

> Predetermined Time System—An organized body of information, procedures, techniques and motion times employed in the study and evaluation of manual work elements. The system is expressed in terms of the motions used, their general and specific nature, the conditions under which they occur, and their previously determined performance times. Syn: Predetermined Motion Time System.

[1] From ANSI Z94.1-12 IE Term. (Footnote 3, Chap. 1).

And the Biomechanics Standard of Z94.1-12 is Z94.1 and its definition from that viewpoint reads:

> Predetermined Motion Time System—Any scheme useful in the prediction of performance times of industrial work tasks. It analyzes all motions into elemental components whose unit times have been computed according to such factors as length, degree of muscle control required, precision, strength, etc. Time standards of several of these are used in bench mark (q.v.) levels for normal performance in biomechanics and work physiology. For these purposes they are divided into systems: derived from taxonomy and kinesiology; nontaxonomic and kinesiological; taxonomic but nonkinesiological.

> Bench Mark—A physiological or biological reference value against which performance is compared, e.g., "normal" heartbeat is 60-80 beats per minute.

These, together, give a good idea of the meaning of the term.

A word of caution is appropriate, however. Neither stopwatch time study nor motion study with therbligs have been superseded. In current methods engineering practice either or both procedures are often used as the soundest approach to the most practical solution of a particular problem involving methods work and the establishment of labor standards. Similarly, informed industrial engineers properly use both the more traditional techniques and predetermined time systems to develop standard data that will efficiently meet the need for rapid standards specification.

It is obvious that only the manual, or operator-controlled, part of work cycles is evaluated by predetermined time systems. The time required by process-controlled elements must still be found with the aid of a timing device. Process time is that consumed when the machine, material characteristics, or other physical or chemical changes (not themselves dependent on an operator's manual motions) control the working time—only time devices will measure it directly. The stopwatch is far from obsolete, it must merely be used in a different manner.

DEVELOPMENT OF PREDETERMINED TIMES

Predetermined time systems were rather slow in development. Only about three of them, notably MTA (Motion-Time Analysis), Work-Factor, and MTS (Motion Time Standards) trace their usage into the 1930-1940 era. Possible, this was in part due to the effects of the depression and its aftermath. The bad taste left by the "efficiency experts" also undoubtedly left its effect. Industrial engineers had to expend much of their energy and thought to overcome this bad reputation, which hindered profitable usage of their time in developing new approaches to surmount the hostile surroundings and in regaining acceptance by managers.

This situation did not continue. The 1940's brought a rapid war mobilization with shortages of material and manpower. There was little time in which to achieve the necessary tremendous increases in productivity. Business was forced to pay closer attention to all types of labor-saving devices and managerial controls. The patient efforts of qualified industrial engineers had also, by then, overcome much of the bad reputation of the past.

Several more of the better known predetermined time systems, including MTM, originated during this period. The pace of production and economic pressure continued after the war, in response to the hunger of demand created by years of non-availability of necessities, luxuries, and investment-type production. During the entire decade, however, management avoided the errors made in the "efficiency era" that thrived under similar economic conditions. This is to the credit of both managers and wiser practitioners of work measurement.

Then, through the 1950's, the predetermined time systems that were already established continued a healthy growth as competition reappeared. The remaining systems also came into existence since then. It is interesting to note that typically human differences in interpretation as to the significance of various aspects of the new concept of predetermined times led to a degree of divergence of systems. Thus arose a new need—the need to attain more universal agreement on one or several of the more prominent systems. Only then could the traditional factional disputes that had plagued conventional time study and motion study be averted. The cycle of development had entered a new phase.

This new phase of developments started in the 1960's and still continues. Just as the early time study practitioners strove to develop standard data and thereby reduce the drudgery of establishing labor standards (a technique still used and covered later herein), today's applicator of work measurement techniques utilizing a combination of time study *and* predetermined time systems tries to make the work of establishing standards more efficient by devising "simplified" or "condensed" predetermined time systems (really generalized *and* specialized standard data) based upon predetermined elemental times.

This trend has been aided by the advent of more and better computers to which work measurement applicators and researchers have turned. While some utilize the computer merely as a system storage device, sorter, and calculator, others have attempted a more sophisticated approach. The most sophisticated approach is to utilize the computer in the interactive mode and to program and store within the computer not only the elemental (detailed) predetermined time system, but also a biomechanical model of the average man, a "library" of standard tools and their usage routines, and the analysis decision rules associated with the predetermined time system so that the proper input instructions for the workplace layout coordinates and work description cause the computer to generate directly a properly detailed and structured predetermined method-time standard. In fact, not only can optimized workplace layouts, together with the work pattern and tool locations, be printed as output, but an animated video display of a worker performing the work as analyzed is possible. Virtually all of this has been demonstrated in the laboratory. Accomplishing the latter approach for a specialized kind of work situation, like conventional assembly and wiring of parts on an electronic chassis, would be more simple than the generalized case.

During the past two decades, especially during the last decade, the emphasis has been to develop simplified and specialized versions of several major predetermined time systems. The trend has also been to develop associated computer programs for several of the more popular types of computers—these latter programs not only contain the "system" details, but also the decision rules and a program to calculate standard and other helpful analysis data. Naturally, a printout in a format acceptable as a standard's document can also be obtained.

Another recent phenomenon is that some of the industrial engineering consultants have taken a basic predetermined time system, almost always MTM-1, and developed a proprietary system(s) that they then use in their consulting work. They, too, have computerized versions that they lease or sell to their clients. While some of these systems work satisfactorily according to some reports, the background material is never made public.

The national MTM associations have a complete array of such systems for which research reports are available. Additionally, these associations have available computer programs, training course materials, training courses, training films, research reports, etc., and they provide certification of graduates of approved courses through examinations graded by the associations. The MTM organizational system will be further described later in this chapter.

CHARACTERISTICS OF PREDETERMINED TIME SYSTEMS

Definition

Two official definitions already have been presented. The principal one conforms closely to the authors' preferred definition originated in the first (1957) edition of this text, as follows:

> *A predetermined time system is an organized body of information, procedures, and techniques employed in the study and evaluation of work elements performed by human power in terms of the method or motions used, their general and specific nature, the conditions under which they occur, and the application of prestandardized or predetermined times which their performance requires.*

Note several salient features of this definition which adequately cover the many systems now known. They specifically concern human energy expended as work, and do not measure process time, as mentioned earlier. It is necessary to confine the attention, for one given study or evaluation, to one specific method at a time. The true essence of a particular motion and its relation to other motions must be clearly understood. The context in which the motions are performed will affect their analysis and performance time. And finally, the time, or range of times, which is assigned to a given motion has been determined in advance; the standardized data promotes uniformity and universality in application.

Predetermined Time Systems—Aims and Ideals

For a field as broad and varied as work measurement, it is difficult to formulate a set of aims and ideals that could be the basis for a technique satisfactory to all concerned. However, ideal criteria are herewith presented which, hopefully, avoid the equal dangers of oversimplification and overelaboration. Since these criteria admittedly are ideal, it is readily conceded that no single system will completely satisfy all of them.

1. The approach and usage of a system must be professional in the best sense of that term. Salesmanship of the system is a necessity, but this should be done with befitting dignity and taste. Industrial engineers, being concerned with the design of procedures and controls for the interaction of people, materials, and machines, have a responsibility in the design sense that is just as real as the honesty of a civil engineer who uses an adequate factor of safety to design a bridge. Human lives are affected, as are the economy and natural resources. For these reasons, the professional aspects surrounding a predetermined time system merit prime attention in evaluating the system.

2. The system must be based on sound concepts with heavy reliance on the fields of physiology and psychology. Competent personnel should follow known and accepted scientific mathematical rules while developing the data which comprises the working portion of the system.

3. The desirability of validating the data both by the developers and independent efforts is self evident. Such validation is not likely to occur unless the methodology and background data are made available to the public. Publicity increases the chance for further validation and for providing convincing evidence of the correctness of the data—both being essential to public trust and confidence in the system.

4. Some form of training control will insure that misuse of a system through partial or improper understanding will not engender a repetition of the "efficiency expert" era. A good predetermined time system provides an

agency or procedure by which training in the system can be properly controlled. Such training should be available to the public subject to limits of
qualification and ability. It could also be further safeguarded by means of
testing, continuing publication of new data, interchange of data between
practitioners, and circulation among practitioners of application experiences with the system.

Such training control is hard to achieve. One approach used to achieve it
is to limit the publication of data and procedural methods. However, this
could produce other bad effects. Therefore, the authors believe that the
agency for training control preferably should be administered separately
from the information and data sources of the system. This independent
agency should work mainly on a level of professional endorsement, the lack
of which would deter use of the system by untrained people. It should, upon
proper request, certify the adequacy of the training courses and the knowledge of trainees to allow interested parties to ascertain the competence of
practitioners seeking to apply the system to their own operations. Admittedly, this is not the strongest form of training control. The individual's
opinion concerning the emphasis this ideal deserves will determine his
ranking of this with the other ideals expressed or which he may desire to
subtract from, or add to, those expressed in this text.

5. The working form of a good predetermined time system must be practical and practicable to make it adaptable to the solution of many types of
work-study problems. It must be applicable to a wide range of variation in
work content, industrial practices, and workers' skills. Lack of these features will tend to hinder beneficial results it might otherwise attain, and
limit its effectiveness as a major tool of work design.

6. A good system is easy to use. Analysts will become discouraged if their
constructive time becomes occupied by an ever-increasing mass of details.
The specific feature of predetermined time systems most directly bearing
on this problem is the fineness of division of the motions classified. These
divisions must be detailed enough to account adequately for all practical
variations without exceeding the tolerance of the analyst for details. On the
other hand, rather fine subdivisions are proper to a good system, especially
if they are amenable to synthesis into larger categories which do not violate
the limits of workability noted.

7. Actual work situations are dynamic, changing with the times and new
technologies. A good predetermined time system should be maintained in a
dynamic condition to meet the challenge of change. Truly valid data, for the
methods of determination used, will not subsequently be shown to be incorrect. A system should, however, be flexible enough to employ improved
means of data determination and to permit being brought up to date as the
occasion demands.

Though a predetermined time system might be soundly conceived and initiated, lack of development will hinder it from maintaining its acceptable status. Even so enduring a structure as a beautiful palace will fall into disuse and ruin without proper maintenance and updating to such modern devices as electricity, running water, central heating, and the like.

8. A good system has a practical emphasis. It must give correct answers for actual work situations. The best means of assuring this facet for predetermined times concerns the use of the two main data sources. Laboratory data results from controlled conditions; it should properly be used only for isolation of the variables, their nature, their interrelationships, and approximate numerical limits they normally exhibit. For the actual application data, however, reliance must be placed on industrial samples of sufficient scope, range, and size to permit valid conclusions. Industrial data inherently show the behavior of motions under actual operating conditions; such data may be confidently applied to achieve correct answers for practical aims.

9. Another ideal of predetermined time systems is universality. Acceptance must be broad, not confined to limits of special industries, or geographical boundaries. It should perform equally well in diverse operations, foreign countries, and in the hands of anyone properly trained. Regardless of the merits of a predetermined time system, limited acceptance for any reason will vitiate its usefulness in solving the many workstudy problems demanding prompt attention. This generally results from inadequate or unfavorable publicity, lack of professional "selling," or custody of the system by persons inept at using the multiplicity of channels for advantageous public communication. A skilled craftsman may be using a superior type of micrometer, but if nobody believes his measurements, the instrument is of little worth in his work.

The data should be preserved in some definite physical form, such as motion pictures, written material, punched cards, punched tape, film, magnetic tape, magnetic disks, etc., to permit reexamination at a later date; thus, anyone who approaches the system without prejudice can evaluate such evidence.

Many of these ideals of predetermined time systems are difficult to attain fully in human situations. These standards, however, do provide a gage which could be used to measure any given system. It is conceded that no existing system is likely to meet all of these criteria.

The fact that a system does not entirely meet a given ideal implies no reflection on its value. Also, it is important to emphasize that systems other than Methods-Time Measurement should not be judged by the limited information included in the space available here. It is readily conceded that valid standards can be developed using predetermined time systems other than Methods-Time Measurement. While they personally believe the Methods-Time Measurement system to be superior from an over-all viewpoint, the authors admit that is probably does not excel in every single aspect.

Under no circumstances would the authors imply or advise that anyone, or any firm, consider the abandonment of one system for another if they are getting satisfactory results from it—since the better known systems generally yield valid and usable labor standards that provide an excellent medium for analyzing and improving manual work.

MAJOR PREDETERMINED TIME SYSTEMS

Space limitations on this text do not permit describing the more popular of the predetermined time systems in detail, only MTM-1 will be so covered—most of the other MTM systems are covered in detail in *Advanced Work Measurement*.[2]

Many predetermined time systems have been formulated since the debut of the first. Some were and are maintained by private industry for internal usage. Others were semi-public and later gained restricted support. Yet others, of which MTM is a notable example, have become widely known and used because all their data have been made public information. The last groups have proved most hardy, simply because human beings tend to react against restriction on their minds or efforts. Obviously, fairer evaluation of their worth is possible because more is known of them. Significantly, the published material on all systems increased greatly subsequent to the first edition (1957) of this text. The better known systems are mentioned below with their common abbreviations, general chronology, and notable features.

Motion-Time Analysis (MTA)

This is the oldest system, developed in 1919 by Mr. Asa B. Segur whose company started actual client usage in 1925. It was constructed on the foundation of the "principle of predetermined times" which he stated. Features are well-defined motion categories (essentially therbligs) based on physiological principles, variable time bases for possible variations in the motion, distinct rules of usage, and an approach to motion combinations. It has been kept secret by agreement between the originator and his clients. Full details have never been published, but some information exists.[3]

Undoubtedly, work measurement is much indebted to Mr. Segur for many of the basic concepts of predetermined times.

[2]Delmar W. Karger and Walton M. Hancock, *Advanced Work Measurement* (New York: Industrial Press Inc.), 1982.

[3]Asa B. Segur, *Industrial Engineering Handbook* (New York: McGraw-Hill Book Company) 1st ed., 1956, pp. 4-101 to 4-118; Wilbert Steffy, 3rd ed., 1971, pp. 5-4 to 5-6.

General Electric (G & P, MTS, DMT)

Always a leader in time study, this famous company also has several private systems of predetermined times. Each system is distinctive in design.

One dubbed the "Get and Place" system—developed in the late 1930's by Harold Engstrom and associates at the Bridgeport, Connecticut plant.[4]

Another, known as "MTS," for Motion Time Standards, was devised about 1950 at the Schenectady, New York plant. Judging by the published data,[5] it roughly corresponds to the detailed type of Work-Factor data.

A later system known as "Dimensional Motion Times" was formulated by Helmut C. Geppinger about 1953 at Bridgeport, Connecticut. The principle of the system[6] is that the time it takes to perform previously standardized operations on parts with standardized designs bears a relationship to the dimensions of the part.

Except for DMT, these systems have not been publicly circulated; but it is common knowledge that usage of time study, data from other systems, and adaptation of data to General Electric operating policies was made a part of them.

Work-Factor (WOFAC)

The Work-Factor systems[7] cover a wide range of work measurement needs. WOFAC has subsystems, computer programs, training materials, etc. However, the underlying material of the work-factor systems are not publically available in the manner that MTM material is available.

The basic Detailed Work-Factor for general analysis also is the basis for three less complicated systems. Simplified Work-Factor is used for less precise work. Abbreviated Work-Factor was designed for fastest analysis and to cover longer cycles with still less precision. Ready Work-Factor is available mainly for simple analysis by persons other than industrial engineers. To assess mental work elements, a special system called "Mento-Factor" has evolved as well. All of these systems are sufficiently complicated so that their application without the prescribed training is not advisable, nor is it recommended by the company.

Methods-Time Measurement (MTM)

This family of systems is derived from the basic data developed between

[4]Ralph M. Barnes, *Motion and Time Study.* (New York: John Wiley & Sons, Inc.) 2nd ed., 1940, Ch. 22 & 23; 4th ed., 1958, Ch. 28.

[5]*Ibid.,* 4th ed., 1958, Ch. 28.

[6]Helmut C. Geppinger, *Dimensional Motion Times* (New York: John Wiley & Sons, Inc., 1955).

[7]Joseph H. Quick, James H. Duncan, and James A. Malcolm, Jr., *Work-Factor Time Standards.* (New York: McGraw-Hill Book Company, 1962); also the same authors in *Industrial Engineering Handbook,* 3rd ed. (New York: McGraw-Hill Book Company, 1971) pp. 5-65 to 5-101. Some information also is in the *Handbook of Industrial Engineering,* 1982, John Wiley & Sons, New York.

1941 and its public disclosure in 1948, plus continuing public research since that time. The universal forms are the basic MTM-1, MTM-2*, and MTM-3*; but a considerable number of additional MTM subsystems now exist that are aimed at specific problem areas such as the following (those subsystems identified with an asterisk are described in detail in *Advanced Work Measurement*[8]):

MTM-C*: A high-level and fast to use system for setting clerical standards—a specialized system.

MTM-V*: A high-level and fast to use system for setting manual labor standards in machine shop work—a specialized system.

MTM-M*: A very specialized functionally oriented system for setting labor standards on work involving magnification for part or all of the work.

4M DATA MOD II*: A computer-aided system based on MTM-1 that is faster to use and more accurate than MTM-1.

MTM-UAS: A universal analysis system for activities having the characteristics of batch production.

MTM-MEK: An MTM system for activities having the characteristics of one of a kind or small lot production.

All of the preceding MTM systems except MTM-3 have associated computer programs. All such computer programs are designed to help speed up the analysis and calculation; however, the computer languages used range widely depending on the system and where the program was developed. The U.S./Canadian association can provide programs for MTM-1, MTM-2, 4M (really MTM-1 that is as fast to use as manual MTM-3), MTM-C, MTM-M, UAS, and MEK.

An additional high-level system has been developed in Sweden and is being considered for inclusion in the U.S./Canada Association approved list of MTM systems. It is identified by the acronym SAM and, according to the Swedish MTM Association, it is more accurate than MTM-3, but it is faster to use than MTM-3.

It is also possible than an operator selection system (MAST), developed in Great Britain and the Netherlands, will be adopted by the U.S./Canada Association.

These latter systems will be included in a revision of *Advanced Work Measurement* if they are adopted by the U.S./Canada Association.

HISTORY OF MTM DEVELOPMENT

Methods-Time Measurement (MTM) is the only predetermined time system for which the entire data *and* research have been made completely and readily available to the general public. It was developed at Methods Engineering Council[9] of Pittsburgh, Pennsylvania, by Dr. Harold B. Maynard,

[8]D. W. Karger and W. M. Hancock, *op cit.*

[9]This firm later separated into Pittsburgh firms called Maynard Research Council and H. B. Maynard and Company, the latter subsequently becoming a division of Planning Research Corporation of Los Angeles, California.

Mr. Gustave J. Stegemerten, and Mr. John L. Schwab following preliminary study at the Westinghouse Electric Corporation. They merit commen-
dation for their high ethics and the unselfish discharge of their professional
responsibilities in making the results of their work completely available.
These men further aided the development of modern management by
founding, with others, a non-profit technical organization known as the
MTM Association for Standards and Research. They assigned all their data
and development rights to this association, which later transferred them to
the International MTM Directorate.

The above factors are primarily responsible for Methods-Time Measurement being the pre-eminent predetermined time system for work study and
measurement. The authors contend that MTM provides a basis of agreement for all practitioners on a single and universally recognized predetermined time system best meeting the ideals for such a system. This opinion,
shared by a substantial majority of predetermined time system users, will
be easily understood after digesting the material in the preceding and following chapters.

All predetermined time systems are far more complex than a merely casual examination can disclose. If the typical reader of a text or handbook
chapter describing MTM, Work-Factor, or another system attempts to
apply it on that basis alone, it will seldom be done correctly and the result
will very likely be substantially in error. It will become apparent why this is
so after the material on MTM-1 in this text has been read. Finally, another
appropriate reason for including great detail on MTM-1 is the belief of the
authors, and of many others around the world, that it is the best *methods*
analysis system in existence. Certainly a study of MTM-1 should afford a
thorough understanding of methods, work simplification, and human work
capability that would be difficult to obtain more efficiently in any other
manner.

An authoritative account of the origination of MTM by its discoverers, in
the first text[10] published concerning any predetermined time system, was in
itself a unique venture in the field of work measurement. It played a leading
role in bringing MTM to its present state of development.

During 1940, a large group of time study men completed a methods improvement program conducted by Methods Engineering Council, a management consulting firm founded in 1934 and headed by Dr. H. B.
Maynard. The organization's name signifies the fact that its founder first
coined the accepted term "methods engineering" and that the firm was primarily doing methods evaluation and methods improvement work. In analyzing the results of their trainees' work the inventors of MTM became con-
vinced that the substantial cost reductions attained were the result of

---

[10]Harold B. Maynard, Gustave J. Stegemerten, and John L. Schwab, *Methods-Time Measurement.* (New York: McGraw-Hill Book Company, 1948).

methods correction rather than an outgrowth of true methods engineering or methods *design;* and they felt such after-the-fact rectification of methods in place of true *methods design* would never end or lessen without a new approach.

Maynard, Stegemerten, and Schwab therefore searched for a means by which good methods might be established *in advance of production.* They deduced that, if operators learned the best method as they began a new task, the later need and possibility of marked improvements would be reduced. Training costs would also be lower. This would assist managers plagued by production problems, labor difficulties, lack of training guides, and a paucity of usable knowledge in establishing correct methods prior to starting production.

They decided to study common industrial operations and endeavor to develop "Methods Formulas." Their initial choice was to investigate sensitive drill press operations. They intended to extend the same approach in other areas if they were successful in finding such suitable formulas in their first effort, so that there would result eventually a body of information which had been desired by even the earliest of work analysts. The findings of the research on sensitive drill press operations surprised even the researchers, since they were able to expand the results into the first predetermined time system to gain general public acceptance and usage. They found that truly basic motions had been isolated and valid times had been established for these motions after trials on other types of work.

It is a cardinal rule of good research that the data should be preserved and compiled in the most convenient, economical form possible that will permit future reference, validation, and extension of the data. To comply with these rules the originators of MTM decided to film the operations, accurately record and explain all data taken from the film, and validate their conclusions by testing under actual production situations. In addition, it was a traditional approach—the Gilbreths had relied heavily on film as a research medium. The major difference in the analytical approach of Maynard, Stegemerten, and Schwab from that of the Gilbreths was that they were also disciples of the work measurement school which believe that a basic connection existed between time and motion and that they fortunately discovered the key by which such equivalence could be demonstrated and evaluated.

Having decided to analyze the drilling operations from the standpoints of both time and motions used, the researchers next reasoned that only one thing prevented the films from being both a time study and a motion study in reproducible, easy-to-review form. This was the question of the performance level of the operator. Obviously, when a constant speed or regulated camera and projector are employed, each frame of film represents a measurable time lapse dependent directly on the speeds involved. However, this time lapse would not necessarily equal the normal or select time for the motion—or portion of a motion—view in a given frame. Frame counts, therefore, are a means of finding the time, provided the performance rating question can be solved.

Performance rating was discussed in detail earlier in this text, especially the LMS system developed by two of the MTM originators and another associate, Mr. Stewart M. Lowry.

The MTM developers took a practical approach to the stumbling block of standard time measurement—the problem of finding what constitutes a standard of normal performance. First, while actual shop runs of drilling operations were being filmed, several seasoned raters used the LMS system independently for the various parts of the operation and for the operation as a whole. Second, the film was later analyzed for the motion content of the operation in question. Third, the consensus of ratings was applied to the frame counts to yield normal motion times. This same general rating procedure also was used later in the MTM Association's continuing research program as long as it was based on cinefilm recording and analysis before the advent of the MTM Electronic Data Calculator (MTM-EDC). This piece of computerized equipment was developed during the 1958–1961 period. It greatly facilitated MTM research carried out at the University of Michigan by Dr. Walton Hancock and associates. The third edition of this text contains more information on this device. It used punched paper tape as a recording medium. Other specialized measurement systems have been used by researchers, including Doppler radar. The experimental procedures and controls have kept new data compatible with the performance base of the original data.

Here it is helpful to quote a passage from the Application Training Course MTM-1 Manual of the MTM Association (Day I-19) as follows:

> MTM time values are, in general, statistical averages, tested for reliability, and developed from samples of industrial motions of the type they are expected to represent. The samples of motions were collected in a random manner from what is hoped to be a balanced cross section of typical industrial operations. They are expected to apply and, most successfully, are applied to operations which can be considered as belonging to a usual or normal range of industrial operations. They are not exact true times for every motion in any operation. Rather, they are approximations and best estimates, which to a greater or lesser degree reflect the correct performance time of a motion in a given operation.

With this understanding and the knowledge that, except in unusual cases, compensating errors within an operating cycle and from cycle to cycle in a repetitive performance tend to reproduce the average condition predicted, it can safely be stated that the given time values will apply to all but abnormal and unusual occurrences of human motions.

Essentially, then, everyone who applies the MTM data times to motions equivalent to the well-defined categories set up in the MTM system tacitly agrees to the ratings of the original observers as constituting a standard of normal. This means that all persons properly applying MTM are using the same yardstick for performance.

Basic Motion Timestudy (BMT)

The BMT system[11] was developed from 1949 to 1953 by Gerald B. Bailey and Ralph Presgrave with associates in the consulting engineering firm of J. D. Woods & Gordon, Ltd, of Toronto, Canada. The data is frankly acknowledged to be a revision of MTM, adding features of several other predetermined time systems.

Modular Arrangement of Predetermined Time Standards (MODAPTS)

This recent system is an ingenious development by an informal research consortium of 17 individuals from 14 companies convened in 1965 at Sydney, Australia, under the leadership of G. Chris Heyde. Upon public notice of basic MODAPTS in 1966, they evolved into a group called the Australia Association for Predetermined Time Standards and Research. Starting in 1967, this group also by 1969 devised Office MODAPTS. Basic MODAPTS is intended for very fast time determination using minimal analysis and data,[12] sacrificing methods analysis capability. The system essentially was derived from the universal MTM systems and certain MTM-based proprietary systems, together with some experimentation.

Other Systems

There are other systems, but they are either not widely known or historically notable. Among these are Holmes,[13] Olsen,[14] Western Electric,[15] and Mundel–Lazarus.[16]

SUMMARY

It can readily be seen that each of the systems discussed has several features to commend it. No single system can be said to be superior in all re-

[11]Gerald B. Bailey and Ralph Presgrave, *Basic Motion Timestudy* (New York: McGraw-Hill Book Company, 1958).

[12]Gerald N. Griffith, "MODAPTS A Predetermined Time System That Can Be Memorized," *Assembly Engineering,* August, 1970.

[13]Walter G. Holmes, *Applied Time and Motion Study* (New York: Ronald Press Company, 1938, rev. 1945).

[14]Joseph H. Quick, James H. Duncan, and James A. Malcolm, Jr., *Work Factor Time Standards* (New York: McGraw-Hill Book Company, 1962).

[15]Wilbert Steffy, *Industrial Engineering Handbook,* 3rd ed. (New York: McGraw-Hill Book Company, 1971) pp. 5-11 to 5-14.

[16]Marvin E. Mundel, *Motion and Time Study,* 2nd ed. (Englewood Cliffs, N.J.: Prentice-Hall, Inc., 1955).

spects to all others. Rather, the thought here has been that the best of these systems, or a new system as yet unfounded, all contain features that approach the ideals laid down in this chapter. It is fairly certain, as well, that the ultimate system will draw its format and features heavily from the systems known today; such as was the case for BMT, MODAPTS, and indeed the new members of the MTM family of systems.

The variety of these systems, and the number of them, all attest in a positive way that the departure from traditional time study and motion study as discrete entities has brought the industrial engineer closer to the ideal form of work measurement.

Surveys and articles numerous in current periodicals usually conclude that predetermined time systems enjoy a steady growth in their general usage. They have solved many work measurement problems either avoided or inadequately resolved with the more traditional techniques.

The attainment of a truly universal, unanimous system, as alluded to earlier, will not be easy nor soon, because legitimately people *are* diverse in their ways and therefore *prefer* variations. However, if these variations are constructed on a sound base—and MTM, the writers believe, does provide that base better than anything else now in view—then the ultimate system is more possible. Continuing research, as explained later, is bringing it every closer to the ideals previously expressed.

CHAPTER 4

More Introductory Material on MTM

DEFINITION OF METHODS-TIME MEASUREMENT

The official definition of MTM has been unchanged since the originators first embodied it in this form:

> Methods-time measurement is a procedure which analyzes any manual operation or method into the basic motions required to perform it and assigns to each motion a predetermined time standard which is determined by the nature of the motion and the conditions under which it is made.[1]

The wording of this definition is precise and carries a number of meanings deserving explanation.

The hyphenation of Methods-Time is a definite, descriptive part of the system because it constantly reminds MTM users of an important fact. This is the connection between the method of doing a task and the time that task will require. It correctly implies that methods work must *precede* the setting of task times, since the time and method are integrally related and time of itself is meaningless for a task unless it is set in conjunction with a definite, identifiable method.

A method, in this sense, consists of a set or sequence of motions logically performed in a definite order to produce the desired effect on the workpiece or the placement of the body members. The physical behavior imparted to objects will depend on the work done by the human mechanism, and many variables exist on which the characteristics of the work and its duration depend. When these factors have all been properly analyzed and any needed corrective action taken, a *method* has been established. Only then can a reliable *time* for the task be assigned. A time study analyst attempts to do the same thing with stopwatch procedures, but the results cannot be nearly as successful as those possible with a fundamental system such as MTM.

[1]Harold B. Maynard, Gustave J. Stegemerten, and John L. Schwab, *Methods-Time Measurement.* (New York: McGraw-Hill Book Company, 1948).

Methods-Time Measurement is an analytical procedure. The proce-durized analysis must be accomplished if the time required for the motions made is to be established. Definite steps must be followed, as shown later in this text, to make successful usage of the time data. Analysis cannot be cir-cumvented if successful results are to accrue from either the method or the time standpoints. The analysis designed into the MTM system inherently aids its user to arrive at better methods and lower times for a given task. The system is thus internally cohesive, in that each motion itself yields clues to methods improvements, in addition to the improvements indicat-ed by the motions considered as a chain of integral links in the method con-cerned. This feature is almost unique in work measurement systems.

The measurement is restricted to manual portions of the method. The term manual is used in a dual sense in MTM. It refers alike to the hands alone and also to other portions of the body and /or body members.

The major body subdivisions, physiologically, are the trunk or torso and all members attached thereto. The hips, legs, and feet are appendages used most for support, carriage, and transport of the trunk. Their actions are in-cluded under Body Motions. The neck and head contain most of the major sense organs, of which the eyes have the most useful function in work per-formance. Eye motions are a separate category of MTM motions that re-quire analysis. The remaining body members—the shoulders, arm, wrists, hands, and fingers are usually categorized as manual members. The major portion of the MTM system has been associated with these latter manual members of the body since they are the most used body members.

The next concept of the MTM definition is that, following analysis of the manual actions into basic motions, a pre-established time value which does not permit interpretation by the analyst is assigned to each motion. MTM provides a true stand for each basic motion encountered. It will consistent-ly produce reliable time measurements when used by competent analysts. This is the main concept of any predetermined time system and meets the test of clarity and practicality.

The time assigned to each motion does not depend on the operator; it is for the average operator in the universally varying distribution. The times are mean values which will gage the performance of any operator, either the average or the exceptional performer at either extreme of the probability curve. The element of analyst judgment is eliminated in this respect and the established times are therefore objective.

Finally, the time does depend, however, on both the true nature of the motion in question and the conditions of performance imposed on the mo-tion by all other agencies besides the operator involved in the situation. This will become clearer as each motion is exhaustively discussed in sepa-rate chapters.

Although, as noted earlier, the primary official definition of MTM has never changed, the MTM Association in 1958 adopted the following subdefinitions to emphasize and clarify it:[2]

1. *Methods*—The specific method must be established before the time can be determined.
2. *Manual* operation: This means the normal operation performed by people, and excludes operations controlled by equipment and process such as welding time and machining time.
3. *Basic* motion: This refers to the finger, hand, and arm motions; eye motions; and the body, leg and foot motions shown on the data card.
4. *Nature* of the motion: The times for specific basic motions are affected by the degree of control or accuracy involved.
5. *Conditions:* The times for specific basic motions are affected by the conditions of the objects, such as jumbled, small, flat; by the restrictions imposed upon the path of the motion; and by the degree of care or precision that must be exercised.

These statements constitute a concise summary of the preceding discussion in official format.

TECHNICAL ASPECTS OF ORIGINAL RESEARCH

With the basic approach to their "Methods Formulas" determined, the originators of MTM were ready to accumulate the needed data on film and then to analyze the film for the motion-time information being sought. At first they tried the conventional micromotion technique of element subdivision using therbligs (Gilbreth elements). The work sequences in the drilling operations were lists in therbligs along with the leveled times (actual times adjusted by the concepts defined earlier) from the films. The trial plots of this data results in a meaningless scatter, showing the need for further element subdivision to permit curve drawing. After checking the actual shop conditions to assure that the data itself was correct, they begin to redefine the elements—thus originating the MTM elements.[3]

The new definitions were in terms of basic motions with readily identified starting and ending points. These motions were classified as to type and another plotting trial showed highly meaningful data has been achieved. Other operations studied in this manner yielded further data which correlated with the original studies. Instead of specialized "Methods Formulas", the originators discovered they now had a predetermined time system documented by many feet of motion picture film with performance ratings coded to the film and record sheets of all the pertinent data: The principal technical factors connected with this work are next discussed.

[2]By permission from *Journal of Methods-Time Measurement,* Vol. 4, No. 4, MTM Association for Standards and Research, Ann Arbor, Michigan, Nov.-Dec. 1958.

[3]The reader should realize that this account of the original MTM research is greatly condensed. A complete chronology can be found in the speech "Dividends from Research" by Prof. D. W. Karger which is part of the *Proceedings* of the 1963 International MTM Conference at New York City, available from the MTM Association, Fair Lawn, New Jersey.

Time Units

The time represented by one motion picture frame naturally depended on the speed at which the film was exposed. Constant speed equipment was used to assure uniform time increments for each frame. By applying the performance rating in percentage terms, it was then possible to find the average time consumed by an average operator in the frame in question.

However, it was of practical necessity to assign time values in units which had the dual advantages of easy usage and yield numerical results that could readily be used to supply the input to cost systems. Most industries rely on decimal minutes or decimal hours for their measurement and subsequent costing of labor. Time units such as these, however, would have been difficult to use because of the fact that many of the basic motions were very short in elapsed time of performance and many zeros would have been needed between the decimal point and the first significant digit.

This is clearly illustrated by the film speed times expressed in commonly used time units. Since the film speed used in the original research was 16 frames per second, each frame covered an unrated elapsed time of 0.0625 second, 0.0010417 minute, 0.00001737 hour.

The obvious way to avoid such unwieldy time units was to recognize that units are arbitrary by nature—the necessity for conversion to other desired units being the only real limitation. Maynard, Stegemerten, and Schwab invented a new time unit known as the Time Measurement Unit and assigned 0.00001 hour as the value of one Time Measurement Unit. Since most wages are in dollars per hour, the TMU can be multiplied by the hourly rate and decimal point then shifted five places to the left to find the cost of labor directly. Also the hours required to produce 100 motions (or pieces) can be found by shifting the decimal two places to the left.

As a result of the unit chosen, the following time conversions will be valid:

1 TMU	equals	0.00001	hour	1 Hour	equals	100,000 TMU
	equals	0.0006	minute	1 Minute	equals	1,667 TMU
	equals	0.036	second	1 Second	equals	27.8 TMU

Under these conditions, one unrated film frame at the research speed was equal to 1.737 TMU. No motion time less than this value could be set from the film. This proved to be the principal limit on the precision of the original data. When other film speeds, such as the faster speeds used in later MTM research sponsored by the MTM Association, are used to get a finer time determination, it is merely necessary to find the new per frame time in TMU to easily fit the new data into the present MTM system. Precision then becomes a matter of film speeds and control of the filming mechanisms; on the other hand, accuracy is inherent in the film frames themselves. The main chance for error is in the performance rating factor applied; but even here consistency is possible if the researchers are experienced in rating techniques and can agree with other raters about the operation on the film. Recently, MTM research sophistication has been increased by the usage of such devices as electronic data collectors in conjunction with extremely sensitive transducers, punched card equipment, and computers.

CONCLUSIONS OF THE ORIGINAL RESEARCH

With the aid of a shorthand coding system for the motions, described fully in the motion chapters to follow, the research results were summarized and condensed into a handy-sized data card. The original data card may be seen in the text by Maynard, Stegemerten, and Schwab. Most of their results, however, form an integral portion of the newer data card information presented in each motion chapter in the present text. An official current data card is enclosed in the pocket at the rear of this text.

Their conclusions also included specific rules for measurement, usage, and appropriate cautions vital to application of the data on the card. Such rules arose because the originators fully recognized that the application of the data would necessarily occur without the wealth of background experience they had gained and habituated from their research work. In other words, they took pains to build practicality into their system so that the benefits of research would not be lost due to poor presentation and indecision during usage.

Verification of the Data

A variety of statistical, mathematical, and graphical techniques were applied to the data and plots to assure that no gross error had occurred during the analysis of films or the subsequent efforts described. It is a tribute to the pains and care taken by the developers that research later conducted by many other people only in rare instances revealed differences greater than one film frame at the film speed of the original research. Such accuracy was due to consistent, professional application of scientific research principles, not to fortuitous chance.

VALIDATION

The research was verified by Maynard, Stegemerten, and Schwab. Validation, however, was possible only by an independent, unbiased source. This quickly followed publication of the original MTM textbook by the inventors. Such an independent investigation conducted at Cornell University was reported[4] for the Management Division of the American Society of Mechanical Engineers.

The Cornell report carefully stated that the information developed from the study was pertinent only to the MTM system, and that the selection of MTM for study should in no way be interpreted as invalidating any other system. It was clearly explained that other systems differ fundamentally in two respects. First, the classification of motions into which elements are subdivided differs both in content and separation points. Second, the performance level on which motion times are based differs between systems. Consequently, correct comparison of one system with another is difficult, if not impossible.

Regarding the second reason given above, traditional industrial engineering always employed two basic concepts of average daily performance. These are the high task and low task standards. The accepted meaning of low task is the maintenance of such a working pace as could be sustained throughout the day under daywork payment conditions. High task, on the contrary, implies that the pace of work, sustained throughout the day would be that usually found under incentive payment conditions. The official standard[5] defines these other related terms as follows:

High Task—Performance of an average experienced operator working at an efficient pace, over an eight-hour day under incentive conditions, without undue or cumulative fatigue. Often stated as a percentage above normal performance. (See normal effort, low task.) Syn.: incentive pace.

Low Task—A term used to indicate that performance rating or production standards are based on daywork levels as contrasted to high task or incentive work performance. Sometimes taken to mean a level of performance below the level expected under measured daywork conditions.

Measured Daywork—
1. Work performed for a set hourly nonincentive wage where performance is compared to established production standards (most frequent use).
2. An incentive plan wherein the hourly wage is adjusted up or down and is guaranteed for a fixed future period (usually a quarter) according to the average performance in the prior period. (Infrequently used.)

Normal Effort—The effort required in manual work to produce normal performance. (See Normal Performance.)

Normal Performance—
1. The work output of a qualified employee which is considered acceptable in relation to standards and/or pay levels, which result from agreement, with or without measurement, by management or between management and the workers or their representatives.
2. An acceptable amount of work produced by a qualified employee following a prescribed method under standard conditions with an effort that does not incur cumulative fatigue from day to day. (See Fair Day's Work.)

[4]Kendall C. White, "Predetermined Elemental Motion Times," *ASME Paper No. 50-A-88*, New York, November 1950.
[5]From ANSI Z94.1-12 IE Term. (Footnote 3, Chap. 1).

Fair Day's Work—The amount of work that is expected daily from an employ-ee. May be established solely by management or through mutual agreement. May or may not be established through the use of various measurement tech-niques. (See normal performance.) Syn: expected attainment.

The MTM data system, for example, is based on a low task standard. The performance rating system used in its generation relies on a concept of nor-mal pace that matches daywork conditions. However, other predetermined time systems employ varying or different normal pace concepts. This is a further reason why direct contrast and comparison of predetermined time systems is technically difficult.

The method of study and analysis used at Cornell paralleled that of MTM's founders. This involved detailed analysis of motion pictures of ac-tual industrial operations. The Cornell films were taken in a variety of in-dustries represented by eight companies on operations ranging from small riveted assemblies up to complex machine handling. Only incentive opera-tions were included. With a minimum of disturbance at the workplace, a small electric camera was utilized at 1,000 frames per minute. Each frame of film thus represented 0.001 minute of elapsed time. In most cases at least two analysts used the LMS rating system to level for skill, effort, and condi-tions.

Cornell's film analysis in terms of MTM elements are summarized below. Their recorded times were leveled to average operator performance. While only a limited amount of data was collected for each element, we quote the conclusions of the report as follows:

> The data collected so far appears to point toward the practicability of defin-ing work elements in terms of fundamental elements of motion common to a wide range of industrial activity and the establishment of standards of time for these elements which are reproducible within smaller limits than those normal to current time study practice. In summary for all elements, the check studies have consistently ranged within approximately plus or minus one percent of the MTM times. For individual elements, time differences between check times and MTM times in the order of 0.1 to 0.5 TMU's (0.00006 to 0.0003 minutes) are not uncommon and, as evidenced in the data in this paper, a few are larger. While some of these time differences do represent five percent to ten percent differences in time for specific elements and in isolated cases as much as twenty percent, they are not considered serious in application since any standard is made up of a composite of a number of elements, the majority of which are well within plus or minus five percent. No reason is now apparent why continued study should not result in reconciling the differences which do exist. In some in-stances further data may alter the check time as pointed out before; in some, modification in the MTM time may be desirable; and in some instances, it is entirely possible that new elements or sub-classes of existing elements may be necessary.

Of equal importance to the subject of validation, but less often recognized, is the fact that the independent research sponsored by the MTM Association at The University of Michigan from 1953 to 1973 effectively reproduced almost all of the original MTM data. While the specific research projects were selected by the sponsor, the university had full control of the projects and published the results in research reports written by its staff. This research caused only minor additions and changes to the MTM studied at Cornell. The fundamentals and most of the system have remained unchanged. In 1973, the MTM Association embarked upon a revised policy of having research groups at outstanding colleges and universities quote on specified projects as well as submitting unsolicited proposals. The first awards under this policy went to researchers at Georgia Institute of Technology.

The mere fact that many of the shortcomings of time study have been overcome in Methods-Time Measurement alone justifies the use of this or a similar system. In addition, such a system has other corollary benefits as follows:

1. The importance of methods is emphasized
2. Methods and equipment can be designed on a scientific basis
3. Time standards will be consistent from job to job
4. Standards can be set in advance of production
5. The ability to set standards in advance of production makes possible more accurate costing of products and the development of more accurate production planning
6. The time required to derive standard data is greatly reduced
7. Methods are precisely and accurately described
8. Grievance cases can be settled on a basis of fact rather than on opinions.

Once the reader begins an active study of Methods-Time Measurement in later chapters of this text, it will become quite evident that the use of the system emphasizes the importance of method. In fact the system cannot be applied without considering various methods and precisely defining the one finally to be used.

Methods-Time Measurement has provided industrial and management engineers with a much needed basic tool. Like any tool, it can be refined and improved through additional research and development. This is being done with MTM through a research program dating from the time of its creation.

MTM ORGANIZATIONS

To meet the problems noted earlier, the originators and their early MTM colleagues sought a whole new approach to implementing the important development of MTM that would extend its benefits to industry while avoiding misuse of the system which might rekindle the earlier "efficiency expert" problems. Something new, organizationally, was essential. Accordingly, a group of 11 practicing consultants met at New York City in January 1951, with Dr. Lillian M. Gilbreth and Prof. Harry J. Loberg of Cornell University, to discuss the condition and to find ways of meeting it. Out of this cooperative approach evolved, in March 1951, the present *nonprofit* MTM Association for Standards and Research representing the U.S.A. and Canada. This body grew into the Cooperating National Associations and their incorporating, in six short years, as the International MTM Directorate. These organizations all will be described generally, and the U.S.A./Canada in more detail.

MTM Association for Standards and Research (MTMA)

The best description of the founding of the MTM Association was given by its first president, Dr. Harold B. Maynard, in the keynote address of the First Annual MTM Conference at New York City, in October 1952, as follows:

Experience has shown that when an MTM practitioner gains confidence in his ability to apply the procedure, he very often develops the urge to improve upon it. When he encounters an unusual motion sequence for the first time, he is quite likely to want to invent one or more new motion classifications and assign tentative time standards to them. If he then checks with a stop watch and finds that they seem to work out all right —which they often do if they are a very small part of the total cycle time—he concludes that he has added to the development of the MTM procedure, and he uses his new motion classifications thereafter without further checking.

This is, of course, a highly unscientific way of developing the MTM procedure. Unless some kind of control is introduced, it can only result in producing as many different varieties of MTM as there are practicing engineers. Furthermore in most cases, the addition of new motion classifications is completely unnecessary. Experience has already shown that the solution to most application problems lies in learning how to apply existing motion classifications and standards properly, rather than in inventing something new.

The matter of control is not difficult to handle, however, once the need for it is recognized. In our own organization in the early days of MTM, we discovered quite unexpectedly one day during a discussion of the application problems which our staff had encountered, that each man had added new motion classifications (all different) or had developed new interpretations of existing motion classifications (also all different) until we were well on our way to having a completely unstandardized procedure.

Obviously this would never do. We, therefore, developed carefully written descriptions of standard motion-classification interpretations and application practices, and we insisted that they be followed to the letter. We admitted that every word in these descriptions might not be correct at first, and we encouraged our engineers to challenge anything which they believed to be wrong. But each point challenged was checked by further research. If it was found to be wrong it was changed, but not until then was the engineer allowed to deviate from the former standard practice. When a new standard practice was approved, all engineers were immediately required to adopt it.

Under this kind of self-imposed discipline, we quickly corrected any initial errors in our application practices and achieved consistency among our own staff. But while this solved our own problem, it did not solve the problems which were rapidly being generated by a growing number of practitioners who were not under our control. We wanted the procedure to be used, for we had a belief in the important contribution which it could make to industry, but we did not want it abused. Too many variations could cause inconsistencies, arguments, weakened faith in the procedure, and all manner of unpleasant consequences. What to do?

The problem was discussed in January of 1951 by a small group of management consultants who were either using MTM professionally at that time or were planning to use it. They recognized the need of bringing both the development and the application of the MTM procedure under control and considered various means of accomplishing this. The result of these discussions was the formation of the MTM Association for Standards and Research whose first Annual Conference we are attending today.

The aims and purposes of the MTM Association stated in its By-Laws are as follows:

> To widen acceptance for the proper use of MTM.
> To conduct basic and applied research in the field of methods-time measurement. To encourage and stimulate members and other organizations—both commercial and academic—and individuals to conduct similar research and to coordinate the work done by them.
> To establish standards and by all possible means to sustain the high quality of work done by all organizations and individuals in their use of MTM.
> To compile all available information pertaining to the development and application of methods-time measurement and to provide members with this information at frequent intervals.

These clearly expressed, dynamic goals have thus far been successful as is attested by the record.

Cooperating National Associations (CNA)

Internationally, the MTM system has enjoyed ever widening acceptance. As noted earlier, its spread to other countries occurred rather rapidly. In a few cases, North American multinational companies introduced it within their overseas operations. However, the chief agency for foreign growth was the management counsel membership of the U.S.A./Canada MTM Association, which actively sponsored provisional groups in an affiliate status prior to the formation of the International MTM Directorate (IMD). Associations achieving CNA status in that period were Svenska MTM-föreningen in Sweden, Schweizerische MTM Vereinigung in Switzerland, l'Association MTM Française in France, and Nederlands MTM Genootschap in Holland.

Once several firms or organizations in a country became involved in

MTM training and application, it was natural to organize a national association; this trend continues today under the aegis of the IMD assisted by the older CNA members. Since the five associations previously named formed the IMD, the number of Cooperating National Associations has more than doubled to the 12 in the following list:

l'Association MTM Française	Paris, France
Deutsche MTM Vereinigung E.V.	Hamburg, West Germany
Japan MTM Association	Tokyo, Japan
Menetelmatekninen Yhdistys MTY	Helsinki, Finland
MTM Association Belge, Belgische Vereniging	Ghent, Belgium
MTM Association for Standards and Research	Fair Lawn, New Jersey, U.S.A.
MTM Association Limited	Warrington, Lancashire, England
MTM Raad	The Hague, Holland
Norges Rasjonaliseringsforbund A/L	Oslo, Norway
Schweizerische MTM-Vereinigung	Zurich, Switzerland
S R F—MTM—Gruppen AB	Stockholm, Sweden
The Israel MTM Group	Tel-Aviv, Israel

Expansion of this list is anticipated by groups in Czechoslovakia and Rumania. Other countries showing some interest are Bulgaria, Poland, and Yugoslavia.

A form of membership in the U.S.A./Canada MTM Association is held by groups and individuals in Brazil, China (Taiwan), Columbia, India, Mexico, Peru, and Republic of South Africa. Similar CNA contact points are Australia, Austria, Brazil, China (Taiwan), Czechoslovakia, Denmark, East Germany, Hong Kong, India, Italy, Mexico, New Zealand, Republic of South Africa, Rumania, Spain, U.S.S.R., and Yugoslavia. All these offer further potential for new CNA groups. All such international activity attests to the faith of the founders of the MTM Association in the universal growth of MTM usage and results.

International MTM Directorate (IMD)

With the primary mission of coordinating the worldwide MTM activity of the Cooperating National Associations, the International MTM Directorate (IMD) was formed on June 25, 1957, in Paris, France, by five CNA founders, as previously mentioned. It was granted *non-profit* corporation status by Ohio, U.S.A., in November 1968, and received federal tax exemption in April 1970. Its objectives were formally adopted in London during May 1969, and read as follows:

> The general objective of the IMD and the national MTM Associations is to develop, spread, and employ knowledge concerning man at work so as to improve

his productivity, his job satisfaction, and his working conditions.

More specifically, the objective of the IMD are to encourage close cooperation among all those interested in the study of man at work, whether it be in research, training, or field of application of MTM and MTM-based systems.

The International MTM Directorate serves as a coordinating body for research activities of the National Associations, for publication of international standards of practice, for membership development, and for communication and exchange of information.

The major areas of activity of the IMD to reach the general objective are:

1. Research . . . . .
2. Standards of Practice . . . . .
3. Membership Development . . . . .
4. Communications . . . . .

With its own officers elected from CNA delegates, the IMD has distinct powers and activities in addition to providing guidance and influence within CNA groups. The first president was Dr. Harold B. Maynard of America, an MTM originator and also the first head of the U.S.A./Canada association. Served by a Secretariat based in Stockholm, Sweden, is the Managing Board of Trustees with executive powers, who meet annually, and the General Assembly holding legislative powers, which meets every three years in conjunction with an International MTM Conference and the CIOS Conference to define policy and ratify interim actions of the Managing Board. Besides the President and Secretary, the Managing Board includes the Treasurer, an Auditor, four General Trustees, two Regional Coordinating Trustees for Europe and the Americas, and five Functional Coordinating Trustees responsible for Research, Application, Training and Qualifications, Human Engineering, and Information Services. The latter five also chair working committees that transact their affairs by periodic meetings and by mail.

Besides its operating policies as they affect CNA affairs and its own publications, the IMD provides direct MTM resources due to its exercise of powers derived from the U.S.A./Canada association. The *Journal of Methods-Time Measurement* was internationalized in 1973 after 18 years of publishing by the original association. While retaining copyright power for the MTM-1 data card, MTM development and revision were delegated to the IMD by the U.S.A./Canada group in 1966. This action stimulated both research and application of the MTM system on an international level, with the IMD sponsoring several new MTM systems[6] and undergirding the research efforts of CNA members. Since then, the U.S.A./Canada association has mainly carried onward with the basic research effort while

---

[6]The basic, original (detailed) MTM system was renamed "MTM-1" by the IMD when later systems were adopted to simplify reference and show priority.

the IMD and other CNA members largely have concentrated on application research and the development of the new MTM systems. As a result of this revised procedure, the approach to MTM data revision and other changes in practice now consists of the following steps:

1. Any CNA decides to undertake programs or projects in response to its membership interests and demands.
2. The proper Coordinating Trustee of the IMD is notified and kept informed to permit international coordination and mutual help between CNA groups with common interests on a continuing basis.
3. Literature searches are made at both the national and international levels. They are subject to publication if deemed appropriate, but always furnish background and guidance to the group carrying on the work.
4. The national group performs laboratory studies and publishes the results. Review permits input by interested parties worldwide.
5. Industrial studies are conducted by the national group and, upon their publication, this may generate additional industrial studies internationally. Validation studies are a mutual CNA responsibility.
6. The form and substance of practical application is decided on the international level after recommendation by the CNA accomplishing the program or project. Upon satisfactory review and reconciliation of the new data or procedures, they are promulgated by official IMD sanctions. Thereafter, CNA groups must conform of the requirements of changes and new developments in publications and practices.

It can be seen, therefore, that the IMD is a practical force in MTM practice worldwide, thus comprising an essential resource for the MTM user.

Membership

The different types of association membership are as follows:

A. *Honorary*—This highest form of membership recognizes persons who have made outstanding contributions to MTM and work measurement. To date, those so honored were the late Dr. Harold B. Maynard and the late Dr. Lillian M. Gilbreth, both of whom were active in association affairs. Honorary membership has also been granted to Jack S. Schwab and Gustave J. Stegemerten.

B. *Individual*—These members are of two classes, as follows:

    1. *Academic*—Such members are persons on academic staffs of colleges and universities whose work relates to industrial engineering or the management sciences.

    2. *Associate*—This form of membership induces individuals who have acquired their "blue card" certification as having taken an approved MTM-1 course and passed the associated examination. They belong, at nominal cost, to obtain all data and literature

pertaining to MTM. They also may participate in the activities and committees of the association. This support mutually benefits the association and the individual associate member.

3. *Instructor*—Any individual qualified by the MTM Association as a licensed instructor.

C. *Organizational*—Two numeral classes are used to group these members:

Class I—This class includes all organizations other than management counsel firms with an interest in MTM and a willingness to participate in the advancement of the association's ongoing programs. Because these organizations most directly represent the application interests of the total membership, their involvement is highly valued in promoting the usage and integrity of the MTM techniques and systems.

Class II—Professional consultant firms providing management counsel that includes MTM practice and services, such as training and installation of MTM systems.

The specific and definitive requirements for membership that emphasize ethical, moral, technical, professional, experience, and performance standards are available from the executive office. Furthermore, the interests of these members extend into all phases and areas of the association.

OFFICES

Organizationally, the U.S.A./Canada association consists of unpaid officers and committee members whose time and expenses toward meeting the needs and goals of the association are generously donated by either the individuals or their employers. Officers are chosen by majority vote of the association membership. The only paid employees are the Executive Director and his staff. There are some regional offices, but the headquarters is now located in Fair Lawn, New Jersey.[7]

COMMITTEES

All of the national MTM associations and the international directorate function through numerous committees cooperating with their executive officers to provide the many membership services and achieve the goals set for MTM development and improvement.

---

[7]The main office address now is: 16-01 Broadway, Fair Lawn, New Jersey 07410. Telephone (201)791-7720.

PROGRAM MATERIALS AND ASSISTANCE

A vast array of published material, services, and other resource material is available from the association. These include training courses for all MTM systems (also available from Class II Professional Consultant Members), training manuals, training materials, data cards, analysis sheets, computer programs, books, training courses, programmed instruction courses, research reports, miscellaneous reports and publications, conference proceedings, and the *Journal of Methods-Time Measurement* (now also the international journal of the IMD). The association also has examinations for testing competency of course graduates and the instructors, who must receive a special certification as an instructor for the course to be approved. The examinations are graded by the association. Certificates and/or certification cards are issued to those who qualify.

RESEARCH

MTM research has shifted in emphasis, locale, and personnel over the years. It started at the Methods Engineering Council in Pittsburgh and then was moved by the Association to the university area, principally the University of Michigan where most of the later fundamental research was either accomplished or else directed by the university research activity. Currently, most of the research is carried out by committees and consortiums of interested companies. This research is primarily of an application or validation nature, rather than basic or theoretically oriented research.

The Research and Project Development Committee, made up of volunteer members, has recently completed and introduced a new Standard Data Course and a course on Electronic Test Data. The committee is currently working on Cost of Work Measurement, guideline for use of various measurement techniques in the white-collar and service areas, a Standard Data Level II course, Gainsharing Programs, and the enhancement of 4M to incorporate features such as time formula and route sheet capability. No primary or basic university research is being sponsored at present, although possibilities are under consideration.

AN MTM SYSTEMS OVERVIEW

For the first two decades of its history, Methods-Time Measurement was understood as only *one system*. In the broad sense, this still is true. Yet the dynamism of the MTM system fostered no only continual improvements in the data and procedures of work study, but also the birth of new MTM-based *systems* to meet the demands of diversity in practical application within economic constraints. One result is that the expanded and refined original system is now called MTM-1 rather than basic or detailed MTM. With MTM-1 as the premier member of a *family of systems* available to attack and solve the diverse range of work measurement problems, today's analyst can attain greatly increased flexibility and effectiveness. Since this trend is likely to continue, this section aims to give orientation and perspective on the MTM family of systems.

However, space in this text will not permit coverage of more than the complete essentials of MTM-1 upon which the other family members are founded. Both the latest advanced topics within MTM-1 and the detailed explanation of other MTM data levels will be deferred to a later text planned for their coverage. Here the attempt is to provide an overview of each existing MTM system and their integrated aspects. To offer more in a balanced presentation of all work measurement would overemphasize the predetermined time techniques, despite the fact that they gradually are replacing many of the historically important approaches. This text provides all the information needed, if augmented with the necessary practical training courses available, to achieve competence in MTM-1. Advanced MTM-1 students and those wishing to extend their analysis capability with the rest of the MTM family will be best served by using this as a first text and then continuing with the future volume.

This chapter is a prelude to Part II, which fully covers the primary MTM-1 data, to keep that section unencumbered by concept and data advancements. This arrangement orients the reader with a familiarization process. It also provides those inclined toward wider study with interest and impetus to thorough learning of MTM-1 as a sound preparation.

SYSTEM PERSPECTIVE

Today's proliferation of "system" and "systems" as applied to many technical subjects and devices can lead to confusion if not carefully grounded. The proper meaning may be gained from Webster's Unabridged Dictionary, which contains these "system" definitions:

"1. A set or arrangement of things so related or connected as to form a unity or organic whole; . . . . .
4. A set of facts, principles, rules, etc., classified or arranged in a regular, orderly form as so to show a logical plan linking the various parts."

In ordinary language, a system is any coordinated collection of things around which an arbitrary boundary is drawn to deal with them as a unified subject. Such a boundary can include a large complete subject or can be used to encompass a sub-system thereof, called more simply a "system" for convenience. Therefore, a number of "systems" belonging to a larger whole system can be described as a "family of systems" when the larger border is meant. This term is especially applicable when the "family of systems" consists of a basic system and all the other systems which have been derived from it; however, the only really basic data remains that of the *original* generic system. To meet widely varying conditions and requirements, a number of versions of MTM-based data were developed into a "family of systems" fitting this concept precisely.

In the inclusive sense, the *MTM system* is a work study and work design technology emphasizing methods description in terms of predetermined basic motion and time data or their derivatives; but, specifically, it is the MTM family of systems. The *MTM family* of systems is a set of *generic* (micro level) standard data systems for evaluating human performance, augmented by sets of *functional* (macro level) standard data systems based on the generic systems but tailored to meet broad application demands of selected activities. Users of the MTM family may extend their application by deriving *specific* standard data systems for MTM-based assessment of particular demands found in selected products, services, organizations, or industries.

Before proceeding, it is well to elaborate those summarized concepts to promote the proper system perspective.

The Measurement Spectrum

A complete measurement system should cover the entire range of cycle time, job frequency, accuracy, method, economics, and any other pertinent requirements encountered in the diverse places and situations justifying labor and cost control. This historical goal has not been met fully by any single approach as yet, but rather has been approximated by adapting the existing techniques and data sets piecemeal, to cover certain portions of this wide spectrum which blends continuously from one extreme to the other. Figure 1 shows one typical scheme[8] of this nature (only an example, not implying author agreement), where various forms of work measurement can be selected based on the cycle time and annual production rate. Recently, however, determined efforts have been made to achieve the basic aim of full coverage by enlarging and expanding the various predetermined time systems, with each "system" designed to fill a certain portion of the spectrum in a fashion compatible within the continuum—the "integrated" approach.

The integrated approach means that during analysis of a job or operation, the motion and time information used directly at different points in the same analysis may be drawn freely from any of the family members of the over-all system without losing full compatibility. This assures flexibility within economical analysis constraints because there will be no need to

---

[8]Adapted from: John C. Martin, "Standards and Wage Systems Audits," *Industrial Engineering Handbook,* 3rd ed. (New York: McGraw-Hill Book Company, 1971), p. 6-161.

have more accuracy than can be controlled, and the methods variation will naturally match the analysis choice within desired or permissible monetary limits. Also, it means that the system parameters will be met with a harmo-

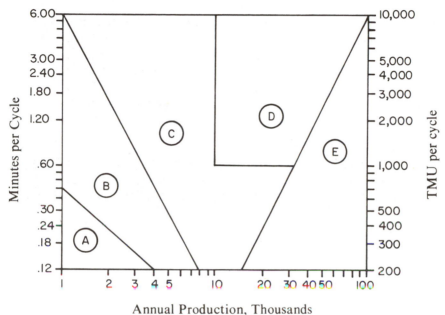

Annual Production, Thousands

Area A = Empirical estimates based on experience
Area B = Rough time checks and job comparisons
Area C = Formal time study and time formulas
Area D = Condensed versions of predetermined times
Area E = Detailed predetermined times

EXAMPLE CHOICES OF TECHNIQUE TO BE APPLIED:

Cycle Time	Annual Production, Thousands								Annual Prod.	Tech. Sel.	Cycle Time, Minutes
	2.5	5.0	7.5	10	25	50	75	100			
TMU: 250	A	B	C	C	E	E	E	E	2,000	A	.12–.23
800	B	C	C	C	C	E	E	E		B	.23–1.59
2,000	C	C	C	D	D	E	E	E		C	1.59–6.00
Min.: .25	B	B	C	C	E	E	E	E	20,000	E	.12–.22
1.00	B	C	C	D	D	E	E	E		C	.22–.60
3.00	C	C	C	D	D	D	E	E		D	.60–6.00

Fig. 1.   Typical scheme to select work measurement technique for establishing operational standards.

nious relationship between the segments of the analysis even if these segments involve switching back and forth between the basic and "higher" level systems.

This kind of approach imposes definite restrictions and trade-offs in the design of the various measurement system levels derived from the basic system. A few of such direct and inverse relationships that affect data system design and selection are depicted[9] in a general way by Fig. 2. Here it is obvious that developing higher level systems involves much more than merely "splicing" or "building" various sizes of elements with the basic data, for the application parameters behave in a far from simple way. As will be seen later in this chapter, much effort and ingenuity are necessary to attain an integrated family of systems anchored to a basic data system.

Data Types and Levels

Before proceeding to describe the members of the MTM family of systems, it is necessary to discuss several important data design terms. These terms are ways of classifying the construction and application of motion and time data as used for operational standards.

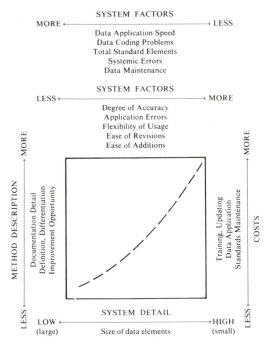

Fig. 2.   General data system design and selection relationships.

[9]Developed from: Roger L. Yard, Sr., "New Methods of Developing Standard Data," *Journal of Methods-Time Measurement,* Vol. XI, No. 3, July/Aug 1966.

All standard data consists of elements put together in various ways to suit certain purposes. For human performance data, the points of interest are the motion description or element content/coverage together with the attendant time value for the element. The size of the elements relative to the amount of time they include is commonly spoken of as being *microscopic* or *macroscopic,* usually abbreviated to micro and macro, to denote whether the smallest subdivisions are very short or relatively longer. As to content, elements are aimed at describing basic motions, motion sequences, tasks, operations, etc. Time and motion descriptors of elements collectively are called the *orientation* of the data. So far, the elemental level is meant.

When the data elements are assembled for application, the mechanism used is called their *construction* and their resulting scope is classified as to *universality.* The two major construction procedures yield *vertical* or *horizontal* data. Vertical data includes elements and element combinations developed for one kind or class of work, or at most a very limited range of work, and therefore fairly specific areas of application. Horizontal data elements are tailored to cover many kinds and classes of work, which means their applicability is quite general at the sacrifice of exact coverage for any given single kind or class of work. The scope of a data set in varying from highly universal to very specialized or restricted, is often termed as being *generic, functional,* or *specific.* Generic data applies almost anywhere due to its generality. Specific data usually applies to rather narrow, or even a single, area of interest or variety of work. Functional data applicability falls between these extremes. These three scopes of data are important enough to justify further elaboration.

A *generic* system is oriented to human behavior, and therefore has maximum universality, with elements recognizable as distinct human actions by the work performer. Such elements can be single basic motions, sequences of serial or averaged motions, or motion aggregates keyed to work actions. All of the motions within the elements are drawn from the basic system, which is itself generic in full detail. Element names are general in that they describe universal human efforts, but do not inherently reveal the nature of the work activity being analyzed. Examples: Reach, Grasp, Walk, Get, Put, Handle, Transport.

A *functional* system is oriented to work actions on and by the parts and tools involved, thus being restricted to similar parts and tools which may or may not be found universally and/or used in the identical way, with elements recognizable as having distinct results produced by the handler of the parts and tools. Such elements can achieve single actions, sequences of serial or averaged actions, or action aggregates keyed to work results. The elements are derived from motions in the basic system and/or elements in one or more of its generic derivatives. Element names are essentially specific in that they use an action verb and an object noun to tie together the operations and tools used, and thus inherently reveal the nature of the work activity being analyzed. Examples: hammer nail, clamp blocks, staple cards, dial telephone.

A *specific* system is oriented to coverage of work goals for particular work tasks, thus lacking universality because all pertinent work conditions must be met for proper application, with elements recognizable as the attainment of the distinct working steps involved. Element derivation may be from basic system motions and/or elements from derived generic and/or functional systems. The size of such elements can vary with the magnitude of the goal, and the element names uniquely identify the narrow or broad sphere of application. Examples: make casting, prepare Form R, machine cam lob, overhaul engine, balance ledger.

In the MTM context, the formal systems are either generic or functional since the users will construct their own specific data from these two kinds of published systems. The existing *generic* systems include MTM 1, MTM-2, and MTM-3.

Finally, one can speak of various data *levels*. A level represents either of two ways of classifying a complete data package. One way is with respect of the magnitude of the number associated with the package. For example, in MTM, MTM-3 is a higher level than MTM-2 and MTM-2 is a higher level than MTM-1. The other way refers to the *generation* history of the data. Being the first developed, MTM-1 would be called first generation as to level. Second generation, being next developed and based on MTM-1, would be MTM-2. MTM-3 is third generation level as regards its order of development. The number thus affixed to the generation being described therefore indicates a level.

Several sources[10] are available for the reader interested in further information and other interpretations of the ideas in this section. Since new data system generation is the dynamic frontier of work measurement, it behooves the aware practitioner to become conversant with the multiplicity of successful innovations appearing in the last decade and ahead.

Motion Aggregates

Regardless of their type and other characteristics, the higher level systems ultimately rest upon the basic motions in the underlying MTM-1 sys-

[10]Daniel O. Clark, "Meet the MTM Family," *Journal of Methods-Time Measurement,* Vol. XVII, No. 3, 1972.

Karl E. Eady, "What is the MTM Family Really Like?," *The MTM Journal,* Vol. I, No. 2, 1974.

William J. Mattern, Kenneth Knott, and Russell W. McDonald. "MTM-GPD and Other Second Generation Data," *Industrial Engineering Handbook,* 3rd ed. (New York: McGraw-Hill Book Company, 1971), pp. 5-102 to 2-121.

Harold B. Maynard and William K. Hodson. "Standard Data Concepts," *Industrial Engineering Handbook* (see last entry) pp. 3-121 to 3-189.

tem. Therefore, to devise larger elements for later generation systems, it is necessary to use a process that can be called "aggregation." With time study, this procedure was severely restricted due to the inability of changing time breakpoints flexibly; but predetermined motions permit complete freedom in selecting breakpoints, so that data developments have mushroomed since their advent. These newer data systems are generated with motion aggregates—new compilations of any desirable portions of the total motions in a work sequence, task, operation, etc., to fit design parameters.

When motion aggregates are developed in various ways, they are thereby defined to a separate existence in terms of method and time information that has been abstracted and amalgamated from the basic source. They can be used for still higher level data generation in the same way as they were themselves derived from the initial data base of the first level system. In addition, they can be applied directly to determine operational standards in the accuracy and analysis speed range for which they were designed.

The general effect of a higher level system as compared to the basic system is to gain speed of application at some acceptable sacrifice of accuracy and loss of method detail, because the motion aggregates are larger units of motion and time data having less flexibility insofar as exactly tailoring the operational standard to the actual occurrence of a work event. Depending on the monetary, scheduling, and other control demands on the standards produced with the aggregated system, these drawbacks may be minimal; indeed, the final operational standards might still be more accurate and descriptive than actually needed to optimize the production, although generally they are less able to meet such criteria than would the basic system if its costs of analysis could be justified.

Many mechanisms or procedures have been utilized to derive motion aggregates. One of the more successful approaches has been called the "building block concept." The concept is that, at any given data level or when changing to higher or lower levels, the data will be compatible between systems if the motion aggregates are:

1. Behaviorally based; that is, defined to include physiological wholes which utilize natural motion breakpoints in the basic system.
2. Traceable back to the motions in the basic system and do no violate any of the definitional rules of that system.
3. In accord with the same requirements for elemental independence as in the basic system, for elemental additivity rest upon this foundation.
4. Unique as to content and description, so that classification of operational occurrences will be possible without the equivocation or undue analysis time; because analysis decisions using the aggregates must be quick, sure, and free of misinterpretation.
5. Coded without ambiguities so that both adaptation to automatic data processing and efficient auditing are possible.

Other sources may express these ideas differently, but the true essentials are as given above.

When using the "building block concept" or any other viable procedure to develop motion aggregates for higher level data systems, a number of distinct strategies are available. Not only must the developer understand these, but the user of the aggregated system can also profit and improve its application by knowing them. Basically, four ways can be used to put together elements aggregated from basic motions or lower level elements:

1. *Sequential.* Using natural breakpoints, the serial motions or elements can be combined with a new name. For example, handling almost any object commonly involves reaching to it, getting hold of it, bringing it to a different location approximately or precisely, and letting go of it. This is the highest frequency of motion occurrence. Whole sequences like this can be easily derived for different ranges of the variables present and sizes of objects involved, then used thereafter as aggregated elements.

2. *Representative.* Motion sequences to do the same job are sometimes highly variable on each occasion with justification. Getting a fair aggregate then means that, after sufficient study of the many possible sequences, a representative sequence can be chosen as an aggregate for all of the occurrences.

3. *Averaged.* When a sequence of motions is desirable for any aggregate and it varies only in key dimensions, an averaging approach can be used. For example, motions can be otherwise identical but require differing time values due to variable distances, weights, etc. If all the points of interest in the range can be timed with the basic data, then an average for the midpoint, mean, mode, etc., of the range can be found to serve as an aggregate value.

4. *Proportional.* This approach is best when a series of motions to be aggregated has, perhaps, one or more of the motions variable as to case, type, kind, other characteristic. If observation or analysis can establish a frequency for each of the variations in the motion sequence, then a realistic usage of proportioning or percentage factoring can be devised to find the aggregate condition.

All of these approaches have been used in developing higher level MTM systems in the MTM family. The important point to note is the really basic contribution the identification of these approaches has made to the standard data practices of industrial engineering, for it is predictable that they will foster other data systems in the future. Especially with electronic computers easily available for simulation and handling of large amounts of data economically, the shackles on analysis and further motion aggregation have melted away!

One result of the use of computers it to extend the use of statistical techniques in the development of standard data. Not only is the approach economical, but it also makes possible the achievement of much greater accuracy combined with the ability to use the computer to do much of the user's searching for applicable standard data elements. A new course, "Principles of Standard Data Construction," developed by the U.S./Canada Association addresses this issue.

MTM ADVANTAGES AND PRINCIPAL USAGES

From its original form, conceived and developed by Maynard, Stegemerten, and Schwab, the basic MTM-1 system has developed and expanded to meet today's work measurement problems through basic research, applied research, and practical refinement in the crucible of worldwide application. It enjoys preeminence among the many current predetermined time systems because it most nearly meets the ideals previously stated, besides having the following inherent advantages over time study:

1. It eliminates the leveling or rating of operator performance.
2. It forces the applicator to concentrate on method rather than the time element of the job.
3. It requires and permits a more exact description of the method used.
4. It permits methods evaluation *prior* to production or trial runs.
5. Its application yields more *consistent* work standards.
6. It reduces the need for a stopwatch to process- or machine-controlled timing and determining delay elements.
7. It shifts grievance resolution from individual performance or rating discussion to factual questions of work content and method.
8. It helps make industrial engineering more professional by providing thoroughly researched and validated data regarding the time needed for an average operator working with average skill and effort to perform well-defined motions.

MTM-1 not only provides superior time data but also is a most efficient methods evaluation tool. It is the *basis for all other MTM family systems,* as well as for many independent standard data systems. This is in addition to formal statements in MTM literature that its principal usages are to:

1. Establish production or incentive time standards
2. Develop many varieties of standard data
3. Promote the rational improvement of existing work methods
4. Provide a basis for developing effective methods in advance of production
5. Estimate the labor content of redesigned or new parts and products

6. Allow the development of superior tool designs based on recognition and usage of "the best method"
7. Aid in equipment selection and facility design by permitting evaluation of the human work content prior to approval, purchase, and use
8. Guide product design in properly considering pertinent human factors
9. Aid in more objective settling of time standard grievances
10. Assist in training supervisors to be methods conscious and to help the technical support staff in designing better working methods
11. Isolate keys to better operator training based on realistic methods data
12. Facilitate research in related areas such as basic bio-mechanics, personnel practices, handicapped worker capabilities, etc.

With this prologue already covering all the more important areas of work measurement, it is indeed remarkable to note the ever-accelerating MTM developments far beyond traditional areas which follow.

METRICATION AND TRANSLATION

The original MTM data and subsequent U.S.A./Canada MTM-1 data all were in customary or English units and the English language. These became and remain the international standard.

Almost immediately after publication, the first technical system development was metric conversion[11] and translation into Swedish. Since then, the increasing number of national members of the IMD have each translated the data into their own language, most of them using the Swedish metric data. The United Kingdom later "went metric" and the Canadian Commonwealth is in process of metrication. The United States is "dragging its heels" with regard to metrication. The attempts to date have been to try to run a dual system, which will not work. As of this writing, we and possibly a minor country or two, are the only nations that have not metricated. Our industries are more and more required to metricate just to compete in the international marketplaces and in order to utilize equipment and components made abroad. Sometimes our politicians will eventually get up enough nerve to do what should have been done a long time ago.

In the case of MTM, there has been a interesting reverse twist with regard to metrication. MTM-2 and MTM-3 systems were developed in the metric system and required conversion[12] and translation into the English units and language. It is not presently known into how many other languages MTM has been translated, but the official U.S.A./Canada data is expected to remain the IMD standard. The mere existence of this technical item attests the MTM-1 dynamic nature!

[11]Franklin H. Bayha, Shlomo Erez, Walton M. Hancock, and Knut Lund. "A Comparison of the Swedish and U.S.-Canadian Data Cards," *Journal of Methods-Time Measurement,* Vol. XV, Not. 4, 1970.

[12]Karl E. Eady, "Translation of the Official MTM-2 and MTM-3 Data Cards into the English System," *Journal of Methods-Time Measurement,* Vol. XVII, No. 2, 1972.

MOTION EQUATIONS

Since the second edition of this text, electronic technology has made readily available to anyone, worldwide, at reasonable prices, the miracle handheld digital electronic calculator with print and direct readout, and portable and desk personal computers with as much memory and more speed as some of the earlier mainframe and minicomputers. Computers are now available to almost all professionals regardless of where they are located.

As a result, it is very probable that the more astute MTM users and researchers will want to have algebraic equations for the MTM motion data;

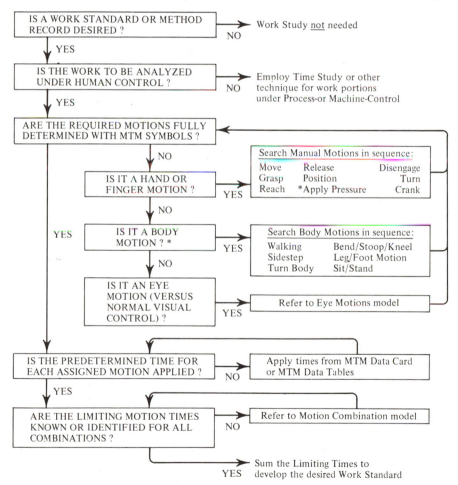

*Apply Pressure can be either a Manual or Body Motion.

Fig. 3. Basic Decision Model for MTM-1 System application.

MODEL	ANALYSIS QUESTION	ANSWER
Basic System	Is a work standard or method record desired?	Yes
	Is the work to be analyzed under human control?	Yes
	Are the required motions fully determined with MTM symbols?	No
	Is it a hand or finger motion?	Yes
	RESULT: Search Manual Motions in sequence (here skip Move and Grasp, going to Reach models)	
Reach* (Sheet 1)	Is the purpose to relocate the hand or finger(s) to a destination?	Yes
	Is there nothing or only a small and/or light object in the hand?	Yes
	Is the destination at an object to be obtained?	Yes
	Is high manual control needed at or near the destination?	No
	Is the destination within 3″ of the body or other hand?	Yes
	Is there a sharp change in direction of 90 degrees or more within a 6″ radius during the motion?	No
	RESULT: Reach Case A Analysis	R__A
Reach* (Sheet 2)	Are the manual members stopped at *both* the start and the finish?	No
	Are the manual members in motion *only* at the start?	Yes
	RESULT: Reach Type II in motion at start	mR $\times$ A
Net* Distance	Is any kind of assistance present to modify the total index knuckle travel?	No
	RESULT: x = L, say 10″, measured knuckle travel	mR10A
Basic System	Are the required motions fully determined with MTM symbols?	Yes
	Is the predetermined time for each assigned motion applied?	No
	RESULT: Find time for mR10A from MTM-1 Data Card	7.3 TMU
Basic System	Is the predetermined time for each assigned motion applied?	Yes
	Are the limiting motion times known or identified for all combinations?	Yes
	RESULT: Analysis is complete, motion is mR10A = 7.3 TMU	

*These models may be found in Chapter 5.

Fig. 4. Example of MTM-1 Decision Model usage.

since these can be used in computing routines and also to develop standard data time values where more than normal precision is justified. If everyone developed such equations on his own, data problems could result for them and when correlating with other practitioners. Therefore, everywhere that they could be developed with justification, the authors have included the equations in this text as a norm for these purposes. It must be distinctly understood that these equations *are not* approved data of the MTM Association for Standards and Research. They *are* based only on approved MTM data and observe the proper statistical cautions, but their authorship is here clearly identified in the event any question of responsibility arises. Whether they are ever used directly, in many cases they help the reader to understand the MTM time variation and thus serve as good instructional material in their own right.

Decision Diagrams

The MTM Decision Diagrams are flow-chart models embodying the system logic in easily accessible format, making the analysis reasoning very clear. Analysis questions on the left-hand side can be answered in binary fashion to lead to further analysis directions or analysis results on the right-hand side of each model. This makes them a good learning device permitting more rapid conceptualizing by the student, thus enabling direct analysis without them after a certain amount of practice and observation. After that, they remain a good reference for analysis problems because of their orderly attention to all factors affecting an analysis. In Part II, where all the diagrams appear, it is good procedure for the reader to refer to them as the text is being mastered. Then, for final review at the end of each chapter, another study of the diagrams it contains is suggested.

The model which ties together the whole MTM-1 system appears as Fig. 3. As are all models, it is used by entering the top left analysis block, answering the question appearing there, and proceeding by the arrows until the desired result has been attained. To afford practice with this top diagram, Fig. 4 has been devised to use it together with several of the models appearing in Chapter 5. The reader is urged to study this example rather closely. Although this artificial example may seem long to arrive at the rather simple result, tracing the logic path through most of the decision trees often yields the answer very quickly. Also, as the text is mastered and more of the diagrams are studied, one soon recognizes their effective usage and thereby minimizes the time needed to use them.

PART II

Detailed MTM-1 Data

Reach

Definition[1]

REACH[2] *is the basic hand or finger motion employed when the* **predominant purpose** *is to move the hand or fingers to a destination.*

1. *Reach is performed only by the fingers or hand. Moving the foot to a trip lever would* **not** *be classified as a reach.*
2. *The hand may be carrying an object and the motion still will be classified as a reach provided the predominant purpose is only to move the hand or fingers and not the object. An example would be the "reach" for an eraser while the performer is still holding chalk in the same hand.*
3. *Short reaches can be performed by moving only the fingers; longer reaches involve motion of the hand, forearm, and upper arm.*

The predominant purpose of the motion is the deciding factor in classifying it as a Reach. Holding or palming a light, small item such as a screwdriver, pliers, or scissors while moving the hand does not justify classifying the motion as a Move rather than a Reach. For example, in the needle trades, a scissors is often held in the palm of the hand while other opera-

---

[1] From "MTM Basic Specifications" prepared by the 1957 Training Committee of the MTM Association. The main definition and the numbered sub-definitions are considered integral parts of the complete specification for each motion. These data are derived from Vol. V, No. 4 of the *Journal of Methods-Time Measurement* for November-December 1958 and are used by permission of the MTM Association for Standards and Research. To save repetition of this footnote in subsequent chapters of this text, reference is made to this source and this footnote by the following notation: From "MTM Basic Specification" (Footnote 1, Chapter 5).

[2] The reader will probably recognize that this is basically the same motion as the Gilbreth Basic Element or Therblig called Transport Empty. The similarity of the other MTM basic elements to Therbligs is also interesting to one familiar with that motion study system.

tions such as reaching from the cloth on the working area in front of the operator to a bin containing pins are performed. The motion to the bin should be classified as a Reach if the predominant purpose is to get the hand in a position to pick up some pins for the next operation.

The analysis and investigation conducted as part of the development of the Methods-Time Measurement procedure, and subsequent research performed under the auspices of the MTM Association for Standards and Research, has definitely shown that the major influencing factor on the time to make the motion is the predominant purpose and not the fact that a small object is carried in the hand.

It is likewise obvious from a practical viewpoint that the palming of a large, heavy, or clumsy object would have an influence on the motion and would, therefore, change the classification to a Move (see next chapter).

This is just one example of the type of distinction that must be made on a practical basis when Methods-Time Measurement is applied. Definite rules to substitute for trained judgment cannot be set down for every specific case that may be encountered. The more experience the well-trained analyst gains, the more likely he is to make correct decisions on the classification of the various movements required in the performance of the work. At best, he must rely on such guiding principles as those in this text.

KINDS OF REACHES

There are five cases or classifications of Reach separately identified in the MTM procedure. They are identified by letters of the alphabet as follows: A, B, C, D, and E.

Case "A" Reach

Case A Reach may be defined as follows:

Reach for an object in a fixed location or for an object in the other hand or on which the other hand rests.

An example of reaching for an object in a fixed location is that of an operator of a punch press reaching for the trip buttons. The example infers that "fixed location" means that the object must be bolted down or be located in exactly the same place each and every time. However, from the required practical viewpoint of the MTM analyst, "fixed" refers to the mental certainty of the object's location and not that it is stationary or bolted down (although such cases could also fulfill the condition).

Practice will help "fix" an object's location in an operator's mind. This is explained by recognizing the function of "kinesthetic sense" which is the *apparent* ability of the muscles to feel the near presence of an object without the visual aid of the operator's eyes. Actually, this "kinesthetic sense" is a capacity of the mind which allows it to almost automatically direct the muscles to return to a location fixed in memory by frequent previous experiences with relative orientations of an object and the body. A popular name for this ability which seems to describe what occurs is "muscle memory."

From a practical viewpoint, all operators who can qualify for factory work can also learn to make Case A Reaches in the time allotted by the MTM procedure. This, of course, will require sufficient practice to reduce the visual or mental control in performing the Reach from the conscious to the automatic level. Because of their greater skill, however, better operators will tend to make reaches to single objects which may vary slightly from cycle to cycle in the time allotted for a Case A Reach. The second type of motion just mentioned is normally identified as a Case B Reach.

The above facts could mean that, on highly repetitive work, a classification decision might sometimes be made which seems to conflict with the definition. For example, in a plant having workers who are generally above average in skill due to selection techniques and where the work is highly repetitive, it may be desirable to classify some motions as Case A, which by rote definition may be strictly Case B, if practical observation shows them obviously to be Case A as proved by a time check. In such event, the Case A time is the correct one to use.

One should not make such decisions contrary to the defined limits unless sufficient data is available to justify such procedure. Perhaps clarity on this point is best achieved by mentioning that the two types of occasion when a motion must be classified are: (1) when the motion is being visualized and (2) when a motion has been observed. Remember then that the classification decision hinges on the fact that the standard method prescription should be for the average operator working with average skill and average effort under the average *prevailing* conditions.

In the instance cited above, the skill and repetition constantly prevailing *would* justify the specification of a Case A Reach.

If the other hand is already holding or is resting on the object being reached for, the latter portion of the Case A definition begins to apply. "Kinesthetic sense" goes into action and, as far as the operator's mind is concerned, he has a mental certainty of the object's location. This mental certainty, however, does not apply in this manner unless the object is being held by the other hand within three inches of the grasping point. Actually, a Reach ending anywhere within three inches of the body (e.g., cigarette in lips) would justify the Case A analysis for Reach. This is an important point to remember, since the data developed to date indicates that three inches is the practical limit for this " kinesthetic sense" to operate properly. Of course, exceptions to the 3-inch rule can be found (e.g. a baseball bat in the hand of a professional player) where the grasp point can be further than three inches from the other hand, yet the Reach will be made consistently.

Case "B" Reach

This case of Reach may be defined as follows:

Reach for a single object in a location that may vary slightly from cycle to cycle.

The Case B Reach is the one most commonly encountered in industry. These Reaches are also usually easy to identify.

Examples of such Reaches are most numerous and some typical illustrations are:

1. Reaching for a paint mask lying anywhere on a stand, prior to painting the part
2. Reaching for a pair of pliers, resting someplace on the work table, which are used for a portion of each work cycle
3. Reaching for a lone pencil loose in one's pocket.

There are two meanings for the word "single" in the definition. As the examples show, only one object is the target. But "single" also means that this object is by itself, isolated or clear enough from other objects so that no barrier to obtaining it is present. In other words, the surface supporting the target object is its only contact or surrounding and therefore it may be easily grasped and/or picked up.

The variance in location normally is understood to be one to two inches; however, the size of the object and other factors need to be taken into consideration from a practical point of view. A variance of two inches for a small part might be of considerably more importance than a variance of three or four inches for a large part. This is because of the need for greater visual and muscular control to direct the hand to the Reach destination.

Mention was made in the previous section that unusual operators often convert timewise a Case B Reach into a Case A Reach if they have had sufficient practice. The analyst should be careful not to classify Case B Reaches as Case A Reaches unless very special consideration indicates that such action is the correct approach. It certainly is not intended to be the usual approach.

Conversely, as discussed under the definition of Case A Reach in the previous section, a Reach to a part on which the operator's other hand is resting is a Case B Reach unless the point of grasp is within three inches of the resting hand.

Case "C" Reach

The Case C Reach is defined as follows:

Reach for an object in a group of objects which are jumbled so that search and select occur.

It should definitely be understood that it makes no difference whether the objects jumbled together are all of the same kind or whether they are different. The main thing is that they must be jumbled into a pile so that search and select occur. Reaching for a single object would rarely be classified as a Case C Reach, by its very definition.

The time for the Case C Reach is longer per distance moved than for any other case of Reach except D, which consumes exactly the same time per inch of Reach. One of the reasons affecting the length of time is the necessity for the eyes and the mind to select which object is to be grasped out of the pile, before grasping action can occur. That is, in addition to the required muscular control, both visual and mental control are necessary. This mind and eye action slows down the rate of hand travel and extends the time before Grasp action begins, which is actually at the end of the Reach time.

Another illustration of the practical decisions which must be made by the analyst is where an operator reaches for a rivet in a pan full of rivets. It may be that the operator usually grasps several with a normal Grasp and then allows all but one to sift through his fingers before the Move motion takes the hand out of the area of the totepan. Practical research has shown that, in many cases, the Reach is not slowed down to the time shown for a Case C Reach but that the actual time is closer to the value for a Case B Reach. This indicates that a Case C Reach may not always apply to all jumbled piles.

The problem of stock depletions will puzzle some analysts. By this it is meant that, as stock is depleted from a tote pan, the operator is reaching toward only a few parts with Reaches logically classified as Case B Reaches. The effect of this can for all practical purposes be ignored; one should always classify the motions for the normal occurrence. If the variations are of enough importance, separate consideration must be given to them.

Case "D" Reach

A Case D Reach is defined as follows:

Reach for a very small object or where an accurate Grasp is required.

An excellent example of a Case D Reach would be reaching for a little piece of broken glass. The glass, being both small and having sharp edges, requires an accurate Grasp to prevent injury to the fingers.

When a very small, hard-to-grasp object is reached to, it is obvious that the time of Reach per unit of distance is extended due to the necessity of controlling the motion accurately, particularly the latter portion of the Reach. The same effect is encountered where a very accurate grasp is required in order to prevent damage to fragile or polished parts.

Because of the above effects, the time for a Case D Reach was found to be greater than for a Case A or B Reach. In fact, it was found to be identical with the time for a Case C Reach. Remember that the MTM time values were established by research and should be used whether we can fully explain each one or not.

If the operator is reaching to a very small part that is also jumbled with many small parts, only the method description is affected by the decision to classify the Reach as either C or D. The analyst should make this decision to best fit the most important characteristic of the operation. However, specification of an R—C directly to one part would never be incorrect where small parts are jumbled together.

In any event, if the above type of problem is encountered, do not allow two Reach times, since one identical value suffices.

Case "E" Reach

Case E Reach is defined as follows:

Reach to an indefinite location to get the hand in position for body balance, for the next motion, or out of the way.

Examples of Case E Reaches are numerous. If a heavy object is lifted with one hand, the other hand and arm are often extended out from the body in order to provide body balance. The arm extension is a Case E Reach. Carrying a chunk of ice at your side with an ice tong is almost always an occasion for extending your other arm in an outward direction to provide body balance.

After an operator feeding a punch press has loaded the material into the die, his next motion, following release of the material, is obviously a Case E Reach—getting his hand and arm out of the way of the ram— unless the Reach were to the activating device (normally Case A). Moving the hand out of a dangerous area is a Case E Reach which could be dubbed the "safety Reach." Such cases could be limiting; that is, subsequent motions could not or should not be performed until the "safety Reach" has been completed. Safety directors often make this motion a required and an automatic one by installing "pull guards" on the press equipment.

Another example of a Case E Reach—that of getting the hand into position for the next motion—would be where a hand is moved in the general direction of the next required motion. This usually occurs while the other hand or some other body member is performing a limiting motion (the one consuming the greater amount of time). After the limiting motion is completed, the hand making the Case E Reach then continues on, usually with some other case of Reach. The practical effect of this is to reduce effectively the chargeable distance of the latter Reach.

It is often true that a Case E Reach is made with one hand while the other hand is performing a limiting operation. In other words, the other hand actually consumes the larger amount of time, and therefore, the Case E Reach need be considered only for methods description. Frequently, this type of Case E Reach is not shown in the motion pattern, as a time-saving convenience.

A final kind of Case E Reach is one where the fingers only are moved off or out of the way of an object. In this case, the distance reached is measured by the travel of the fingertip(s). Normal MTM Case 1 Release motions (see Chap. 11) actually are very short Case E Reaches by the fingers. However, if the fingertip travel is an inch or more, the Reach should be recorded as a Case E Reach of the proper distance and the time allowed, unless it is limited by other motions. Sequences of such reaches frequently occur when dropping a handful of small parts over a tote pan.

General

The analyst may find the following word association will assist in identifying the five cases of Reach:

> Case A Reach—Automatic
> Case B Reach—By Itself
> Case C Reach—Crowded and Choose
> Case D Reach—Diminutive or Dangerous
> Case E Reach—Ease or Extricate

TYPE OF MOTION

In making start-stop motions, such as normal Reaches or Moves, MTM research has revealed that the body member or members go through a period of acceleration to a maximum velocity, that this velocity is maintained through a portion of the motion, and that the body member or members are finally decelerated to a stop. This is quite logical by analogy to the action of an automobile between starting and stopping points. However, as some critics of MTM point out, the time (Time = Distance ÷

Velocity) to perform a given motion *can be* influenced by the motion preceding or following it. That this is not a valid criticism of MTM is easily seen by (1) the motion classification descriptions and also by (2) the provision in MTM for consideration of the Type of Motion as given in this section.

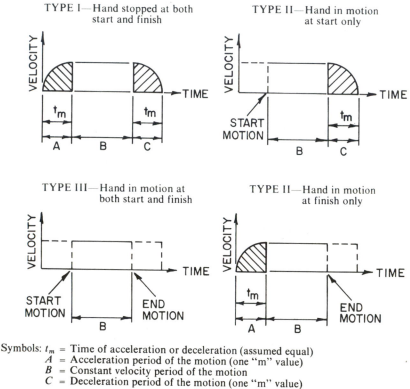

TYPE I—Hand stopped at both start and finish

TYPE II—Hand in motion at start only

TYPE III—Hand in motion at both start and finish

TYPE II—Hand in motion at finish only

Symbols: t_m = Time of acceleration or deceleration (assumed equal)
A = Acceleration period of the motion (one "m" value)
B = Constant velocity period of the motion
C = Deceleration period of the motion (one "m" value)

Fig. 1.   Type of Motion for hand movements.

The three Type of Motion categories are graphically depicted in Fig. 1 and described as follows:

I. Type I Motion is the condition where the hand is moving at neither the beginning nor at the end of the element. This is the normal start-stop motion and is by far the most common. In the figure, it is idealized to show a smooth acceleration to peak velocity, a period of travel at this constant velocity, and a smooth deceleration to a stop. While actual "ballistic" motions may depart considerably from this picture, the main purpose in the simplification is to give the concept a form and to show the basic assumption that the acceleration and deceleration periods are practically equally divided for most conditions. Either of these is considered an "m" value in the MTM procedure.

II. Type II Motion is identified as having the hand in motion at either the beginning or at the end of the work element. It is obvious that by eliminating either a period of acceleration or a period of deceleration from the Type I Motion, the distance traveled by the hand will be greater for a given amount of time. Conversely, for a given amount of distance to be traveled, the time required will be less for a Type II Motion than for a Type I Motion because either an acceleration or a deceleration period has been eliminated.

III. Type III Motion is where the hand is in motion at both the beginning and the end of the work element. The velocity of hand travel is for all practical purposes constant, so the curve relating velocity and time is a straight line. Thus, the time is minimal, even less than for a Type II Motion. Type III Motion is rarely found in practice.

It should also be apparent that for Case C or Case D Reaches, the hand cannot be in motion at the end of the element. This is because during a Case C Reach, search and select occurs, and for a Case D Reach a very accurate Grasp is required at the end. For these reasons, Case C and Case D Reaches cannot be Type III or Type II in motion at the end; these cases can only be Type I or Type II in motion at the start.

MTM research has evaluated the effect of acceleration and deceleration on the various cases of Reach. This information is shown on the data card.

The effect of eliminating one period of acceleration or deceleration is presented in the data card information for Case A and Case B Reaches. In order to establish the time values for Type II Motions for other than Case A Reaches and Case B Reaches, and for all Type III Motions, a set of formulae are given later in this chapter. Basically, those formulas depend on subtraction to find the time required for one acceleration period—or its equivalent, one deceleration period. Table 1 shows these

Table 1. Value of "m" in TMU for Case A and Case B Reaches

CASE OF REACH	REACH DISTANCE, INCHES																				
	f	1	2	3	4	5	6	7	8	9	10	12	14	16	18	20	22	24	26	28	30*
	Value of "m" in TMU																				
A	.4	.2	.5	.8	1.2	1.2	1.3	1.3	1.4	1.4	1.4	1.5	1.6	1.7	1.8	1.8	1.9	2.0	2.1	2.2	2.2
B	.4	.2	1.3	1.7	2.1	2.8	2.9	2.8	2.9	2.9	2.9	2.8	2.9	2.9	2.8	2.8	2.8	2.7	2.7	2.7	2.6

* This value applies for all distances exceeding 30 inches.

values of "m" for both Case A and Case B Reaches.   It also shows the range through which this value varies with distance.

MTM research showed that the "m" value for the Case B Reach is applicable for all other cases except the Case A Reach.   This latter Reach is the fastest possible, and it also has shorter acceleration and/or deceleration periods than are required for a Case B Reach.

When developing times for Type II and Type III Reaches exceeding 30 inches, it is assumed that the "m" value listed in Table 1 for 30 inches will govern beyond that distance.   This is because the time curve has become essentially a straight line beyond that distance, with a uniform slope for extrapolating values, in the plot of Reach times.

An example of Type I Motion would be where a punch press operator has just finished placing the part in the die and reaches toward the side to the trip button.   The hand is stationary at both the beginning and the end of the Reach.   (See Fig. 2.)

A typical example of a Type II Motion would be where drop delivery into a totepan is used and the operator keeps his hand in motion while reaching for a part in another totepan to prepare for the next assembly cycle.   (See Fig. 3.)

An example of a Type III Motion would be where an assembly operator employs drop delivery into a totepan, then reaches his hand to a part lying on a smooth workbench surface and employs what is known as a Contact Grasp when his hand touches the part, then slides the part along the workbench in the same continuous arc in which the hand was moving at the moment of contact.   For this Type of Motion, there would be no period of acceleration from the moment the part was dropped nor any deceleration up to the point where the part was contacted with the hand or fingers.   Obviously, the part contacted in this example cannot have such a size, weight, or shape nor any other feature that would injure the hand; for all of these would affect the velocity at the point of contact.   The motion after contact normally is classified as a Move.   (See Fig. 4.)

CHANGE IN DIRECTION

If a very sharp change in direction occurs during a Reach motion, it will affect the time for any Reach.   The change must be abrupt enough to necessitate an appreciable amount of muscular control.   A change in direction will affect the time for the Reach if a turn of approximately 90 degrees or more is made in a radius of no more than six inches.

An example of this type of situation would be where an assembly operator reaches in over the edge of an assembly frame preparatory to doing work at the bottom corner nearest the operator.   This is illustrated in Fig. 5 .

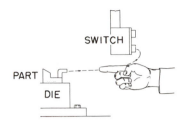

Fig. 2. Type I Reach.

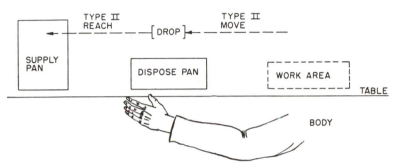

Fig. 3. Type II Reach.

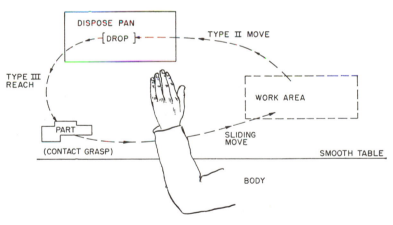

Fig. 4. Type III Reach.

The Case A Reach is the only one slowed enough by a change in direction to consider the time effect as a separate item in the application of the MTM data. This is accounted for by assigning such Case A Reaches the time value for a Case B Reach of the same length. All other cases of Reach are affected by other factors to such an extent that the change in direction effect can be neglected.

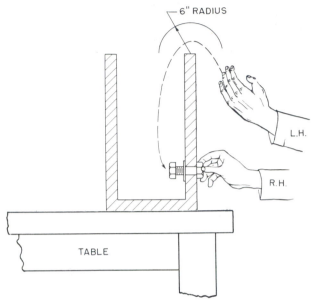

Fig. 5. Change of direction. If the nut were within 3 inches of the right hand, the left hand would be making an R—Acd.

DATA CARD INFORMATION

As previously indicated, the MTM analyst utilizes a data card containing all of the required information. The MTM-1 data for Reach is reproduced in Table 2, which shows the Reach time data (Table I on the card) in the form currently approved by the MTM Association.

The time values for Reach Case A in motion and Reach Case B in motion at distances of 3 inches or under, for Reach Case C at ¾ inch or less, and for Reach Case B in motion at distances of 20 inches and greater were obtained by extrapolation from existing research data. Because of this, they were shown in light-faced type on the MTM-1 data card prior to 1972, when this practice was abandoned. The Hand in Motion values for Case A and Case B Reaches can be used directly for Type II Motions, as the "m" values are already subtracted from the basic time.

A research report on *Short Reaches and Moves*[3] is available for help where needed. It includes data which was fully validated and then incorporated on the revised MTM-1 card in March, 1955. The information given in this text agrees with that data. It represents closer evaluation of the shorter Reaches by running film at four times the speed used in developing the original data. This research revised the values for Reach

[3]Available from the U.S./Canada MTM Association.

Table 2. Time Data for Reach
(MTM—I Data Card Table I—Reach—R)

Distance Moved Inches	Time TMU				Hand In Motion		CASE AND DESCRIPTION
	A	B	C or D	E	A	B	
3/4 or less	2.0	2.0	2.0	2.0	1.6	1.6	A Reach to object in fixed location, or to object in other hand or on which other hand rests.
1	2.5	2.5	3.6	2.4	2.3	2.3	
2	4.0	4.0	5.9	3.8	3.5	2.7	
3	5.3	5.3	7.3	5.3	4.5	3.6	B Reach to single object in location which may vary slightly from cycle to cycle.
4	6.1	6.4	8.4	6.8	4.9	4.3	
5	6.5	7.8	9.4	7.4	5.3	5.0	
6	7.0	8.6	10.1	8.0	5.7	5.7	
7	7.4	9.3	10.8	8.7	6.1	6.5	
8	7.9	10.1	11.5	9.3	6.5	7.2	C Reach to object jumbled with other objects in a group so that search and select occur.
9	8.3	10.8	12.2	9.9	6.9	7.9	
10	8.7	11.5	12.9	10.5	7.3	8.6	
12	9.6	12.9	14.2	11.8	8.1	10.1	
14	10.5	14.4	15.6	13.0	8.9	11.5	D Reach to a very small object or where accurate grasp is required.
16	11.4	15.8	17.0	14.2	9.7	12.9	
18	12.3	17.2	18.4	15.5	10.5	14.4	
20	13.1	18.6	19.8	16.7	11.3	15.8	
22	14.0	20.1	21.2	18.0	12.1	17.3	E Reach to indefinite location to get hand in position for body balance or next motion or out of way.
24	14.9	21.5	22.5	19.2	12.9	18.8	
26	15.8	22.9	23.9	20.4	13.7	20.2	
28	16.7	24.4	25.3	21.7	14.5	21.7	
30	17.5	25.8	26.7	22.9	15.3	23.2	
Additional	0.4	0.7	0.7	0.6			TMU per inch over 30 inches

Case A at 3 inches or less, Case B at 4 inches or less, Case E at 1 inch, Case A in motion at 3 inches or less, and Case B in motion at 2 inches or less. It is also the original source for the values at ¾ inches or less.

Note that the times are 2.0 TMU for all Reaches ¾ inches or less. In the research on Short Reaches and Moves, it was discovered that this is apparently the minimum performance time for any human motion. It is apparently due to the physiological limits of human reaction time. This time minimum also seems to be independent of the degree of control involved in the motion in question. Isolation of this interesting fact is another original contribution by MTM research to basic motion data.

The data card, plus the information previously given, plus practical working rules which follow in the later portions of this chapter will enable one to establish time values for the various kinds of Reaches encountered in the home, office, and factory.

THE SHORTHAND OF REACH

In order for the analyst to write down efficiently the various motions required, he must utilize a type of shorthand. This shorthand for the element Reach is outlined in the following material.

Reach

Each time the work element Reach is encountered, the engineer will indicate on his study sheet the capital letter R denoting Reach.

Distance of Reach

The distance in inches that the operator Reaches is noted following the letter R.   All distances are expressed in inches when using the official data card values.[4]   Decimals or fractions are to be rounded off to the nearest whole number, unless they are less than one inch.   For indicating distances under one inch, fractions may be used when the distance has been accurately measured and notation of this fact is desirable.   In the more general case where the analyst knows the distance is ¾ inch or less, but the fraction is not known or easily determined, use of a small case f will suffice.

> *Example:*   R12 denoting a Reach of 12 inches
> R⅝ showing a ⅝-inch finger Reach
> Rf indicating a short Reach under ¾ inch, precise distance not known.

Case of Reach

The case of Reach is identified by the capital letters A, B, C, D, and E, and is noted following the distance.

> *Example:*   R12A denoting a Reach of 12 inches to a single object in a "fixed" location.

Type of Motion

Type I Motion, where the hand is not moving at either the beginning or the end of the Reach, requires no particular identification since it is the standard or most often encountered situation.   The example of R12A illustrates this.

Type II Motion, where the hand is moving at either the beginning or the end of Reach, does require special identification.   The lower case letter "m" denoting motion is placed either before the R denoting Reach or after the letter denoting the case of Reach. If it is placed before, it means that the hand is moving at the start of the motion; if it is placed after the case of Reach, it denotes that the hand continued moving at the end of the Reach.

> *Example:*   mR12A denotes that a Case A Reach of 12 inches occurred when the hand was already in motion at the start of the Reach.

[4] This, of course, applies to the U.S.A./Canada data card.   When using the metric data cards also available, the distances and symbol numbers are in centimeters.   The analyst usually will know if the units are metric or English.

Type III Motion, where the hand is moving at both the beginning and the end of Reach, is identified by placing the lower case letter "m" both before the R and after the letter denoting the case of Reach.

Example: mR12Bm indicates the hand was in motion at both the beginning and the end of the Reach.

Change in Direction

A change in direction is denoted by placing the *small* capital letters CD to the right of the letter denoting the case of Reach. This is used for Case A Reaches only, to show that the time value used is that for a Case B Reach (see previous explanation).

Example: R12ACD

For other cases of Reach, this symbol is not needed because the time is not materially affected by change in direction.

PRACTICAL WORKING RULES

The previous information has defined Reach, the various cases and types of Reach, special conditions encountered, etc. The material which now follows contains additional practical rules of application, an interpretation of the data card, and other information which will be of value to the MTM analyst.

Measurement of Distance

The distance reached should generally be measured with a flexible steel rule. If one makes several sample Reaches and observes them closely, it will be noticed that the hand does not normally travel in a straight line but rather in an arc or with a curved motion. The curved path represents the distance that should be used in applying the MTM data. It is for this reason that a flexible steel rule is better for measurement purposes than, for example, a yardstick.

Distances should be measured and not estimated,[5] unless one has had a great deal of practice in estimating distances and can do so quite accurately. All fractions or decimals should be rounded off to the nearest inch except those of ¾ inch or less. For these extremely short distances the minimum time of 2.0 TMU applies and the distance is normally indicated as an f (stands for fractional). The value for 1 inch applies to all distances over ¾ inch up to 1½ inches. This is a "rule of thumb" for short distances.

[5] This is true for MTM-1, but the rules may differ for other members of the MTM Family of Systems.

A practical reference point of measurement on the hand is the knuckle of the index finger. Figure 6 illustrates the proper method of making a measurement of Reach for an ordinary situation. In measuring short finger Reaches, the finger nail is an easy point from which to obtain dimensions of movement.

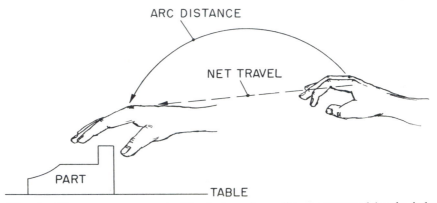

Fig. 6. Distance measurement. The inches of arc distance traversed by the index knuckle, as shown, is the chargeable distance of the Reach. It is improper to use the straight line net travel as Reach distance.

It has been found that the average person can Reach a maximum length, *horizontally,* of only 24 inches directly away from the body without making assisting body movements. If a radially forward horizontal Reach exceeds this distance, it can generally be assumed that some body movement was employed. The symbol and time for a 24-inch Reach would, therefore, be used in this event. Maximum arc distance for a move involving lateral motion with the arm extended could, naturally, exceed this; the same is true for *vertical* hand displacements. Shorter persons, of course, cannot Reach even 24 inches, whereas tall persons with long arms could Reach an even greater distance without supporting body movements. In defining the length of Reach, one should specify the distance proper for the average operator, not the exceptional one. This principle of accounting for the average operator holds for specifying *any* MTM motion or motion combination.

Figure 7 illustrates the areas that can be covered by an average-build operator with arms comfortably extended and also for the condition where the elbow is close to the body. As shown in the figure, the distances covered are:

Arm Comfortably Extended	*Elbow Close to Body*
18″ Forward	12″ Forward
23″ Sideways	16″ Sideways
30″ Arc Swing	20″ Arc Swing

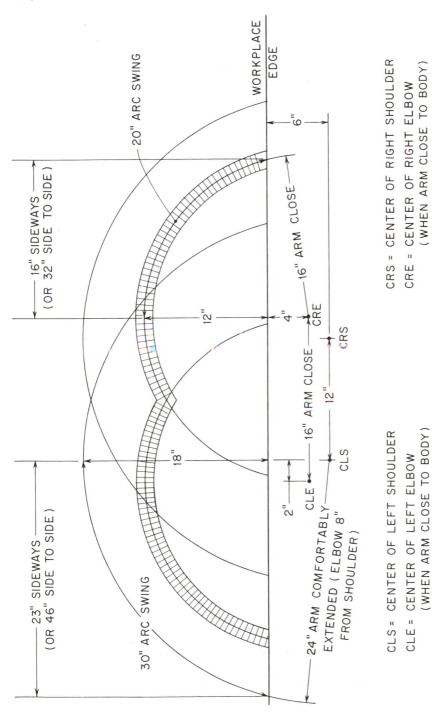

Fig. 7.   Areas covered by *horizontal* Reaches and Moves without body assistance.

CLS = CENTER OF LEFT SHOULDER

CLE = CENTER OF LEFT ELBOW
(WHEN ARM CLOSE TO BODY)

CRS = CENTER OF RIGHT SHOULDER

CRE = CENTER OF RIGHT ELBOW
(WHEN ARM CLOSE TO BODY)

The shaded areas near the inner arc boundaries represent the optimum location of materials, parts, and tools for the most efficient use of a worker's hands with minimum fatigue. To promote motion economy in tasks, *standard workplace* layouts should be based on this factor. Another noteworthy point is that, as a matter of regular policy, the operator should not be required to work for extended periods either beyond the outer limits shown or within the triangular area directly in front of his body. This figure thus gives valuable clues to guide the motion analyst when he prescribes the *workplace layout* to which any assigned working method is directly connected.

If the workplace requires the operator to Reach beyond the areas indicated, this is a general clue that it should be closely studied to see if arrangements could not be made to bring the material and the work area closer to the operator. Then assisting body movements which require more energy are not needed.

Body Assistance

The MTM data found on the data card is all based on motions made without the aid of body assistance. Therefore, if the body moves and thereby shortens the effective distance that the hand or arm must travel in making a Reach, or for that matter a Move, this assistance must be subtracted from the total distance traveled in order to determine the proper distance to use on the MTM data card. This net distance would also be shown in the symbol. You will recall, for example, that it has previously been stated that the average person is not able to Reach more than 24 inches radially from the body in the horizontal direction without utilizing a body movement to assist the arm and wrist movements.

The first case of body assistance is a simple one where the shoulder or torso moves simultaneously with the arm and in the same direction as the hand is traveling. To calculate the effective length of the arm motion, it is merely necessary to subtract from the total distance traveled by the hand the distance that the shoulder moved.

Figure 8 illustrates pictorially the situation just described. The allowed distance, and therefore the distance shown in MTM symbolism, is equal to the total distance moved by the index knuckle from P to Q *minus* the distance that the shoulder moved, from X to Y. This rule automatically adjusts the time allowed for the motion to a value which MTM research has shown to be justified for the kind of situation in question. An alternate method of measurement of perhaps greater accuracy and ease would be to place the body in final position, while holding the hand and arm in the original location relative to the body and then measure the hand travel to get the allowable distance directly without extra measurement and calculation.

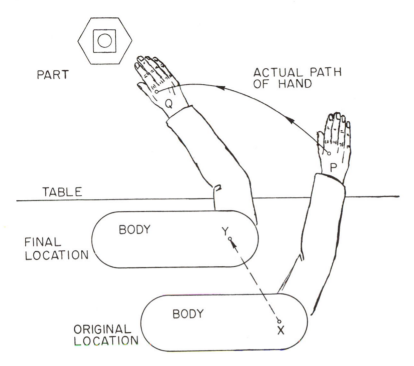

PART

ACTUAL PATH
OF HAND

Q

P

TABLE

FINAL
LOCATION

BODY

Y

ORIGINAL
LOCATION

BODY

X

Fig. 8.   Simple body assistance to Reach.

All MTM Reach and Move data is based on the distance moved without the help of body assistance.

When measuring shoulder movement, it is suggested that a point be selected about six inches from the spine, which is about the tip of the shoulder for most people.

A more complicated situation exists when body assistance is given the arm motion by means of a turning action due to pivot of the body about the spinal axis.   This is known as radial assistance, but actually consists of a combined assistance partly due to lateral assistance of the type described above and partly due to the pure radial assistance effect.   The amount of lateral assistance can be ascertained by noting the shoulder movement as already described.   The amount of radial assistance can best be found by the theory of concentric circles as shown in Fig. 9 and discussed below.

The normal length of the *fully extended* arm being about 30 inches, a forward travel of 1 inch by the *shoulder* (*BB'*) with the spine as a pivot will cause the *hand* to travel 6 inches (*EE'*).   To get the allowable distance of Reach, then, would require a deduction of 6 inches from the total Reach distance; however, 1 inch of the deduction is lateral assistance

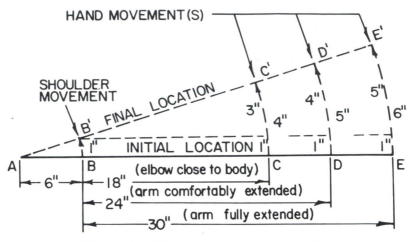

Fig. 9.   Radial body assistance to arm motion.

(*BB'*) and the remaining 5 inches is due to radial assistance.   The length ratios on the diagram should make this clear.   Note that 5 inches of radial assistance was caused by 1 inch of shoulder movement under these conditions.

In case the arm is only *partly extended,* which would normally mean a length of 24 inches from the *shoulder* to the hand, a forward travel of 1 inch by the shoulder (*BB'*) with the spine as a pivot will cause the *hand* to travel 5 inches (*DD'*).   Therefore, 5 inches must be deducted from the total Reach distance to get the allowable Reach distance without combined body assistance.   Note again that, although 1 inch of this distance is lateral assistance (*BB'*) and the other 4 inches is radial assistance, the net effect is to gain 4 inches of radial assistance from 1 inch of shoulder travel.

If the elbow is held close to the body, pure radial assistance to the Reach will be 3 inches; and the combined assistance will be 4 inches (*CC'*) after adding the 1 inch (*BB'*) of lateral shoulder assistance.

To summarize the above information, at least insofar as the *radial assistance* itself is concerned, the following formula can be stated:

$$\text{Allowed distance} = \text{total distance hand} \\ \text{moved} - X \times \text{distance shoulder moved}$$

where: $X = 5$ if the arm is fully extended, or
$X = 4$ if the arm is comfortably bent in the normal working position, or
$X = 3$ if the upper arm is close to the body.

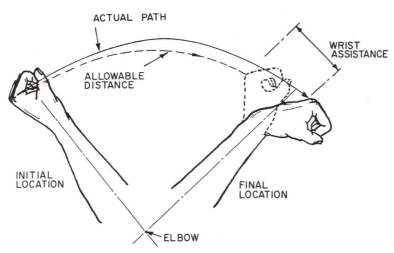

Fig. 10.   Wrist assistance to arm motion

Another case of body assistance occurs when the wrist assists the arm motion as illustrated in Fig. 10.   The object was actually moved a distance equal to the distance moved by the arm plus the distance moved by the wrist.   However, the wrist motion is not to be allowed because it was done simultaneously with the arm motion, which was, of course, the limiting factor and took up the total amount of time required.

Allowed distance = total distance moved by the index knuckle — distance attributable to wrist motion.

This last situation is mostly encountered in motions such as hammering.

When finger movements are used with arm movements similar to the above wrist assistance, the method for determining allowed distance is:

Allowed distance = total distance moved by the index knuckle — distance attributable to finger movement.

One of the best and most commonly used ways of determining the amount of any kind of body assistance is to make the arm and body movements separately, so that the amount of each can be measured.   To do this, first move the body to its final position while holding the arm stiff in its original position, then move the arm along to its final position and note the effective arm movement.   This technique works especially well for radial assistance.   In using this method one is not required to use the formula for calculating allowed distance, because the measurement is directly the allowed distance.

It is well to remember in applying MTM that the average worker will usually tend to use body assistance in his motions, and only rarely will it be found that the worker uses body action which increases the chargeable distance—what could be termed body resistance. In some rare cases, body resistance is legitimate to the job. When this situation is encountered, it is obvious that the body resistance to the movement should be added to the effective length of the motion, rather than being deducted as is the case with body assistance.

Type II and III Reaches

Practical formulas for computing a TMU value for cases and types of Reach *not shown directly on the data card* are as follows:

FORMULA NUMBER	CASE AND TYPE OF REACH	SYMBOLS FOR THE MOTION*	FORMULA FOR TMU VALUE*
(1)	Case A, Type III	mR (Dist.) Am = A Std. − 2 (A Std. − Am)	
(2)	Case B, Type III	mR (Dist.) Bm = B Std. − 2 (B Std. − Bm)	
(3)	Case C, Type II	mR (Dist.) C = C Std. − (B Std. − Bm)	
(4)	Case D, Type II	mR (Dist.) D = D Std. − (B Std. − Bm)	
(5)	Case E, Type II	mR (Dist.) E *or*	
		R (Dist.) Em = E Std. − (B Std. − Bm)	
(6)	Case E, Type III	mR (Dist.) Em = E Std. − 2 (B Std. − Bm)	

* "Std." represents the TMU value for a Type I Motion of the same case and distance as listed directly on the data card. An "m" in either the symbol or the formula designates the "m" value involved when acceleration and/or deceleration is absent from the motion, while Bm refers to the Hand in Motion B column time value from the data card. Am represents the time value from the Hand in Motion A column of the data card. Also see Table 1 for the "m" values.

As an example of the use of these formulas, consider a Reach symbolized by mR12Am. Using Formula (1) and the data in Table 2,

$$\text{TMU value} = 9.6 - 2\,(9.6 - 8.1) = 9.6 - 2\,(1.5) = 6.6\ \text{TMU}$$

The basis for these formulas becomes self evident with a little study of the Type of Motion section. They are merely algebraic expressions for the rules cited there. In essence, they show how the values in Table 2 are used to remove the time value of an "m" (see Table 1) in order to account for the absence of either an acceleration and/or deceleration period during the motion. The symbols themselves show the number of "m" values to be considered.

Odd Distance Reaches Over 10 Inches

The data card distances are listed at each inch up to the 10-inch value. Beyond this point, only the even number distances to 30 inches are shown. It can be shown from a plot of time versus distance that the curves permit use of straight-line interpolation to obtain inbetween values to sufficient accuracy. By the time the 10-inch value has been achieved, the curve becomes very close to a straight line for the remaining distances.

To find the time values for Reaches of odd distance between 10 and 30 inches, therefore, merely add to the next lower "even inch" time half the difference between that time and the time for the next higher "even inch." For example, to find the time for R19C the procedure is as follows:

$$R20C = 19.8 \text{ TMU}$$
$$R18C = 18.4 \text{ TMU}$$

$$\text{Diff.} \quad = \quad 1.4 \text{ TMU}$$

So that R19C = 18.4 + 0.5 (1.4) = 18.4 + 0.7 or *19.1* TMU

Reaches Over 30 Inches

This situation is encountered, for both Reaches and Moves, even though it was previously stated that Reaches over 24 inches radially from the body do not normally occur without body assistance in the *horizontal* plane. If the operator is reaching straight out in front of the workplace in a horizontal dimension he normally cannot exceed a distance of 24 inches if body assistance is eliminated. However, if a sweeping motion is made from the right side of the work area to the left side with the arm extended, the total allowable distance traveled can be well over 30 inches. Reaches exceeding 30 inches can also occur when the motion is vertical.

To calculate TMU values for Reaches over 30 inches, use the following formula:

$$R x'' \text{ (Case of Reach)} = R30 \text{ (Case of Reach)} + (x'' - 30'') \times \text{(factor}$$
$$\text{for case of Reach shown in Table 3, below)}$$

where: R = Symbol for Reach;
         x = Distance reached, inches.

Examples: A. TMU for an R42B:
              R42B = 25.8 + 12(.7) = *34.2* TMU

         B. TMU for an R39A:
              R39A = 17.5 + 9(.4) = *21.1* TMU

Table 3.   TMU Values for Reach Increments Over 30 Inches*

Case of Reach	A	B,C,D	E
Additional TMU per inch of Reach over 30 inches	0.4	0.7	0.6

* Since the 1972 revision of the MTM-1 data card, these data have been included as the bottom row of the Reach Table (see Table 2).

The above formula is based on the theory that the MTM data curves for distances over 30 inches have become straight lines; however, it is not correct to extrapolate with the Hand in Motion columns on the data card. The factors in the table therefore were found for each Type I Case of Reach by subtracting the TMU value for 28 inches from the TMU value for 30 inches and dividing the result by 2 to get the per inch rate of movement. It is not necessary to memorize this procedure since the factors now are listed on the data card. It is necessary only to remember that the slope of the curve has become a straight line with the per inch rate listed. Therefore, to calculate any time above 30 inches, using the extension factor listed, merely set down the TMU value for a 30-inch movement for the case of Reach involved and add to it an amount equal to the factor multiplied by the number of inches over 30 inches that has been traveled.

How to Identify and Analyze MTM Motions

In making MTM studies, it is suggested that first the *kind* of basic motion be identified—such as Reach, Move, etc., and that the proper main symbol, such as R, M, etc., be written on the analysis form. No attempt should be made to analyze the complete symbol for each motion in turn until all the main symbols have been listed for the element or cycle; attention should be focused on one thing at a time as discussed here.

The second thing for the observer to determine is the *classification* of each Reach, Move, etc. For example, when Reach is observed, it is, of course, necessary to determine which of the five *cases* of Reach are involved and this part of the symbol is next written down.

Third, the analyst should identify the *type* of motion and indicate it with "m" symbols on the form. The other MTM variables and factors for Move, Grasp, Position, Turn, etc., are handled in a similar manner.

The fourth step is to fill in the *distance* involved. In doing this, one must be careful to set down the correct allowed distance, taking into account the effect of body assistance, wrist assistance, finger assistance, etc. Similar practical rules will be given for each MTM element.

The fifth and last thing to put down is the TMU values and their sum. This can be done in the office, as it is a clerical routine. TMU values from the data card are to be shown to the nearest tenth.

For the benefit of those not having previous experience with MTM, it might be mentioned that the problem of simultaneous motions, how to analyze these motions, etc., will be given as part of the material in later chapters and the reader should not concern himself with such questions at this point. They will be discussed only when all of the MTM elements have been explained in sufficient detail to make such consideration worthwhile.

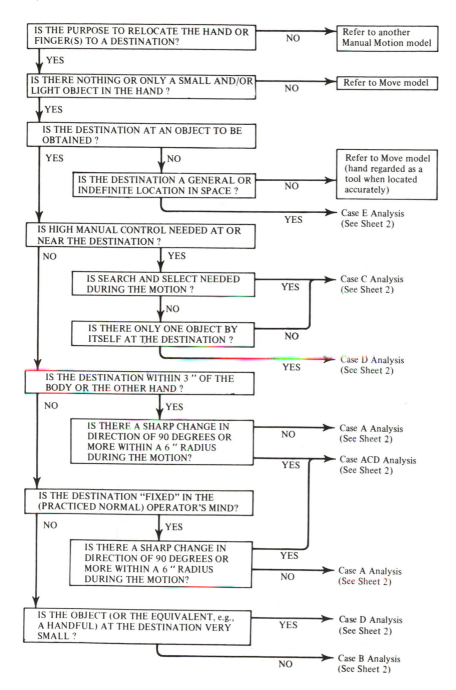

Fig. 11A.   Decision Model for Reach.

(from Sheet 1)

			ARE THE MANUAL MEMBERS STOPPED AT <u>BOTH</u> THE START AND THE FINISH ?			
			YES	NO		
				ARE THE MANUAL MEMBERS IN MOTION <u>ONLY</u> AT THE START ?		
				YES	NO	
					ARE THE MANUAL MEMBERS IN MOTION <u>ONLY</u> AT THE FINISH ?	
					YES	NO
CASE ANALYSIS	Relocating the manual members	TYPE I		TYPE II in motion at start	TYPE II in motion at finish	TYPE III in motion at <u>both</u> start and finish
A	to an object in a "fixed" location <u>or</u> in the other hand <u>or</u> within 3 " of the body	R <u>x</u> A		mR <u>x</u> A	R <u>x</u> Am	mR <u>x</u> Am
* ACD	by Case A with a sharp change in direction	R <u>x</u> ACD		mR <u>x</u> ACD	R <u>x</u> ACDm	mR <u>x</u> ACDm
B	to an object by itself	R <u>x</u> B		mR <u>x</u> B	R <u>x</u> Bm	mR <u>x</u> Bm
C	to one object in a jumbled group	R x C		mR <u>x</u> C	Not Possible	Not Possible
D	to a very small object <u>or</u> to an object requiring an accurate Grasp	R <u>x</u> D		mR <u>x</u> D	Not Possible	Not Possible
E	for body balance <u>or</u> for next motion <u>or</u> out of the way	R <u>x</u> E		mR <u>x</u> E	R <u>x</u> Em	mR <u>x</u> Em

* Case ACD is assigned the same time values as Case B.

NOTE: x = Refer to Net Distance for Reach and Move model.

Fig. 11B.   Decision Model for Reach.

LET: x   = NET DISTANCE, inches (except use "f" for 3/4″ or less)

L   = length of <u>index knuckle</u> travel path, total including any arc
       or curve length

Lss   = length of <u>shoulder</u> travel during <u>simple</u> body assistance

Lsr   = length of <u>shoulder</u> travel during <u>radial</u> body assistance

Lks   = length of <u>index knuckle</u> travel due to <u>simple</u> wrist assistance

Lkr   = length of <u>index knuckle</u> travel due to <u>rotary</u> wrist assistance
       (as occurs in Reach-Turn or Move-Turn combined motions)

Lff   = length of <u>finger tip</u> travel during <u>finger</u> assistance

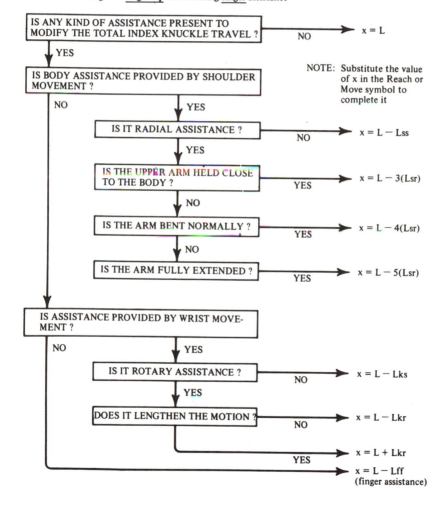

Fig. 12.   Decision Model for finding Net Distance for Reach and Move.

DECISION MODELS

For each motion chapter, all of the practical working rules and other guides to analysis of the motion will be displayed—usually at the end—in the form of a Decision Model. These models (see Figs. 11A, 11B, and 12) are derived from "MTM Decision Diagrams" discussed in Chapter 4.

It is suggested that the reader study these analysis models *immediately after* reading and studying the written rules, since they are the best possible review of the material. Also, if questions arise during application regarding the correct analysis, these models also provide answer most quickly.

CHAPTER 6

Move

Definition[1]

*MOVE[2] is the basic hand or finger motion employed when the **predominant purpose** is to transport an object to a destination.*

  *1. Move is performed only by the fingers or hand. Pushing an object with the foot would **not** be classified as a move.*

  2. The hand must exert control over the object during the motion. In tossing an object aside, for example, the move motion ends when the fingers or hand releases the object.

  3. The fingers or hand may be pushing the object or sliding it; it is not necessary to carry the object.

  4. Using the fingers or hand as a tool is classified as a move. The fingers or the hand itself would be considered as a tool being carried by the hand.

As was true for Reach motions, the deciding factor in classifying a motion as a Move is its predominant purpose. An example of this distinction from the needle trades was given in the Reach chapter. In that case, Reaches were made while a pair of scissors were palmed, their presence in the hand being ignored if the predominant purpose was to locate the hand in a given area—this being a Reach motion. Now suppose that, after pinning some pieces of cloth together, the operator unpalms the scissors and brings them to the edge of the cloth to begin cutting. This latter motion is definitely a Move of the scissors to bring them to cutting position, and the motion would be so classified. Generally, carrying such hand tools as

---

[1]From "MTM Basic Specifications" (Footnote 1, Chapter 5).

[2]The reader will recognize this motion as basically equivalent to the Therblig called Transport Loaded. Along with Reach, Therblig Transport Empty, it is the only other manual motion in which large distances of movement can be involved. All other MTM motions are of relatively small displacement, and this fact is a definite key to motion recognition.

pliers, tweezers, tongs, screwdrivers, etc., to a destination for the purpose of picking up or working on a part would properly be classified as a Move.

The predominant purpose of the motion is the deciding factor in classifying it as either a Reach or a Move. Another mode of expression should make this matter crystal clear. If the hand is empty, the motion will obviously be a Reach, except in the special case where the hand is used as a tool as described below. If the hand holds an object which must be located in a different place, the motion is definitely a Move. Finally, if the hand holds an object as an incidental fact to the reason for the motion, a Reach will be performed. Basically, then, the decision as to predominant purpose hinges on the reason for moving the hand—was it to relocate the hand, or was it to relocate the object? Answering this question in any given case will leave no doubt as to the kind of transport motion involved.

The use of the hand as a tool will be classified as a Move, not a Reach. In other words, *when the hand performs useful work on the workpiece,* the intent of the Move definition is fulfilled. The case of Move involved will naturally depend on the conditions of operation observed. Examples of using the hand as a tool are pounding with the fist to seat an object in a packing carton or using the fingers to crease a piece of paper.

As the nature and variables of Move are studied, it is helpful to remember that the Decision Models pertaining to Move (at the end of this chapter) are the best guide and summary when any questions arise. (See also Ch. 5, Fig. 12).

KINDS OF MOVES

At present, the MTM Association recognizes three different cases of Move, as identified by the following letters of the alphabet: A, B, and C. It is important to mention at the beginning that these are not equivalent categories to those given for Reach motions. Such assumption would lead to errors in recognition, classification, and combination of Move motions. Rather than attempting to match the cases in his mind, the reader should rely on the definitions for each case of Move independently of Reach cases which happen to be identified by the same letter of the alphabet.

Case "A" Move

Case A Move may be defined as follows:

Move an object to the other hand or against a stop.

The same time value was found to be appropriate for both parts of this definition. It is believed that this is due to the degree of automaticity imparted to the motion by either the "kinesthetic sense" of the operator or the presence of an object which will automatically cause motion to cease. At any rate, the degree of muscular control required by a Case A Move is at a minimum for Move motions.

An instance of the first type of Case A Move occurs when a bridge player moves the shuffled deck from tapping it on the table toward the other hand to begin dealing the cards. The dealing hand, in this example, often performs a Reach, Case A toward the deck during the same time interval. This could, therefore, be one example of the fastest types of Reach and Move being performed simultaneously.

The point to notice about this part of the definition is that, because the destination of the object is the other hand, little doubt exists in the operator's mind as to the final location due to its being within the limits of proximity to the body where kinesthetic sense will function accurately. In this respect, the "fixed location" portion of the definition for Case A Reach is fulfilled; and remarks made about automaticity of motion under that discussion apply with equal force here.

There is an inference in the last paragraph that mental certainty of the object's location at all times during the Move will promote the likelihood that a Case A Move is occurring. This justifies classifying another frequently observed Move as a Case A—when the motion path of an object being moved is mechanically constrained or controlled.

Typical occurrences are lowering a drill spindle with the feed lever, operating the gear shift lever of an automobile, and sliding a coat hanger along a rack bar to obtain clearance for hanging additional clothing. In each of these events, there is almost no hesitancy in the motion, since the moving object is guided mechanically along a path of travel while the hand essentially supplies only the motive power. Measurement of the line or circular distance of knuckle movement in such cases greatly simplifies distance determination as well, the hand imparting almost negligible arc travel. All lateral motion of the object will depend on the mechanical constraint, the only deterring factor to excessive Move speed being the exertion of a minimum degree of control to prevent damage to or by the object moved. In most (but not all) cases of mechanically constrained or controlled Moves the MTM analyst will, therefore, specify a Case A Move. The main exception to this is when the object must be approximately or precisely located somewhere along the direction of travel at the end of the motion.

The second part of the definition, Move an object against a stop, is typified by many common examples such as moving a book end against a group of books. The motion needed to close an open desk drawer that offers only minor resistance to sliding shut fulfills both parts of the definition for a Case A Move. Opening the same drawer fully could also exemplify a Case A Move if the travel was limited by another set of stops. Partial opening could be justified as a Case A Move on the basis of mechanical constraint by the slide rails, as discussed above.

The Case A Move is the fastest Move up to a distance of 14 inches, at which the Case B Move becomes faster. This behavior probably is due to a change in the degree of muscular control required for the Case A Move as the travel distance increases. Some effort is then needed to avoid building up so much momentum (which depends on the weight and speed of the object) that the impact against the stop will endanger the part, the stop, or the operator. The resulting damage, injury, or loss of grip to any of

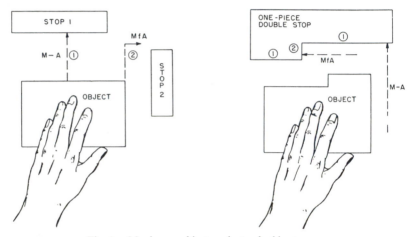

Fig. 1.   Moving an object against a double stop.

these would not be worth a small time reduction in performing the motion. Apparently, this tendency is no problem at distances short enough to prohibit the attainment of excess velocity.

Moving an object against a double stop such as is shown in Fig. 1 is an additional action which sometimes occurs. Film analysis during MTM research indicated that the average worker will first bring the object with a Case A Move against one stop, and this will be followed by a very short Case A Move of the object against the second stop. In the figure, the motion lengths have been exaggerated for pictorial purposes. The second Move actually would be only two inches to fractional in length. However, a very skilled operator will essentially bring the object diagonally against both stops in a single Case A Move, deleting the second motion.

The angle between the stops or their configuration will affect the motion analysis in an interesting way. As the included angle between two straight stops becomes less than ninety degrees (within reasonable limits), the possibility of sliding a properly fitting piece down into the final location with a minimum of extra motion increases. As the included angle goes above 90 degrees, however, the tendency to make two separate motions to seat the object in the stops increases. For example, in the left sketch of Fig. 1, by rotating Stop 2 to the right until the included angle becomes, say, 140 degrees, the probability of hitting both stops in one motion would be decreased.

The MTM analyst must determine at the time of observation whether a given configuration will require one or two Moves by the average operator to place the object against both stops. All will be Case A Moves, however.

The preceding discussion on double stops affords an opportunity to present some more basic and general points of MTM theory. When the reader attempts to actually observe the second short Move against the double stop, he often becomes dubious about the above analysis. There is a perfectly logical reason for such confusion as to whether the second Move actually takes place. This reason is that the motion time is less than the time for him to actually "see" a motion occur. In fact, an Eye Focus time of 7.3 TMU (refer to Chapter 14) is the minimum motion time that can actually be observed by the naked eye, and any motion taking less than this time must be recognizable by some other means. This would be with the aid of one of the MTM "rules of thumb," such as was given above for double stops.

The justification for many of the working rules of MTM is the fact just noted. Such rules are not wishful thinking or crude approximations. They result from the fact that the researcher is viewing films and/or analyzing other recorded data, much of it in great detail, to establish MTM times and motion rules. He is thereby armed with far more information than an analyst working on the factory floor or in an office without such aids. In practical terms, based on the Eye Focus time value, the research man can see the motions in anywhere from four to sixteen times as much detail, considering possible film speeds, as the MTM analyst can observe in the same motion sequence. The obvious import is that the MTM analyst must accept the "rules of thumb" in good faith with the research until such time as further research indicates a need for revision in the rules.

His attention should be on correct usage of the rules rather than on venturing to discover flaws and errors in them. If an apparent mistake in rules is believed to exist, the proper procedure is to send a request for further research to the MTM Association, since this is one of their most important and vital functions.

In practice, then, the MTM analyst must consider all of the factors and working rules involved prior to his decision as to what constitutes the correct identification for the motions in a given situation.

Case "B" Move

This Case of Move may be defined as follows:

Move an object to an approximate or indefinite location.

Moves of this kind are frequent in industrial operations and are relatively easy to identify in practice. Examples are:

1. Laying aside a paint mask
2. Moving a pair of wiring pliers to a clear area on the assembly bench prior to dropping them
3. Moving a pencil to another place on a desk to make room for work papers.

As was true for a Case B Reach, approximate location implies that location of the object within an inch or two of a certain point is sufficient. Obviously, which point in a given area might make little difference so long as the object's location was satisfactory at the end of the motion.

The second portion of the definition refers to an indefinite location. This covers such instances where at any convenient point in the path of travel of the object, optional within sensible limits of measurement, the object is tossed aside by letting go of it. Since the point at which the toss aside Move ends will vary from piece to piece, a definite measurement cannot be made. However, a sensible average distance for the normal length may be estimated even though it is an indefinite quantity.

Clarity on the Case B Move is best served by a little historical elaboration. As would seem quite logical, the inventors of MTM presumed the existence of a definite similarity in the cases of Reach and Move and hence the need for similar case classifications. Reference to their book shows:

Move Case B—Move object to approximate location[3]
Move Case D—Toss object aside[3]
Move Case E—Move object to indefinite location.[3]

Interestingly enough, the time values listed for the B and E Cases were identical, while Case D times equalled those for in-motion Case B.

The subsequent research and examination of the records showed that these similarities were, in fact, true identities. In other words, there is not of necessity any one-to-one connection between the motions of the hands during reaching and moving, and this fact obviates the need for the original manner of classification of Move. The old Case E Move category was therefore added to the present definition for Move Case

[3] By permission from *Methods-Time Measurement* by Maynard, Stegemerten, and Schwab. Copyright, 1948. McGraw-Hill Book Company, Inc.

B, and *tossing movements* (old Case D) *are properly identified as an M—Bm.* The latter Type of Motion is a Type II as discussed below. Symbols D and E are therefore no longer used to designate Move cases.

A distinguishing feature of the Case B Move is that a low degree of control exists, and no special care or caution is needed to accomplish the Move. It might be termed in everyday language as an "ordinary" Move.

Case "C" Move

The Case C Move is defined in this manner:

Move an object to an exact location.

Examples of such Moves are transporting a screw to a threaded hole or bringing the point of a screwdriver to the slotted head of a screw.

The term exact location usually implies that the hand will transport the object within half an inch or so of the destination point. Because of this closer distance limitation, it is evident that Case C Moves require appreciable control of the object, which causes greater time consumption. Case C Moves are the slowest Moves at all distances, although the differences are not as great as one might expect. Besides being due to higher mental and muscular control requirements, this is even more directly caused by the general requirement for the eyes to provide a high degree of visual control during the performance of Case C Moves. The operator must either be looking at the destination point, or there must be some way in which equivalent information can reach his mental centers via kinesthetic sense. He must certainly know where to stop the motion in order to exert the proper degree of muscular restraint during the movement.

The analysis of assembly motions permits another example of the practical type of decision required of an MTM analyst. Since the reader does not know about the MTM motion called Position at this point, it should be mentioned that it basically consists of all assembly motions which follow movement of the parts to within the final limits for a Case C Move given above.

Now, looking at the question of assembly, suppose an object held in one hand must be assembled closely to another object in the other hand. This could occur in several ways. A highly skilled worker might assemble them under highly repetitive conditions with a Case A Move, since he is capable of developing a high level of automaticity where the hands end up close together. Even an average assembler with sufficient practice could perform certain such assembly using only a Case C Move. The normal occurrence, however, would be a Case C Move followed by a Position. Keeping these alternatives in mind, the MTM analyst must resolve each occurrence to the proper motion categories, since time differences are noticeable in the three working methods.

Case C Moves will not necessarily be followed by Position motions. It should be clear, however, that when objects must be positioned, the prior performance of a Case C Move is essential to bring them within half an inch of each other. This leads to the important MTM rule that *Positions are always preceded by Case C Moves.* Only rare exceptions have been found to this rule.

General

The three cases of Move recognized in the MTM procedure should now be clear. As was done for the Reach categories, a memory device, or mnemonic, to assist the MTM analyst in identification is given by the following word associations:

Case A Move—Automatic
Case B Move—Basic or Broad
Case C Move—Control or Careful

TYPE OF MOTION

Most of the comments made in the Reach chapter concerning the nature and theory of Types I, II, and III motions apply equally well to the motion defined as Move.

Case A Moves of Types I, II, and III motions are all possible. One example of this would be a Case A, Type III Move used as part of a sequence to activate an automatic feed lever of a machine using a cupped hand as in Fig. 2. Such a sequence would consist of a Reach-in-motion-at-the-end, a Contact Grasp (explained in a later chapter), the Case A Type III Move, a Contact Release, and, finally, a Reach-in-motion-at-the-start. Examples of Type II Moves may be seen in Figs. 3 and 4 in the chapter on Reach.

All three motion types can also occur with Case B Moves. For instance, the now obsolete Case D Move was found to be a Case B Type II Move. It is the Move involved in tossing aside an object, symbolized (for the selected distances) in MTM as shown:

Toss object aside	M12Bm	)	10.0 TMU
Let go of object	~~RL1~~		—
Hand to clear	mR4E		4.7 TMU
	Total		14.7 TMU

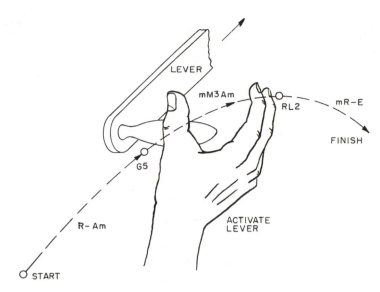

Fig. 2. Striking a machine feed lever with a cupped hand. Here a Case A, Type III Move is used as part of a sequence to activate an automatic feed lever of a machine.

The dash (—) in the TMU column for RL1 shows the time is limited out; the meaning of this term will be given later in this text.

Disposing of objects in this manner obviously saves time, provided no injury to the part occurs from such handling.

The nature of the destination and control involved would naturally eliminate the possibility of Case C Moves in motion at the end. This leaves Type I and Type II in-motion-at-start as the only categories for C Moves.

As was true for Reaches, the MTM data card provides a means of determining the effect of acceleration and deceleration on Move motions. It was found that the effects of all cases of Move were essentially the same as for Case B Moves. The data card listing for Hand in Motion B shows the result of eliminating only one period of either acceleration or deceleration from the Type I Motion and therefore gives the times for Case B, Type II Moves.

A set of formulas is given later in this chapter for all other Type II and Type III Move motions. They all depend on subtraction to obtain the time value of one "m" or its equivalent and it is instructive to note the variation of these "m" values throughout the data-card distance range in Table 1. As in the case of Reach, the assumption is made again that, when developing times for Type II and Type III Moves exceeding 30 inches distance, the "m" value listed in Table 1 for 30 inches will govern beyond that distance.

Table 1.   Value of "m" in TMU for all Cases of Move versus Distance in Inches

Inches	f	1	2	3	4	5	6	7	8	9	10	12	14	16	18	20	22	24	26	28	30
TMU	0.3	0.6	1.7	2.1	2.6	3.0	3.2	3.2	3.4	3.6	3.6	3.4	3.2	3.0	2.8	2.6	2.4	2.2	2.0	1.9	1.6

The values of "m" as shown in Table 1 can be copied on the data card and are then always available when needed.

It will be noticed that the effect of one period of acceleration or deceleration gradually increases to a maximum of 3.6 TMU and then decreases back to 1.6 TMU.   It is believed that this effect is caused by the relatively lower degree of motion confinement in longer distances, the use of heavier and stronger muscles for longer Moves, and the characteristics of velocity and acceleration which change to some degree at distances longer than about one foot.

Whether the suppositions just noted are valid or not, the "m" values given represent validated results of careful analysis of motion picture films. Recall the earlier explanation of the significance of this fact near the end of the discussion covering Case A Move above.   Reporting values based on adequate research is not the same thing as validating reasons why the values vary as they do.   In other words, the reasons behind the variation of validated data are highly desirable and should be determined; but this requires more complex and exhaustive research than that necessitated by determination of the values alone.   This latter type of research as yet has not been done regarding the variation of "m" values; perhaps it might be at a future date.

CHANGE IN DIRECTION

Variation in moving time does occur due to change in direction, but the magnitude of this effect is so small as to justify neglect of it.   Velocity compensations in the motion effectively cancel out the slight time increase caused by the small additional control in the Moves.

EFFECT OF WEIGHT OR RESISTANCE

That an object's weight or the resistance it offers to motion will affect the time required to Move it is a recognized fact.   The originators of the MTM technique tentatively evaluated the extent of this factor's influence on moving time, and the result was a set of data utilizing percentage increases in time values as the weight being moved increased.   These factors were used until March of 1955, when the MTM Association made official a new set of data by which the effect of weight on the movement of objects could be found.

The new factors are based on MTM research which may be found in detail in MTM Research Report No. 108.[4] The latest data for the evaluation of time effects of weight may be seen in Table 4.

The research was basic and exhaustive concerning the effect of weight on Move time and provides answers to almost any question which could be raised concerning it. Two main factors were determined to be major influences on the increased time needed for weighted Moves, and they apply for all Cases of Move. One factor usually, but not always, present was identified as the *Static Component*. The other factor, which is *always* present, is known as the *Dynamic Component*. Both of these factors, along with other major considerations involved with moving weighted objects, are summarized next.

Factors in Weight Control

1. TYPE OF MOVEMENT

An object may be moved in two general ways, depending on conditions surrounding the motion. It may be transported from one location to another *spatially*. This means that the only control and restraint of the object is by the hand(s) and that the object travels through space with no attachment to or guidance from any other agency. The other means of movement depends upon or is influenced by the contact or attachment of the object with other objects; for lack of a better term, this is denoted as *sliding* movement. In this case, part of the control comes from the hand(s) and the remainder from the support or guidance of all contacting or attached objects along the path of motion.

When an object has sufficient weight or resistance to movement to influence the time of movement, it then becomes important to identify which type of movement—spatial or sliding—is being evaluated. In the case of spatial Moves, the full weight of the object must be overcome and controlled by the worker. For sliding Moves, however, the effects of friction must be considered as well when determining the actual amount of force and control required of the operator. In the case of horizontal non-spatial (sliding) Moves, the operator must merely overcome the sliding friction; and this will be the product of the weight and the proper coefficient of friction, because the weight (gravity) force vector is normal to the sliding surface. When the Move is non-horizontal, e.g. on a

[4]Available from the U.S./Canada MTM Association.

raised incline, there is a combination or resultant force (between the sliding frictional force and that of gravity acting upon the object being moved) which must be overcome; so such force combinations must be found by Applied Mechanics analysis, because the force vectors are affected by the angles involved.　Stated in another manner, for all non-spatial Moves of an object, the manual control or effort must merely over-balance the restraining force offered to performance of the motion; and this force always is something less than the weight of the object due to the pull of gravity alone.

2. STATIC COMPONENT

The prior relationship between the object to be moved and the hand will directly influence the extent of manual control which must be gained and the amount of time involved in the total moving time.　Two conditions are possible:

(1) Either the hand already had enough control to start the movement, as when the object has been held spatially during earlier motions in a cycle; or

(2) the hand has not yet assumed sufficient control to permit the Move to start, even though the fingers already are closed on the object.

In the first situation (1), the purpose of the Static Component—to gain control of the object *weight*—already was achieved prior to the motion in question.　Thus, no Static Component time is needed prior to object movement.

For the other situation (2), the preceding Grasp motion has been completed before Move time begins; only the object, not its weight, is under control. Reference to Chapter 10 on the MTM element Grasp will show that motion to be completed, by definition, as soon as the fingers have closed on the object to be moved.　However, the Grasp times do not include the time to overcome other than minimal weight (less than 2.5 pounds).　As a practical approach, therefore, all the extra time due to controlling the weight of the object has been included in the Move weight factors.　This avoids complicating the analysis for Grasp with weight factors in addition to those required for analyzing the Move of the object.

To Move an object that has been grasped, the worker must first achieve *full* control of the weight of the object.　This effort to gain control or "make ready" for moving is the *Static Component* of a weighted Move.　The time required to perform this portion of the Move depends primarily on the weight lifted or resistive force to be overcome.　The MTM procedure provides a set of TMU values that are constant *additive time factors* for any given amount of weight or force application involved in a given Move motion.

The data card table is based on a TMU time constant of: TMU = 0.475 + 0.345 × weight (in pounds). The value given by this formula[5] may be used directly for very precise work. This time factor accounting for the Static Component of a given weight must be added to the total Move time in all cases *except when the object was under full control of the hand prior to the start of the Move.*

3. DYNAMIC COMPONENT

The movement of an object having negligible weight requires the expenditure of a certain amount of time and energy which basically depend on the degree of control and the distance through which the object is moved. These factors are accounted for by the Case of Move and the net distance traveled. However, during the movement of a weighted object, a number of *additional* variables increase the time needed and tend to induce greater fatigue. These will be discussed below. Their effect upon performance time is covered in MTM practice by the use of a *multiplying factor* which depends on the effective net weight (ENW) of the object being moved. This multiplier accounts for the time increase that is attributable to the *Dynamic Component* of a weighted Move. It is always applied to the basic motion time *before* any addition is made for static component requirements. In other words, the Move of a weighted object, besides requiring the basic motion time and the time required for the static component where it is applicable, also involves or requires more time for the "travel" portion of motion.

It is easy to understand the dynamic component when one remembers that work measurement procedures basically evaluate the amount of work an operator performs. In physics, work is defined as the product of force or weight and the distance of action: Work = Force × Distance. For a given distance, then, the energy needed to do the work of moving a weight will increase as the amount of weight increases.

Moving a greater weight, in addition to requiring more energy will require more *time* to make the Move. This fact emerged from the MTM research on weighted Moves which showed that the average individual is capable of a certain power or rate of energy expenditure. Physics defines power as a given amount of work done in a given amount of time:

$$\text{Power} = \frac{\text{Work}}{\text{Time}} = \frac{\text{Force} \times \text{Distance}}{\text{Time}}.$$

Since the power of a worker is limited, he will logically be governed

---

[5] The metric equation is: TMU = 0.475 + 0.7606 × weight (in kilograms).

by a certain time limit to perform a given amount of work (the work to Move a weighted object). Therefore, the time consumed by the dynamic component of movement will vary with the weight. The net result of this is a tendency for workers to Move larger weights through shorter distances (less arc travel—more straight line travel) than they would for smaller weights.

The MTM-1 Move data accounts for the Move *travel* time except for the effect of weight, which is logically and easily compensated for by an appropriate multiplier. The exact increase, usable for precise work, is 1.1 per cent per pound over the basic time. This makes the multiplier factor $1.000 + 0.011 \times$ weight (in pounds).[6]

There are two things which the analyst must be cautioned against regarding the movement of heavy objects. The first is that, when visualizing Moves for MTM analysis, the engineer must be careful to specify reasonable distances which the normal operator can be expected to Move the weight. During observation analysis, this is not a problem because the actual distance involved is directly measured. The second point is that mechanical handling with hoists, cranes, conveyors, and other material handling aids is indicated and used in most modern plants when the power output of the worker to Move a given weight might be exceeded or where the power required will induce undue fatigue.

A suggestive per hand figure for the upper limit of weight to be handled is 35 pounds for women and 50 pounds for men. This will vary greatly with the class of worker, nature of the job, state of health, and similar considerations. In many states, legal limits are specified by law (usually in terms of total weight of objects *to be lifted*) and the reader should check the applicable laws in his locality to avoid making errors in work content specification.

4. Dynamic Component Variables

By the nature of the film analysis used, the effect of the following variables on ordinary unweighted Moves was fully included in the basic data times. However, when appreciable weights are being moved, *additional* time is consumed over the basic values because the variables here discussed assume greater effect on the velocity and acceleration characteristics of the movements. Several of the variables have a negligible effect, while others significantly influence the time to Move a weighted object.

A. Manner of Hand Usage

There exists a number of ways in which the movement of a weight by the hands can occur. When an operator is right-handed, his right

---

[6] The metric equation is: $1.000 + 0.02425 \times$ weight (in kilograms).

hand is termed the preferred hand; the opposite is true for a left-handed worker.   It was found in the research that the use of preferred or non-preferred hands by a given operator made no great difference in the time to Move a weight.   Handedness thus appears to be more directly related to habit than to the time to perform work.

When transporting a weight either spatially or by sliding, the worker can use either one or both hands.   The basic time values and the new weight factors were developed directly for the use of one hand.   During the study, however, investigation was made into what effect was produced by the use of both hands to Move a weighted object.   This effect depends on the manner in which the two hands hold the object.

When both hands are actively supporting the weight, as for instance when they are both underneath a boxlike object, half of the weight will be supported by each hand.   The *effective net weight* (*ENW*) for each hand in this case is one-half of the gross weight of the object for spatial Moves, or one-half of the net friction force to be overcome in the case of sliding Moves. This is the *usual* assumption for average analysis of two-handed weighted Moves.

It also sometimes happens that one hand fully supports the weight of the object or overcomes frictional resistance, while the other hand merely aids in guiding the object.   For example, a box may be held during a Move with the right hand beneath it and the left hand contacting the left side of the box to steady it.   In this case, the effective net weight for the right hand is the gross weight of the object, while the left hand essentially exerts zero effort to overcome the weight of the box.   In general, then, when both hands are used, the relative loading of either hand can vary from zero to the full weight of the object; but, in every case, the sum of the loads for each hand must equal the total object weight or total restraining force being exerted.

The procedure is the same for either manner of two-handed Moves.   The *effective net weight* (*ENW*) is shown with the symbol for each hand motion in accordance with the actual situation for the hand in question.   When time is assigned in accord with the rules governing "limiting" given in a later chapter, the hand carrying the same or greater ENW, whichever is applicable, will govern the time assigned.   In essence then, when both hands are equally dividing the weight, the time assigned will amount to the same value as would be assigned in the case of one hand carrying a weight half as large as the weight actually being moved.   The same procedure holds for cases of two-handed sliding except that friction must also be included in the calculation as discussed below.   If the hand loading is unequal, the principle still applies and only the proportions of force and time between hands is something other than half and half.

B. *Angular Direction of Movement*

The research showed that the angular direction of the Move path with relation to the vertical plane of the body as a reference does have more effect when weights are moved than is true for negligible weights. However, the time increase is not significant enough to require accounting for it in practical applications.

No appreciable time differences appeared for up and down motions (vertical path) or left to right movement and vice versa. This lack of effect was not dependent on the weight of the object moved.

Radial directions of motion in the horizontal planes, in or out from the body cause small time differences. Motion toward the body seems to take slightly less time than movement of weights away from the body; but, as mentioned above, this need not be considered in setting practical standards.

When the direction of motion is intermediate to the vertical and horizontal, the motion path will be at some angle between zero to ninety degrees from the horizontal. There is again a small Move time increase for weights at the larger angles, but practically this can be neglected. Further research into the methods value of direction of motion may, however, be worthwhile.

C. *Effect of Gravity*

The time difference to Move a weight straight up or straight down is extremely small. This was not the anticipated answer for the research. The data showed that the only difference is in the velocity and acceleration characteristics of the Move. The velocity varies at different stages of the same path length, but the same end point for total time taken is achieved for either up or down movement. Also the maximum velocity attained is equal for both directions although the velocity peaks are at different length points in the path of travel. In any event, though, the length of path permitted by the arms is relatively so short that neither velocity nor acceleration affect the total time significantly.

D. *Sex of Operator*

The ability of male or female workers to handle weights in respect to the time required differs chiefly when larger weights are being moved. The proportional difference was found to be relatively constant regardless of the weight moved or the distance of movement, although the numerical difference increased with added weight.

For methods value, it is instructive to know that female operators are slightly more dextrous with weights of five pounds or under, the reverse is true for weights of over five pounds. The prevalent practice of using female operators for light hand assembly is thus verified, while the use of male workers for assembly of heavier parts is also suggested. For very large weights, a female will require from 10 to 18 per cent more move time than a male worker.

Sex differences are not therefore important from the time standpoint so long as the male workers are properly assigned to heavy jobs and legal or sensible limits as to weight are respected in specifying job content for either sex. The weight factors as given are directly applicable to male operators. For either sex, fairness to the operator demands any reasonable revision in the methods of handling, such as mechanical devices, with heavier weights.

E. Effect of Friction

The static and dynamic factors were developed directly for spatial Moves and are to be applied in accordance with the ENW for the Move in question. When sliding Moves occur having friction or equivalent resistive effects, the same factors will apply provided the true ENW (which will ordinarily be less than for a spatial move of the same weight object) is determined by the laws of friction.

The object's weight will be the primary cause, for example, of frictional resistance to sliding a large packing case along a concrete floor. The differentiating point about such cases of friction or similar weight-induced resistances to Moves from the spatial type is the fact that part of the weight is borne by the supporting or guiding surface along which the object is moved. The hands must exert only enough extra control to overcome the friction force component in the direction of motion, and this is almost always smaller in terms of ENW than for spatial movements. However, one may occasionally encounter a case where the worker must overcome the friction forces caused by a mechanically constrained system. Also, in some cases he must also overcome, as mentioned earlier in this chapter, some or all of the weight of the object, the amount being dependent upon the angle to the horizontal of the movement involved. In the latter case the weight component and friction forces must be combined vectorially to find the ENW. In all cases involving friction, the easiest way to determine the ENW may often be to use a spring scale and observe the amount of pull required to Move the object.

Obviously, the engineer applying MTM can utilize his knowledge of Applied Mechanics effectively in determining the net value of the resistance in such occurrences. When measurements are difficult or impossible to make for the data needed to apply this knowledge to a given problem, the sound engineering judgment of the MTM analyst should enable a reasonable estimate of the friction effect. Note also that when adhesives are present, special motion analysis should be made, since this would not exemplify the ordinary concept of sliding friction.

To determine the ENW when friction is present, the ENW as found for a spatial condition must be multiplied by a factor, always less than one, known as the Coefficient of Friction. For all practical purposes, a good average figure for all conditions is 0.4, in other words the ENW for sliding will be 40 per cent of the ENW for spatial moves of the same object weight. Actually, the coefficient of friction varies with the composition of the surfaces, the nature of surface finish, the velocity of motion, and there is even a difference between static and dynamic coefficients.

The earlier remarks about relative hand loading, of course, apply equally well for friction or resistive forces as for purely weight (gravity) forces. In either case, the ENW per hand—whether equal or different—fully expresses the weight control exerted.

Research has shown that, for the complexity and range of velocities encountered in manual motions, the effective change in coefficient of friction is very small. As to the surface composition and nature of finish, as well as state of lubrication, it may be desirable to account for the change in the coefficient in some instances. For this purpose, the commonly accepted values of friction coefficients given in Table 2 should prove adequate.

Table 2. Coefficients of Friction (Average Values)

Surface	Friction Coefficient	Average Value
Wood on Wood	0.3 to 0.5	0.4
Wood on Metal	0.2 to 0.6	0.4
Metal on Metal	0.3	0.3

It can be seen from this table that use of 0.4 as an average value is a relatively safe procedure.

Effective Net Weight

When all of the variables discussed above are summarized, a practical procedure for the evaluation of the extra time involved in moving weights above the basic time value will obviously depend on the ENW for the case in question. The current weight data provides a constant factor for the *Static Component* and a variable factor for the *Dynamic Component,* but both of these factors are based on the ENW of the Move being assigned a time value. The procedure in any particular case is as follows:

1. Decide the case of Move and the net distance involved for the hand(s)
2. Determine the ENW, per rules given below for the average analysis or by actual study for unequal hand loading
3. Write the complete motion symbol, as described in the section on THE SHORTHAND OF MOVE below, for each hand (either one or both hands may be involved)
4. Find the TMU value for the basic motion of each hand from the data card
5. Find the dynamic factor to be applied to the basic time value for the ENW of each hand from the data card. For precise work, the exact value of 1.1 per cent increase per pound may be used instead
6. Multiply the basic time for each hand by the dynamic factor
7. Read the constant factor to be added to the time for each hand for the static component, *if* the static component is needed. It is needed if the Grasp was just completed. It is not needed if the object was already under firm control by the hand in question. For precise work, the exact value of [0.475 + (0.345) × (weight in pounds)] TMU may be used
8. Add the appropriate constant value to the time for each hand found in Step 6
9. When both hands are involved, allow the time for the hand taking the greater amount of time as found in Step 8.

Although it takes a number of words to describe this process of finding the time for weighted Moves, the calculation is very easy and quick after a little practice. The rules for determining the Effective Net Weight (ENW) are given in Table 3, as they appear on the U.S.A./Canada MTM-1 data card since 1972. When using this table, remember that an average value of 0.4 for the coefficient of friction may be used with acceptable results in most cases. Additional analysis help is available in the Decision Model for ENW. The reasons behind these rules have been fully explained

Table 3. Rules for Determining Effective Net Weight (ENW)

EFFECTIVE NET WEIGHT			
Effective Net Weight (ENW)	No. of Hands	Spatial	Sliding
	1	W	$W \times F_c$
	2	W/2	$W/2 \times F_c$

W = Weight in pounds
F_c = Coefficient of Friction

in the section on variables in movement of weights.   A little practice will make the computation almost automatic.

THE SHORTHAND OF MOVE

To enable the analyst to record the various Move motions in a minimum amount of time, the conventional symbols outlined in the following paragraphs are used.   These are similar to the shorthand for Reach motions.

Move
Each time the work element Move is encountered, the analyst will record the capital letter M denoting Move on his study chart.

Distance of Move
The distance in inches that the operator moves the object is written after the letter M.   Rounding off to the nearest whole inch is the practice except for distances less than 1 inch.   Fractional dimensions (refer to Reach chapter) will be indicated by either the known fraction or a small case f.

> *Examples:*   M12 denotes a Move of 12 inches.
> $M\frac{5}{8}$ shows a $\frac{5}{8}$-inch Move.
> Mf indicates a short Move under $\frac{3}{4}$ inch, precise distance not known.

Case of Move
The Case of Move is identified by the capital letters A, B, and C appearing after the distance specification.

> *Example:*   M12A denotes a Move of 12 inches to the other hand or against a stop.

Type of Motion
This variable is symbolized exactly as was done for Reach.

Weight Factor
Following the convention for the case of Move, there is written the necessary numerals for the effective net weight (ENW) corresponding to the weight factor used from the data card.

> *Examples:*   M12B25 shows that a 25-pound object was moved 12 inches to an indefinite location.
> M12Bm5 shows that a 5-pound object was moved 12

inches and then dropped or tossed aside as the hand opened. (This kind of motion is often encountered when an operator drops an object in a tote pan and continues on to reach for another object or when he tosses aside an object.)

Notice that the measured or estimated weight or resistance to the nearest pound is written in the symbol to show what weight factor was used to determine the Move time. Refer to the "up to and including" specification mentioned in a paragraph following Table 4.

If the MTM analyst wishes to make a record of how an exact measured weight or resistance was determined for the particular Move motion, he should do this in the word description column of the work study form. When a weight was accurately determined during observation, the method used does deserve mention on the study form. In fact one MTM convention (accepted rule) uses this idea to distinguish between two differing possibilities. The two would otherwise be indeterminate by the usual method of recording motions on the study form. Assume that, on the same line of an analysis, the left-hand symbol column shows an M12B5 and the right-hand symbol column an M12B15. If no other MTM symbol appears, this means that simultaneously the left hand moved a 5-pound object and the right hand moved a 15-pound object. If actually a 20-pound object was moved using both hands, with the left and right hands supporting 5 and 15 pounds respectively, the convention also requires writing of the symbol "20/2" in both *description* columns. This shows that one object was handled by both hands together, with each hand supporting the amount of ENW shown in the symbol recorded for that hand in the *symbol* columns.

When the weight moved is so minor as to not require mentioning, no numerals after the case letter need be shown. To indicate token resistance for any special reason, the symbol may show a "2" for weights not exceeding that amount; since the dynamic weight factor would then be 1.00 and the static factor would be zero, the time would be the same as for the basic Move motion.

Example: M16B2 illustrates the presence of more than minor resistance but not enough to affect the Move time.

DATA CARD INFORMATION

The MTM-1 data for Move from the MTM-1 data card (Table II on the card) is reproduced fully in Table 4.

Table 4.   Time Data for Move (MTM-1 Data Card Table II-Move-M)

Distance Moved Inches	Time TMU				Wt. Allowance			CASE AND DESCRIPTION
	A	B	C	Hand In Motion B	Wt. (lb.) Up to	Dynamic Factor	Static Constant TMU	
3/4 or less	2.0	2.0	2.0	1.7				
1	2.5	2.9	3.4	2.3	2.5	1.00	0	
2	3.6	4.6	5.2	2.9				A Move object to other hand or against stop.
3	4.9	5.7	6.7	3.6	7.5	1.06	2.2	
4	6.1	6.9	8.0	4.3				
5	7.3	8.0	9.2	5.0	12.5	1.11	3.9	
6	8.1	8.9	10.3	5.7				
7	8.9	9.7	11.1	6.5	17.5	1.17	5.6	
8	9.7	10.6	11.8	7.2				B Move object to approximate or indefinite location.
9	10.5	11.5	12.7	7.9	22.5	1.22	7.4	
10	11.3	12.2	13.5	8.6				
12	12.9	13.4	15.2	10.0	27.5	1.28	9.1	
14	14.4	14.6	16.9	11.4				
16	16.0	15.8	18.7	12.8	32.5	1.33	10.8	
18	17.6	17.0	20.4	14.2				
20	19.2	18.2	22.1	15.6	37.5	1.39	12.5	
22	20.8	19.4	23.8	17.0				C Move object to exact location.
24	22.4	20.6	25.5	18.4	42.5	1.44	14.3	
26	24.0	21.8	27.3	19.8				
28	25.5	23.1	29.0	21.2	47.5	1.50	16.0	
30	27.1	24.3	30.7	22.7				
Additional	0.8	0.6	0.85		TMU per inch over 30 inches			

As was true for the Reach data, certain time values were obtained from extrapolation of curves rather than direct observation. These are Move Case B in motion at 3 inches and under, Case C Moves for 26 inches and greater, and in-motion B Moves at 26 inches or more distance. Remarks on this subject in the Reach chapter are equally valid here. The research report on *Short Reaches and Moves*[7] suggested several modifications of the data in question.

The data card information, combined with the information previously given, plus practical working rules which follow in this chapter will enable the MTM analyst to establish time values for the various kinds of Moves met in ordinary work situations.

It is often the practice for MTM analysts to weigh or estimate objects to the nearest five pounds. This fact influenced the way in which the data card factors for weight allowance shown in Table 4 were tabulated.

The first column of the Weight Allowance section of Table 4 is headed "Wt. (lb.) Up to." The values above 2.5 pounds in this column are the end points of five-pound weight intervals. The factors given in the two adjacent columns apply to all weights in the respective five-pound interval up to and including the weight value alongside the factors. The factor

[7]Available from the U.S./Canada MTM Association.

of 1.11 and 3.9, for example, apply to any ENW above 7.5 pounds up to and including 12.5 pounds.

The values of the factors listed are actually correct only for the middle of the weight intervals shown, which would be at the five-pound breakpoints. Thus, for example, the factor 1.11, which applies for weights in the interval from 7.5 pounds up to and including 12.5 pounds, is, strictly speaking, correct only for an ENW of 10 pounds and is only approximately correct for all other ENW values in the interval. When more precise answers are warranted by study conditions, the exact formulas given in the weight factor discussion should be applied. The data card values are, however, adequate for the usual motion analysis and were therefore presented in the form shown to eliminate the need for interpolation when weights slightly more or less than the midpoints in the five-pound intervals are encountered.

Complete clarity on the usage of the Move weight factors should be afforded by studying Fig. 3, which is self-explanatory.

Because body motions or both hands usually are employed when very large weights are lifted, there seldom is need to extrapolate the data card weight factors for loads in excess of 47.5 pounds. In most such cases, the use of mechanical handling equipment or adequate exploration of body motion application is strongly indicated. However, when on occasion pure Moves for loads in excess of 47.5 pounds must be evaluated, the static and dynamic weight factors can be found from the following extrapolation formulas which are based on *the exact formulas previously given*:

Static: $16.0 + 0.345 \, (ENW - 45)$

   This extrapolation formula says that for an ENW greater than 45 pounds the number of TMU is equal to 16.0 TMU for the first 45 pounds plus 0.345 TMU for each pound of ENW in excess of 45 pounds.

Dynamic: $1.495 + 0.011 \, (ENW - 45)$

   This extrapolation formula says that the dynamic factor for an ENW greater than 45 pounds is equal to 1.495 for the first 45 pounds plus 0.011 for each pound of ENW in excess of 45 pounds.

For example, the weight factors for an 80-pound ENW would be:

Static: $16.0 + 0.345 \, (80 - 45) = 16.0 + 12.1 = 28.1$ TMU for 80 pounds

Dynamic: $1.495 + 0.011 \, (80 - 45) = 1.495 + 0.385 = 1.88$ for 80 pounds.

To demonstrate usage of the Dynamic Component (DC) multiplier and the Static Component (SC) constant in weighted Moves, this example shows handling of a 25-pound object in various ways, with a coefficient of friction of 0.4, while making a basic M5B = 8.0 TMU transport of the object.

HANDS and LOADING	Type of Movement	Prior Weight Control?	MOVE TIME USING DATA CARD WEIGHT FACTORS												Allowed Time, TMU	USING EXACT WEIGHT FACTORS
			Gross Load, pounds LH	RH	ENW, pounds LH	RH	DC multiplier LH	RH	Travel Time, TMU LH	RH	SC const., TMU LH	RH	Total, TMU LH	RH		
ONE (RH only)	Sliding	Yes*	—	25	—	10	—	1.11	—	8.9	—	0	—	8.9	8.9	8.9
	Spatial	Yes			—	25	—	1.28	—	10.2	—	0	—	10.2	10.2	10.2
	Sliding	No			—	10	—	1.11	—	8.9	—	3.9	—	12.8	12.8	12.8
	Spatial	No			—	25	—	1.28	—	10.2	—	9.1	—	19.3	19.3	19.3
BOTH (Equal Loading)	Sliding	Yes*	12.5	12.5	5	5	1.06	1.06	8.5	8.5	0	0	8.5	8.5	8.5	8.5
	Spatial	Yes			12.5	12.5	1.11	1.11	8.9	8.9	0	0	8.9	8.9	8.9	9.1
	Sliding	No			5	5	1.06	1.06	8.5	8.5	2.2	2.2	10.7	10.7	10.7	10.7
	Spatial	No			12.5	12.5	1.11	1.11	8.9	8.9	3.9	3.9	12.8	12.8	12.8	13.9
BOTH (RH 3/5)	Sliding	Yes*	10	15	4	6	1.06	1.06	8.5	8.5	0	0	8.5	8.5	8.5	8.5
	Spatial	Yes			10	15	1.11	1.17	8.9	9.4	0	0	8.9	9.4	9.4	9.4
	Sliding	No			4	6	1.06	1.06	8.5	8.5	2.2	2.2	10.7	10.7	10.7	11.0
	Spatial	No			10	15	1.11	1.17	8.9	9.4	2.2	5.6	10.7	15.0	15.0	15.0
BOTH (RH 4/5)	Sliding	Yes*	5	20	2	8	1.00	1.11	8.0	8.9	0	0	8.0	8.9	8.9	8.7
	Spatial	Yes			5	20	1.06	1.22	8.5	9.8	0	0	8.5	9.8	9.8	9.8
	Sliding	No			2	8	1.00	1.11	8.0	8.9	0	3.9	8.0	12.8	12.8	11.9
	Spatial	No			5	20	1.06	1.22	8.5	9.8	2.2	7.4	10.7	17.2	17.2	17.2

With only these 16 of many possible handling methods, the time ranges from a low of 8.5 TMU to a high of 19.3 TMU; an average for this methods sample is 11.5 TMU with the data card weight factors and 11.6 TMU with the exact weight factors. Such averages could be used as standards for desired mixes of hand loading and handling methods, illustrating one form of detailed methods study possible with MTM-1.

*NOTE:　It would be very infrequent that prior weight control would exist for sliding conditions.　This is possible *only* if the contacting surface is *not below* the object; that is, when no gravity force reaction diminishes the amount of weight under manual control before the sliding starts.　In this case, the frictional forces must act only on the side(s) and/or top of the object being slid.　The bottom then would be the only part under full manual control and therefore able to meet the criteria of prior weight control.

Fig. 3.　Example Methods Study of Weighted Moves.

PRACTICAL WORKING RULES

Most of the information contained under this heading in the Reach chapter will apply equally well to the motion defined as Move. The main exception would be the practical formulas used to compute TMU values below. Therefore, only reference listing will be made here unless additional factors affect the particular topic listed.

Measurement of Distance

This is the same as for Reach motions. Note especially that Fig. 7 of the Reach chapter applies specifically to Moves without body assistance.

Body Assistance

One extra factor not listed in the Reach discussion merits mention here. Since large weights will be moved with either body assistance or extra body motions, rather than with the hands and arms alone, the MTM engineer should carefully notice during analysis of such Moves just what the true facts are. Being alert to this idea will help him to classify his motions correctly.

Types II and III Moves

Practical formulas for figuring types and cases of Move not shown directly on the data card are as follows (where the symbolism is the same as in the Reach chapter for similar formulas):

Formula Number	Case and Type of Move	Symbols for the Motion	Formula for TMU Value
(1)	Case A, Type II	m M (Dist.) A or M (Dist.) A m	= A Std. — (B Std. — Bm)
(2)	Case A, Type III	m M (Dist.) A m	= A Std. — 2 (B Std. — Bm)
(3)	Case B, Type III	m M (Dist.) B m	= B Std. — 2 (B Std. — Bm)
(4)	Case C, Type II	m M (Dist.) C	= C Std. — (B Std. — Bm)

The basis for these formulas becomes self-evident with a little study. The point to remember is that the acceleration or deceleration time for a given length of Move is a constant, regardless of the case of Move. See also Table 1 for the "m" values.

Odd Distance Moves Over 10 Inches

This is the same as for Reach motions.

Moves Over 30 Inches

This is the same as for Reach motions except for the extension TMU values. These are given in Table 5.

Table 5.   TMU Values for Move Increments Over 30 Inches*

Case of Move	A	B	C
Additional TMU per inch	0.8	0.6	0.85

* Since the 1972 revision of the MTM-1 data card, these data have been included as the bottom row of the Move table (see Table 4).

How to Identify and Analyze Move Motions

The general analysis procedure for Move motions is the same as was described in the Reach chapter.   However, several additional items are presented here which broaden the reader's analysis techniques.

Occasionally an analyst observes Moves which do not exactly fit any of the standard MTM definitions.   Examples of such Moves are those found in spray painting, welding, metal filing, buffing, hand rubbing, polishing, etc.   In practically every such case, an object being used to produce useful work is in the operator's hand.   However, the control of the Move is not strictly manual due to process factors.   There seems to be some room for analyst judgment here.   If most of the control is by the operator, some analysts accept a slight loss of accuracy and assign MTM Moves to the task.   Others "tinker" with the situation to make up the time deficit from a normal MTM Move by assigning, say, a Case C Move instead of Case A or B or by adding synthetic weight factors.   While both of these practices often yield a workable answer, they are a *misuse of the MTM procedure*.   The proper analysis is to recognize them as process time and use a stopwatch or other timing instrument to more correctly and accurately determine the required time.

However, as previously mentioned, MTM Moves are correct for the usage of many hand tools.   This does not imply that the application is simple in all cases, as is illustrated by considering hammering motions. Raising the hammer to get ready for striking definitely is a Case B Move. If the hammer weighs more than 2.5 pounds, the raising motion always would involve the dynamic weight factor; and the first time it was picked up and raised would also require the static weight factor.   To this point, the analysis has been routine with respect to MTM motions.

Analysis of the striking motion, however, presents some challenge to the determination of hammering movements.   To establish the first fact, the striking motion definitely is operator controlled rather than process controlled.   But does it conform to the usual behavior reflected by the MTM Move motions with respect to acceleration and deceleration?   At the start of the striking motion, the acceleration does build up in the normal manner. However, the stroke does not achieve a relatively stable velocity and then gradually decelerate (for a period of time regarded equal to the acceleration period) to the end point as does a normal spatial MTM move. Indeed, the whole purpose of the striking motion is to increase the momentum of the hammer head to assure that the required impact force is generated. This means that the aim is to continue speeding up throughout the stroke. Only at the point of impact does any deceleration appear, and then it is extremely abrupt. Therefore, the striking motion must be similar to a Type II Move in motion at the end; but it probably requires less time than does the normal MTM Type II Move. Thus, either the striking motion should be timed for accuracy or else the analyst can specify an MTM time with the knowledge that it may be too "loose." If the hammering is a relatively small part of the total work cycle, the latter approximation may well be reasonable. To carry this illustration forward, let it be assumed that such is the case.

With this assumption, the striking motion can be analyzed as follows. The dynamic weight factor will apply for all hammers over 2.5 pounds in weight. Also, wrist assistance will tend to shorten the net Move distance, as well as to impart some of the required momentum. An easy way to measure the wrist assistance is to first locate the wrist in its final orientation and then determine the length of the remaining striking path at the index knuckle without the complication of wrist assistance. As to case of Move, the destination accuracy required will decide this. Case A will apply for rough pounding blows, Case B for routine hammering, and Case C for accurate aiming of the hammer head. The net analysis therefore could be M—Am, M—Bm, or M—Cm together with the appropriate net distance and ENW. Note that the M—Cm is not an exact motion description, because Type II Case C Moves are not possible; this is merely a way to show deduction of the deceleration time for this analysis. Actually, the M—Am and M—Bm also would not denote in-motion at the end either. This is an approximation of the real event—an "equivalent" motion—for purposes of assigning a reasonable time value by eliminating the deceleration time in these two cases as well.

To complete the hammering analysis, consider next the return stroke after striking has occurred. Usually this will be an M—B with any required ENW for the dynamic weight factor. However, in certain types of hammering, the struck object imparts a definite and significant bounce to the hammer head. To shorten this analysis, the reader should recognize that this would justify an "equivalent" motion of mM—B, a Type II in-motion at the start. Of course the ENW and wrist assistance would again enter into the symbol and time assigned.

With this discussion of process time and the "equivalent" motions as illustrated by the hammering analysis, the reader should be able to apply the Move data with greater awareness of both its utility and limitations. Also, this knowledge will aid in understanding practical usage of all other MTM motion data.

SPECIAL LATE CHANGE

As this edition was nearing completion, a basic change was being considered by the U.S.A./Canada MTM Association regarding the symbolization of weighted Moves. Since the proposed change actually would adopt a practice used for many years in Europe, it merits explanation as a special item even if the U.S.A./Canada revision action is not taken. However, the existing text is not now being changed except by this special section.

As previously discussed, objects weighing more than the minimum of 2.5 pounds justify an increase in Move time allowance. This increase is analyzed by applying the static and dynamic weight factors, either from the data card or by using the exact equations also presented earlier. When appropriate weight time must be allowed, the *Dynamic Component* multiplier always applies against the basic Move time; however, the *Static Component* time constant may or may not be needed as already clarified. None of this practice is to be changed.

However, the symbolization of weighted Moves as shown in this chapter had two deficiencies which can be solved by revising the symbol convention. Since the numerals for ENW in the convention were the only symbolic recognition of extra time for weight allowance, it was not always evident whether the pattern time did or did not include the Static Component unless the pattern reader performed a checking calculation; and this could be further complicated by no indication whether the data card or the exact equation had been used. Also, the same weighted Move convention could—as a result—mean more than one thing and legitimately have more than one time assigned to it within the existing rules. The proposed change would remove these ambiguities from the practice regarding weighted Move symbols.

The solution is a separate convention for the Static Component in addition to the existing weighted Move convention. The Static Component symbol is SC and is followed by the ENW to complete the convention. The time value is thus unique in respect to each convention. For example, the analysis for 12 pounds ENW in one hand could be:

$$SC12 = 3.9 \text{ TMU for data card usage, } or$$
$$SC12 = 4.6 \text{ TMU for exact formula usage.}$$

In either case, the convention and time would be written on the analysis line *preceding* the weighted Move symbol; this truly represents the actual physical method. Also, the SC convention would be present only if needed.

With this revised SC convention, the weighted Move convention would not be changed; but its meaning is slightly altered, because it now shows *only* the Dynamic Component presence and time effect. For example, again using a 12-pound ENW in one hand, the *new* result would be:

M10B12 = 13.5 TMU for data card usage, *or*
M10B12 = 13.8 TMU for exact formula usage.

This would differ from the previous practice where a 12-pound ENW with *both* weight factors applied has been correctly shown with the *same* convention as:

M10B12 = 17.4 TMU for data card usage, *or*
M10B12 = 18.4 TMU for exact formula usage.

The net result is unchanged as to both method and time analysis, since the combined *new* conventions would be:

SC12 = 3.9 TMU
M10B12 = 13.5 TMU } giving 17.4 TMU for data card usage, *or*
SC12 = 4.6 TMU
M10B12 = 13.8 TMU } giving 18.4 TMU for exact formula usage.

As can be seen, the proposed new (for U.S.A./Canada; not Europe) practice is identical in total result, but with *no* ambiguity or mystery as to either the method or time allowance on direct reading from the analysis.

Accordingly, the reader should readily understand and follow this proposed revised practice even though the rest of this text and illustrative materials are unchanged except for this section.

DECISION MODELS

The Decision Models for Move analysis follow as Figs. 4A, 4B, and 5. These cover the basic motion and effective new weight data. Of course, Fig. 12 of Chapter 5, Reach, on finding net distance is also part of the Move models and should be reviewed as well.

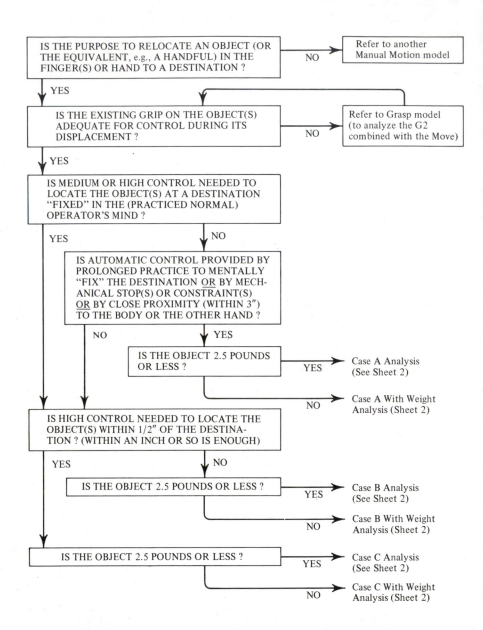

Fig. 4A.   Decision Model for Move.

(from Sheet 1)

			IS THE OBJECT STOPPED AT <u>BOTH</u> THE START AND THE FINISH ?		
				NO	
				IS THE OBJECT IN MOTION <u>ONLY</u> AT THE START ?	
			YES		NO
				YES	IS THE OBJECT IN MOTION <u>ONLY</u> AT THE FINISH ?
					YES NO
CASE ANALYSIS	Relocating the object	TYPE I	TYPE II in motion at start	TYPE II in motion at finish	TYPE III in motion at <u>both</u> start and finish
A	to the other hand <u>or</u> against a stop	M <u>x</u> A	mM <u>x</u> A	M <u>x</u> Am	mM <u>x</u> Am
A With Weight	Case A for object over 2.5 pounds	M <u>x</u> A <u>#</u>	mM <u>x</u> A <u>#</u>	M <u>x</u> Am <u>#</u>	mM x Am <u>#</u>
B	to an approximate or indefinite location	M x B <u></u>	mM x B <u></u>	M x Bm <u></u>	mM x Bm <u></u>
B With Weight	Case B for object over 2.5 pounds	M x B <u>#</u>	mM x B <u>#</u>	M x Bm <u>#</u>	mM x Bm <u>#</u>
C	to an exact destination	M <u>x</u> C	mM <u>x</u> C	Not Possible (Analyze Case B)	Not Possible (Analyze Case B)
C With Weight	Case C for object over 2.5 pounds	M <u>x</u> C <u>#</u>	mM <u>x</u> C <u>#</u>	Not Possible (Analyze Case B)	Not Possible (Analyze Case B)

x = Refer to Net Distance for Reach and Move model.

= Refer to Effective Net Weight for Move model.

Fig. 4B.   Decision Model for Move.

LET: W   = total weight, pounds, of the object moved

    #   = the <u>per hand</u> Effective Net Weight, pounds

    ˑDC   = <u>Dynamic Component</u> multiplier of the Weight Factor (for added time to control the weight during Move)

    SC   = <u>Static Component</u> additive of the Weight Factor (for added time to gain full control of the weight—not the object—before the Move can begin)

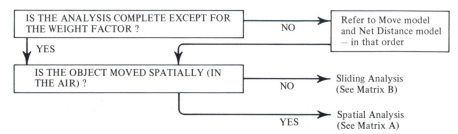

MATRIX A

SPATIAL ANALYSIS

WAS THE OBJECT ALREADY SPATIALLY CONTROLLED PRIOR TO THE MOVE ?		IS THE OBJECT MOVED BY ONLY ONE HAND ?		
		YES	**NO**	
			IS THE LOADING EQUAL BETWEEN HANDS ?	
			YES	NO
	YES	# = W (Using <u>only</u> the DC)	# = 0.5 W (Using <u>only</u> the DC)	# = 0 to W <u>and</u> #(LH) + #(RH) = W (Using <u>only</u> the DC)
	NO	# = W (Using <u>both</u> DC and SC)	# = 0.5 W (Using <u>both</u> DC and SC)	# = 0 to W <u>and</u> #(LH) + #(RH) = W (Using <u>both</u> DC and SC)

MATRIX B

SLIDING ANALYSIS

IS THE MOVE A TYPE III OR A TYPE II IN MOTION AT THE START ?		IS THE OBJECT SLID ON THE SURFACE BY ONLY ONE HAND ?		
		YES	**NO**	
			IS THE LOADING EQUAL BETWEEN HANDS ?	
			YES	NO
	YES *	# = 0.4 W (Using <u>only</u> the DC)	# = 0.2 W (Using <u>only</u> the DC)	# = 0 to 0.4 W <u>and</u> #(LH) + #(RH) = 0.4 W (Using <u>only</u> the DC)
	NO	# = 0.4 W (Using <u>both</u> DC and SC)	# = 0.2 W (Using <u>both</u> DC and SC)	# = 0 to 0.4 W <u>and</u> #(LH) = #(RH) = 0.4 W (Using <u>both</u> DC and SC)

* In these cases, the inertia of the object to starting will be overcome by the impact of the hand in motion.

SLIDING NOTE: The Coefficient of Friction used here is 0.4; if appropriate, other values may be substituted and the equations adjusted.

GENERAL NOTE: Substitute the # value in the Move symbol to complete it.

Fig. 5. Decision Model for Effective Net Weight for Move.

CHAPTER 7

Turn

Definition[1]

*TURN[2] is the basic motion employed to rotate the hand about the
long axis of the forearm.*

  1. The hand may be empty or holding an object.
  *2. Turn cannot be made while holding the wrist firm. Turn involves
  the two bones in the forearm and a pivoting motion at the elbow.*

While pure Turns as defined above do sometimes occur in industrial
operations, they occur more frequently in combination with Reaches and
Moves. This fact has led to statements by some sources that Turn is
a special way of performing a Reach or Move. Such is not the case.
While it is true that Moves and Reaches may include performance of the
basic motion Turn, this does not preclude the ability of the worker to
accomplish pure Turns alone.

Clarity on the distinction just made may be afforded by resorting to
the reader's knowledge of mechanics and physiology. When the hand
reaches or moves, with no shoulder or body assistance being employed,
the manual members (hand, wrist, forearm, and upper arm) all exert their
muscular influence on the motion with essentially *hinging* actions of each

---

[1]From "MTM Basic Specification" (Footnote 1, Ch. 5).

[2]The motion defined here is not a Therblig, but was first identified and isolated as to time
during the Methods-Time Measurement research. That it is a distinctly different motion
from Reach or Move as to performance, time, and measurement will be evident from this
chapter. This discovery afforded ready explanation of many previously dubious motion occurrences, and the results derived from its use in a practical way has justified separated consideration.

or several skeletal joint(s) guiding the performance. When, however, the manual members must accomplish a Turn, the required rotating effect of the hand must be produced by additional muscles not used during Reach or Move. The bone joints must now be subjected to *torsional* actions by the employment of less efficient muscles, and this clearly indicates a distinct performance time from that described above. Logic thus amply supports the isolation of this Turn motion from other manual actions.

Turn as just described may occur as a separate motion or in combination with others. When it occurs by itself as a pure Turn, the forearm axis (axis of rotation) is not displaced laterally. When in combination

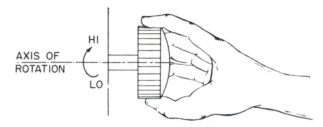

Fig. 1. Example of pure Turn. A pure Turn illustrated by the rotation of a control knob about its fixed axis. That the axis of rotation does not change location is evident.

with Reach or Move, lateral displacement of the forearm axis does occur. Examples of Pure Turn, Reach-Turn, and Move-Turn are shown in Figs. 1, 2, and 3.

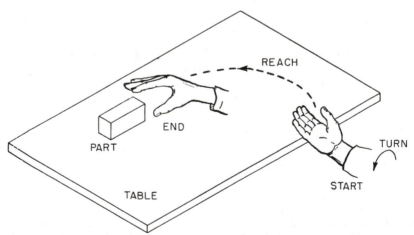

Fig. 2. Example of Reach-Turn. With the palm initially facing upward, a Reach for an object is made while the palm revolves to face downward. The axis of rotation for the turning has obviously shifted almost as great a distance as was reached.

A pure Turn motion practically always involves the use of only one hand. This is because the long axis of the forearm must be nearly perpendicular to the plane of rotation to permit the Turn to occur. This is an important recognition fact to remember, and is pictured in Fig. 4. There are occasions, however, when simultaneous Turn motions will be made by both manual members. These may be simultaneous independent

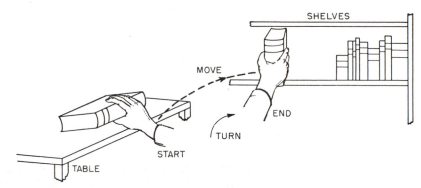

Fig. 3. Example of Move-Turn. Suppose a book was lying flat on a table just prior to being grasped by the operator. Should it next be moved to a library shelf nearby in such a manner that it will be inserted upright, a right-angle Turn must have taken place in the distance of movement. During this Move-Turn, of course, the axis of rotation again shifted almost as much as the Move distance.

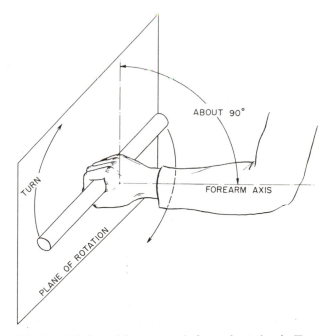

Fig. 4. Relation of forearm and plane of rotation in Turn.

Turns or a coincident Turn where both hands are on the object being turned. Taking a shower bath provides a simple example of a simo Turn situation. The hot and the cold water handles are often turned using both hands simultaneously.

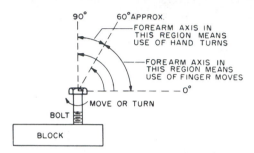

Fig. 5.   Location of forearm as it affects Turn or Move.

The MTM analyst must be careful to avoid confusion sometimes caused by the ordinary expression of "Turn the object." This instruction will not of necessity imply the use of a Turn motion as defined in MTM since objects can also be turned with Move motions of the fingers or hand, and by means of a special motion called Crank (see Chapter 8).

The criterion is the location of the forearm long axis. Study Fig. 5 which shows a loose bolt that is to be turned into a threaded block. As long as the arm axis is at an angle of from zero to approximately sixty degrees to the surface of the block, the fingers and hand will perform a series of Move, Release, Reach, Grasp, Move, etc. But when the axis of the arm is located between about sixty to ninety degrees with the surface of the block, the fingers and hand will execute a series of Turn, Release, Reach, Grasp, Turn, etc. Thus, the motion analysis requires clear distinction if proper classification is to be achieved.

In the MTM element known as Position (see Chapter 12) objects must often be turned slightly to provide orientation prior to insertion. This kind of Turn need not be indicated as a separate motion during analysis, because its presence and time value has been integrated into the Position motions, which also involve other basic MTM motions.

TURN VARIABLES

Two major variables, *Degrees of Turn* and *Resistance to Turn,* have been evaluated for Turn motions. These are discussed in detail in the following material. A possible third variable is also discussed briefly.

Degrees of Turn

The further the action of the forearm muscles causes the wrist and hand to be rotated, the longer will be the time required. The most rational manner in which to express the amount of Turn performed is in terms of the number of angular degrees involved. Experimentation by the reader will show, however, that the rotation cannot exceed 180 degrees due to the physical structure of the forearm. The two bones of which the forearm is composed limit the rotation to this maximum amount.

As a practical matter, there is also a lower angular limit for Turn (about 15 degrees) below which it becomes difficult for the analyst to distinguish the Turn motion from ordinary minor twisting of the wrist present in all normal Reaches and Moves. It would, in fact, be correct to state that small amounts of Turn are present in almost every Reach and Move motion of the hand. To attempt to isolate and measure such Turns, however, would be impractical.

Resistance to Turn

The rotation of the *empty* hand will obviously not require any effort or work above that normally associated with such muscular movements.

When a Turn is applied to an object having appreciable weight or offering resistance to rotation, however, the net result is to increase the time required to produce the rotation desired.

For a given amount of rotation, existing experimental data indicates that the time for Turn due to resistance generally increases with the weight of the object as a smooth curve. The amount of data available did not permit the originators of MTM to develop a complete weight curve, so therefore they classified objects being turned into three weight categories:

> Small — 0 to 2 pounds, inclusive
> Medium — Over 2 to 10 pounds, inclusive
> Large — Over 10 to 35 pounds, inclusive

From a practical viewpoint, no substantial loss in the accuracy of time values has resulted from using this simplified classification procedure.

When weights larger than 35 pounds must be rotated, motions other than Turn will be involved. This fact explains why the upper limit of the classifications was set at 35 pounds. Until additional research is made (a need for which was indicated by the originators of MTM) these classification ranges should suffice. Successful use has validated this conclusion.

The first type of situation in which the weight of the object can prolong the Turn time occurs when the object is turned while essentially free of attachment in space, that is, the entire weight is supported by the unaided hand of the operator. The weight will retard the Turn motion according to the magnitude of the weight; light objects will slow the Turn but little, while heavy objects will create a noticeable reduction in the speed of Turn. This behavior parallels that of weight in Move motions because the worker's muscles must overcome the pull of gravity in much the same way to maintain control of the object.

Not only does the weight of an object influence the time of Turn, but the distribution of the mass of the object also has an effect. This can be verified simply by picking up a steel bar, say 1 inch in diameter and 12 inches long, holding it horizontally with one hand, and turning it back and forth about its long axis. If the same bar is next grasped at the center of its length, and twisted back and forth about an axis at right angles to the axis of the bar (with the forearm aligned to the perpendicular axis), it will be found that considerably more effort and time is required to make, say, ten Turns back and forth than in the first case. The explanation is simple: The moment of inertia of the object about the given axis of rotation is greater when the bar is held in the second position; hence, greater turning effort is required in accelerating and decelerating the bar during the Turn motion.

Under certain conditions, resistance effects may actually exceed the weight of an unattached object. Study of the laws of levers in physics will show that minimum operator effort in turning will prevail when the object is being held at a point close to its center of gravity. When the object is held at a point sufficiently displaced from its center of gravity and with its axis of rotation perpendicular to the force of gravity, the unbalanced condition could easily cause an increase in the control effort required of the arm muscles and, therefore, an increase in Turn time. This unbalance factor will require evaluation during future research since no data now exists on it.

For the present, the actual weight of the object is used in determining which of the three weight categories previously defined apply to evaluate the Turn resistance of free objects as just discussed.

A second type of Turn resistance occurs when the object being rotated is attached to or part of another object, or when mechanical constraint is present. In this case, the object and/or its attachments exerts a resisting torque in opposition to the turning effort of the hand and, if motion is to ensue, this resisting torque must be more than equalled by the product of the force applied by the hand and the distance from the point of holding to the axis of rotation. When the resisting torque is known, the hand force may be found from the following equation:

$$\text{Resisting Torque (inch pounds)} = \text{Hand Force (pounds)} \times \text{Lever Arm (inches)}$$

In using this equation, the lever arm is the measured distance from the axis of rotation to the point of application of the force as seen in Fig. 6. The force desired is that component perpendicular to the lever arm. It is this force that must be used as the weight factor in the three categories previously defined. The control effort the hand must exert to produce turning against a resisting torque is the same as it would exert to Turn a free object weighing the same as the force applied.

A torque wrench or torque balance may sometimes be used to obtain the value of the resisting torque directly without any need for Applied Mechanics calculations. In such cases the measured resisting torque and the length of the lever arm used by the operator's hand (measured from

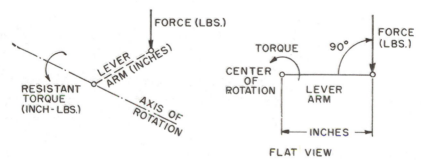

FLAT VIEW

Fig. 6.　Measurement of resistant torque.

between the thumb and the most remote finger as actually gripped) are all that is needed to determine the desired hand force from the formula:

$$\text{Hand Force F in pounds} = \frac{\text{Measured Resistance Torque T in inch pounds}}{\text{Lever Arm L used by operator's hand, in inches}}$$

A direct method for obtaining the force or weight factor present when turning against a resisting torque is to use a spring scale applied to the object being turned at the point where the hand would hold it. The scale must, of course, be held perpendicular to the lever arm when the force reading is taken and only enough pull should be exerted to barely turn the object.

In the majority of cases none of the foregoing procedure will be necessary since the hand force applied by the operator to cause rotation can usually be estimated and classified as being in the small, medium, or large category with sufficient accuracy.

Three examples of torque application during pure Turns are:

　　Revolving a door knob when the lock mechanism is rusty.

　　Driving a wood screw into a piece of green or hard wood, with the arm almost perpendicular to the surface of the wood.

　　Setting a rheostat knob that is constrained by stiff detent springs.

Type of Motion

The third Turn variable is type of motion.   Research has not yet shown whether Type II and Type III motions described in the chapter on Reach are possible during turns, but the inventors of MTM surmised the possibility in their textbook.   They felt that, for isolated observations of what appeared to be Type II Turns, reduction of the table time value by 1.4 TMU would be a reasonable procedure.   This may, therefore, be *used with caution* pending further research.

DATA CARD INFORMATION

The time values for Turn given on the MTM-1 Data Card are shown in Table 1.

Table 1.   Time Values for Turn (MTM-1 Data Card Table IIIA—Turn—T)

Weight	Time TMU for Degrees Turned										
	30°	45°	60°	75°	90°	105°	120°	135°	150°	165°	180°
Small — 0 to 2 Pounds	2.8	3.5	4.1	4.8	5.4	6.1	6.8	7.4	8.1	8.7	9.4
Medium — 2.1 to 10 Pounds	4.4	5.5	6.5	7.5	8.5	9.6	10.6	11.6	12.7	13.7	14.8
Large — 10.1 to 35 Pounds	8.4	10.5	12.3	14.4	16.2	18.3	20.4	22.2	24.3	26.1	28.2

Turns with the empty hand are assigned times as given for the S (small) category.

Notice that time values are given for degrees of rotation in 15-degree increments.   The 15-degree increments represent a practical division of the angular measure because of the difficulty of closer measure in most application cases.   In view of this, it is logical to eliminate consideration of any turn 15 degrees or under; and the table therefore begins at 30 degrees.   Also, the time assigned from the data card should be that for the 15-degree interval nearest the measured angle, unless demands for increased accuracy prompt the MTM analyst to interpolate in the table.   For example, a Turn measured at 62 degrees under normal accuracy requirements would be symbolized and given a time value for the 60-degree turn.   For more stringent accuracy, the symbol and time for 62 degrees would be assigned.

The originators of MTM suggested that even the 15-degree increment may sometimes be too fine for identification of observed Turns.   As an unofficial approach they suggested the use of the "45-degree step rule" which indicated that the analyst should only attempt to isolate angular values of 45, 90, 135, and 180 degrees in such cases.

Perhaps further research and better means of measurement will result in a change of such rules as are given above, but this type of practical approach combined with validated research and successful application has done much to advance MTM ahead of other predetermined time systems.

THE SHORTHAND OF TURN

This discussion presents the shorthand symbols, or writing conventions, for the MTM element defined as Turn.

Turn

The MTM analyst will show a capital T on his study sheet to indicate the occurrence or necessity of the work element called Turn.

Degrees of Turn

The amount of Turn expressed in degrees is noted on the study form following the letter T. This should normally show the relative amount of rotation to the nearest 15-degree interval in accordance with the data card limits. Thus, a turn of 36 degrees would be noted as 30. When greater accuracy is justified and desired, however, the symbol should show the true angular measure, in this case T36. (Refer to the previous section.)

In a manner similar to the handling of distances for Reach and Move motions, the amount of Turn will generally be determined and noted after all of the basic motions have been listed in sequence on the study sheet. That is, the analyst will first decide the kind of motions present or required and will then proceed to classify them.

Resistance Factor

This is denoted by the capital letters S, M, or L as needed (not needed for an empty hand) to show the relative size of the object being turned or the resistance it offers. Examples are:

T90L indicates a 90-degree Turn of an object weighing between 10.1 and 35 pounds; or it may also show that the force required to overcome the torque resistance of the object ranged in this equivalent weight category.

T150 would denote a 150-degree Turn of the empty hand, since this involves zero weight.

Type II Turns

It was indicated that Type II Turns might occur in some industrial operations. In such cases, it is suggested that the analyst use the conventional symbolization as given above, and then add a note opposite the symbol that it is a Type II Turn. Remember this is not official data, however.

The use of complete and correct symbolization will eliminate confusion as to whether the motion variables were properly considered.

COMBINED TURN MOTIONS

After the definition for Turn, it was stated that pure Turns do not occur as frequently in industrial operations as do Turns in combination with Reach and Move.   Clarification of this statement was delayed to now because consideration of the combined motions presumes prior familiarity with the pure Turns.   The first step now is to achieve understanding of the specialized meaning of the term *Combined Motion* as used in MTM.

Up to this point, material on motions has discussed each MTM element without regard to the sequence in which it may appear during a task. Obviously, one way to perform the motions is to do them consecutively. Combining is another.

An example from everyday life illustrating consecutive and combined methods of performing a common task occurs when a man rides the bus to his place of employment and then reads his morning newspaper before the starting whistle blows.   This exemplifies consecutive performance and would require addition of the time for each task to find the elapsed time. Both tasks might also be completed, obviously in less elapsed time, by the worker reading his newspaper on the bus while he is riding to work.   In this way, he has *combined* both tasks to save himself time.   This statement, extended to motion times rather than task times, should make readily apparent the meaning of *Combined Motion.*

A definition for Combined Motions is:

Combined motions are two or more non-consecutive elemental motions performed during the same time interval by the same body member.

The average industrial worker will naturally tend to combine Turns with Reaches and Moves as a matter of ease and convenience.

In respect to Turn motions, several questions (answered below) are raised by the definition of *Combined Motions:*

1. How is the time determined for combined Turns?
2. How are combined Turns symbolized?
3. What effect on the distance of Reach and Move is caused by their combination with a Turn?

A further question, covered in later chapters, concerns the kinds of motions which can be combined and the ways in which such combinations can occur with average operators.   For the present, it is enough to realize that Turn can be combined with Reach and Move.

Limiting Time

When several motions are combined, the elapsed time allowed for the combination is the time to complete the motion requiring the greatest amount of time. A shorter way of expressing this is to say that the longest-time motion is "limiting" and the shorter-time motions are "limited." The allowed time is the "limiting time." The reason for this is obvious from the previous discussion.

To obtain the time for a Turn combined with Reach or Move, it is merely necessary to compare the Turn time with the Reach or Move time to see which is "limiting."

As an illustration, a T135 will "limit" an R3B because the Turn requires 7.4 TMU and the Reach takes only 5.3 TMU; a combined Turn containing these motions would thus be assigned the time value of 7.4 TMU, since the R3B is "limited out." In another instance, an M6B6 (11.6 TMU) will be "limiting" in a combined motion with a T75M Turn (7.5 TMU), which is the "limited" motion.

There are no special rules of thumb to find out which motion in a combined Turn is limiting. The analyst must examine the data card in each case. However, the likelihood of the Turn being limiting is greater with short Reaches or Moves; while for long Reaches or Moves, the Turn will generally be limited.

Combined Conventions

The symbolism for combined motions requires that the separate motions be written on two or more lines of the analysis form. The limited motions are shown by drawing a slanted line through them, as shown in the example at right. The bracket is used to denote the combination, as it helps show whether the limiting motion is the one written on the line above or below the limited motion. Note that a dash (—) in the TMU column signifies that the T135S time is not allowed.

Motion	TMU
R18B	17.2
G1A	2.0
M12B⟩	13.4
~~T135S~~⟩	—
G2	5.6
M6C	10.3
P2SSE	19.7
RL1	2.0
Total Allowed TMU	**70.2**

The method of writing both the consecutive and combined motions should be evident from the example, even though the reader may be unfamiliar with the motions other than Reach, Move, and Turn at this point. The manner of adding times for a task is also illustrated here, this being what is called a motion pattern. The time for the Turn, it being limited out, is not allowed in the total for the task.

Wrist Assistance in Combined Turns

A combined Turn will usually assist the Reach or Move; that is, it effectively shortens the allowable distance involved. In a few cases, a combined Turn can have the opposite effect—what may be called wrist resistance—it lengthens the effective Reach or Move distance. The idea of resistance would be the same as discussed in the Reach chapter—it increases the time needed to complete the motions.

An example of wrist assistance is shown in Fig. 7 along with the

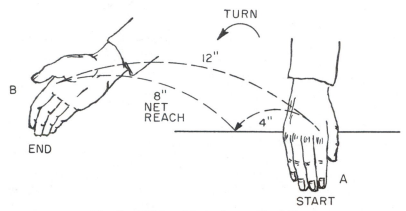

Fig. 7.   Wrist assistance in combined Turns.

distances involved. In view A, the right hand starts palm down on a table and makes a 12-inch Reach to the right while the palm is turned upward with the thumb outward so that a combined 180-degree Turn has occurred. The Turn assisted the Reach by about 4 inches, about palm width for an average male operator. Thus, the knuckle-to-knuckle distance of 12 inches should be decreased to 8 inches to give the allowable distance of Reach for reference to the data card time value. Other cases of combined Turn assistance would be measured in a similar manner.

The best way to measure a Reach-Turn or a Move-Turn is to first make the Turn and then complete the Reach or Move. Then the degrees turned and the amount of assistance or resistance can easily be measured separately and distinguished from the total index knuckle travel.

PRACTICAL WORKING RULES

For all of the Turn variables isolated above, practical means of recognition and measurement of the quantities are necessary.

Angular Measurement

The long axis of the forearm was identified as the axis of rotation for the manual members during Turn motions. This axis intersects the hand through the middle finger. That intersection therefore serves best as the origin of the angle of Turn. Point A in Fig. 8 is this origin.

To observe and measure the angle turned, the analyst needs another point to watch. The line connecting this latter point and the origin will

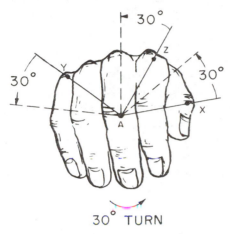

Fig. 8. Angular measurement points on the hand.

generate the angle of Turn as the hand revolves about the Turn axis. The further this second point is located from the origin, the easier will be the angular observation.

Another convenient point on the hand is often selected for this purpose. See Fig. 8 again. Point X, at the thumb knuckle, is the furthest from the middle finger and provides the best "measure point" on the hand. The knuckle of the little finger, at point Y, is next in desirability. Least desirable, because it is closest to the middle finger, is the knuckle of the index finger at point Z. None of these points travel more than a few inches during the Turn, which makes this means of measuring the angle difficult.

A better way to observe this variable is to extend the principle given above. Observe a point further away from the middle finger than the points on the hand just mentioned. This point may be located at some distance on the object being turned, somewhere on a prominent projection of the part, or on an object palmed for the purpose of providing an observation point. A rod or rod-like object such as a pencil palmed between the middle and index fingers provides the best means of finding the angle during a pure or combined Turn. This is especially true if the Turn is made with an empty hand that provides no object for selection of an observation point.

For MTM analysts with enough practice, the most practical way to obtain the Turn angle for normal accuracy is to estimate it to the nearest 15-degree increment shown on the data card. Notice that this suggestion differs from that given for distances in Reach and Move. For these distances, it is preferable to measure, although accurate estimation is permissible. For Turn angles, however, estimation is *usually* adequate.

Size Classification

Due to the relatively large differences in the weight categories on the data card for Turn, the MTM analyst can usually judge closely enough the Small, Medium, or Large size classifications. For borderline cases, however, he may resort to a measuring device such as was discussed under resistance to Turn in this chapter.

A combined Reach-Turn would always be in the S time category because the empty hand carries zero weight. Combined Move-Turns may fall in any of the three size ranges, depending on the weight of the object or the resistance offered.

An interesting relation between the size or weight categories is the relative increase of performance times with increased weight. The data card values for Medium are 1.57 times the Small category, and those for Large are 3.00 times the Small value. In other words, the time required is increased 57 and 200 per cent in the respective cases.

Earlier, it was stated that in *extremely* rare cases one might wish to indicate Turn symbols to the nearest degree of rotation. With the size time ratios just mentioned above and a time equation for the Small category, exact Turn times could be allowed for each degree in these very unusual cases. By plotting the Small time values, it can be seen that the time variation is linear; and the time equation is: TMU = 1.48 + .044 × degrees. Therefore, exact Turn times can be generated for every degree of angular rotation (from 30 to 180 degrees) by the following formulas:

$$T\_S \text{ , TMU} = 1.48 + .044 \times \text{degrees}$$
$$T\_M, \text{TMU} = 2.32 + .069 \times \text{degrees}$$
$$T\_L \text{ , TMU} = 4.44 + .132 \times \text{degrees}$$

DECISION MODELS

The Decision Model for Turn is given as Fig. 9. Note also, in Fig. 12 of Chapter 5, that wrist assistance (or wrist resistance) such as occurs in

combined Reach-Turn and Move-Turn motion sets are covered by the
Decision Model for Net Distance; therefore, that model also is a part of
the review for Turn motions.

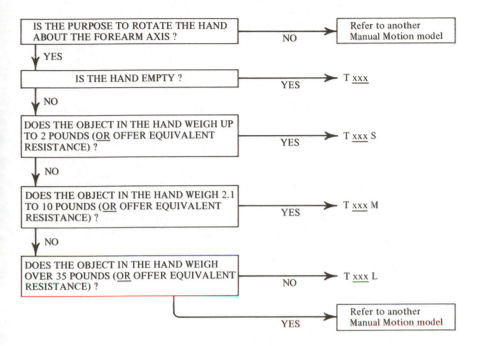

NOTE: Fill in angle of rotation, xxx, to the nearest 15-degree
      increment to complete the Turn symbol; but with a 30-degree
      minimum and a 180-degree maximum value.

Fig. 9.   Decision Model for Turn.

CHAPTER 8

Crank

___

Definition[1]

CRANK is the motion employed when the hand follows a circular path to rotate an object, with the forearm pivoting at the elbow and the upper arm essentially fixed.

1. *Crank is defined in terms of manual action, not by object movement.*
2. *With small cranking diameters and light cranking loads, the wrist undergoes a whipping action while the forearm pivots only slightly.*
3. *With large cranking diameters and heavy cranking loads, the wrist stiffens to rigidity and the forearm pivoting generates a conical surface.*
4. *If the upper arm begins to move, cranking action has ceased and the object is being rotated by a move.*
5. *The axis of rotation of the object may be at any angle to the forearm pivot axis if the manual action meets the cranking criteria.*

Among the numerous examples of the motion are these:

1. Cranking a hand wheel to locate the carriage on the ways of a lathe
2. Starting a small gasoline motor by means of *cranking* a common crank
3. Rotating the stencil cylinder of a mimeograph machine to produce copies.

[1]Although it is a generally recognized MTM-1 element, no definition for Crank was included in the "Basis MTM Specifications" (Footnote 1, Chap. 5) because of its status at the time. All Crank data are tentative and did not appear on the official MTM-1 Data Card for U.S.A./Canada until added as Supplementary Data in 1972, since the research behind it is not fully approved. Also, due to its relatively low frequency of occurrence, it is not high on the priority list for further research. Therefore, this definition is continued, as delineated by the authors in their 1966 edition (as enlarged from the 1957 edition), based on all available literature including the training course manuals approved by the MTM Association.

At the outset, it is stressed that the motion Crank denotes a particular kind of muscle usage by the worker; it should not be confused with the employment of a common crank in a manner which may or may not coincide with the definition as stated.   Crank may be used for many other purposes besides rotating a common crank, although a crank may be actuated with a Crank motion.   MTM motions identify and assign proper time values for actions performed by the *operator*; the motion of an *object* being observed is not necessarily described.   In other words, Crank motions are recognized mainly by the defined muscular action rather than by the predominant purpose of the motion.

The muscular action in cranking involves the hand, wrist, and forearm in a compound series of motions.   However, note that the definition of Crank specifies that the upper arm must remain relatively fixed in space during the motion.   If the operator's elbow changes location significantly during movement of a crank, a Move rather than a Crank is being performed.   Of course, the muscles of the upper arm do act during Crank to increase the manual force applied, even though the upper arm is displaced very little during this action.   The contribution of the other manual members during cranking may be understood by considering the extreme cases described by the second and third sub-definitions, within the limitations of the fourth and fifth sub-definitions.

Light cranking, which means small cranking diameters and low cranking resistance, is illustrated in Fig. 1.   The fingers and hand necessarily hold an object, because the purpose to "rotate an object" is given in the Crank definition.   Close scrutiny of light cranking reveals that the hand hinges about the wrist with a whipping action.   The whipping action tends to give momentum to the object, since the hand hinges toward the direction of rotation.   At the end of each revolution, the hand will have returned to the same angle it had at the start.   During this hand and wrist motion, of course, the axis of the wrist is itself being displaced by the forearm action.

Examining Fig. 1 more closely, it is seen that in Sketch B the relative forearm location and grip are most favorable for light Crank performance, provided the crank can be rotated without stiffening the wrist.   In Sketch A, for a true light Crank motion to be made, both the lack of wrist stiff-

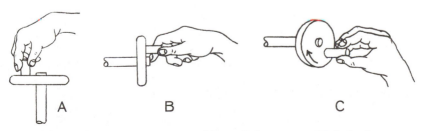

Fig. 1.   Examples of cranking with small diameters and light loads.

ness permitting "whipping" and the requirements of the fourth sub-definition would need checking by the analyst. Even more care in evaluating *all* the restrictions on light Crank motions would be necessary when considering a condition as shown in Sketch C, especially the fourth sub-definition. However, all of the situations in Fig. 1 are within the limitations of the fifth sub-definition and, therefore, could be designated as light Crank motions unless one of the other criteria was violated. Note especially that light Crank does not utilize the heavier arm muscles as is true for heavy cranking.

Heavy cranking, shown in Fig. 2, occurs when the cranking diameter is large and higher cranking loads are being rotated. The whipping action of

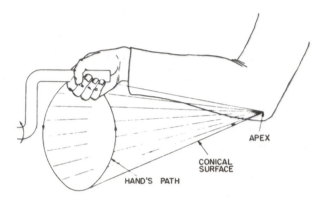

Fig. 2. Path of forearm motion in cranking for large diameters and heavy loads.

the wrist tends to disappear when the wrist becomes almost rigid during heavy cranking. Motive force is exerted on the object being cranked by the heavier muscles of the worker's arm. The forearm generates a conical surface in space as it pivots at the elbow with a rotary twisting. The apex of the cone at the elbow remains relatively fixed because of the high mobility of this ideally equipped joint.[2] Recognition of heavy cranking motions is greatly aided if the analyst has a clear concept of the conical forearm action.

Of prime importance in the identification and recognition of cranking is the relation between the *axis of cranking* and the *axis of rotation* of the object being cranked. The axis of cranking is the centerline of the pivoting action of the forearm during cranking. For the object being cranked, the axis of rotation is the line about which turning takes place. The manual

---

[2] Biomechanically, the elbow is recognized as a hinged articulation (joint) that allows for humerus (upper arm bone) rotation and thus appears to provide ball and socket movement. Therefore, while the axis of the upper arm does not change, its muscles and bone structures are directly involved in cranking action.

action required for light cranking is frequently possible at almost any angle between these axes. However, the likelihood of such freedom of forearm location during heavy cranking is greatly reduced. Usually the axis of cranking must, as shown in Fig 3, be perpendicular, or nearly so, to the plane of rotation. The axis of cranking is an imaginary line passing through the elbow joint which would correspond to the axis of the conical surface shown. The plane of rotation is analogous to the base of the cone. In a true right cone, the base would be perpendicular to the cone axis; and this condition must be reasonably approached before heavy cranking is successful. That is, the axes of cranking and rotation usually coincide.

Since cranking causes an object to move with a turning action, it is interesting to compare the muscular actions it involves with those required

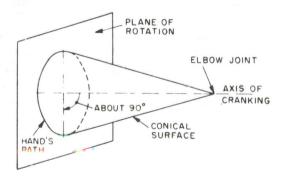

Fig. 3.   Relation of axis of cranking to the plane of rotation during heavy cranking.

by Move and Turn. There are fundamental differences which appreciably affect the time consumed.

Objects commonly called cranks can, under certain conditions, be motivated without employing cranking motions. If the elbow is displaced during the crank's operation, Move instead of Crank motions will have been used. A case of this is shown in Fig. 4 depicting a worker's arm turning a crank which lowers the cutting tool of a shaper to the work surface. Note that the criteria determining whether Move or Crank motions are performed is the movement of the upper arm.

On two counts, the forearm action during cranking differs from that found in Turn motions. First, the forearm does not *twist* about its own axis as is true in Turn, rather it maintains a fixed relation to its own axis all during cranking. Secondly, the forearm axis itself is relatively stationary during pure turning, while the requirement for the forearm axis to rotate during cranking was clearly shown above. Obviously, the differing muscular action during cranking—as compared to Move or Turn—will necessitate different time values.

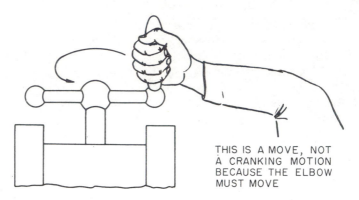

Fig. 4.   Lowering the cutting tool of a shaper to the work surface.

The Crank definition stipulated that the hand follows a circular path. Except for a slight discrepancy due to the whipping action of the hand at the wrist noted earlier, the path will ordinarily be nearly circular.   However, the motion path need not be a pure circle to classify the motion as a Crank.   It may be elliptical as when, for instance, an operator wraps twine around a relatively large rectangular package.   A sensible limit to this idea is approached as the path becomes more nearly linear, say square or rectangular.   The worker would then use a series of four Moves rather than a continuous cranking motion.

KINDS OF CRANK

Cranking motions may occur in two basic ways.   The discussion that follows also includes one kind of motion resembling cranking that is an exception to cranking rules.

Continuous

The first basic way to Crank an object is without pause or interruption. The time required obviously will be less than for intermittent cranking for a given path length of the hand.   A further discussion of this is given in the Type of Motion section of this chapter.

A continuous cranking motion starts from rest, proceeds for the required revolutions, and only then stops completely.   This is perhaps the most commonly encountered form of cranking.   The operator's muscles develop a rhythm and force during the constant motion that results in appreciable momentum during the middle turns.   Such cranking occurs when a fireman actuates a manual type warning siren long enough to caution the parties being notified.   The momentum build-up mentioned above is obviously present in this example.

Intermittent

The second basic type of cranking includes noticeable pauses during its performance which cause it to require more time for a given number of revolutions of an object than continuous cranking. Accounting for the effect of alternating periods of acceleration and deceleration is made in the Type of Motion section.

In intermittent cranking, one single revolution from a start to a stop is completed before a succeeding turn is made. This means that each and every revolution will include an acceleration and a deceleration. The resultant pauses are the recognition cue for intermittency. Because the interval of continuous movement is shorter, the momentum given to the object cranked is less than for continuous cranking. Control requirements may also be present to affect the time requirements.

Encountering high resistance or inertia is a basic cause of intermittency in cranking motions and explains the slower movement.

Note that nothing about intermittent cranking, as defined, precludes either: (1) proceeding directly from the end of the completed revolution to another; or (2) extending the pause between revolutions for longer periods. High resistance cases often occur in the first way; the latter situation often arises when other operator motions or process time intervene. In any event, the time for an intermittent cranking motion is based on a single, complete revolution; so the extent of the pause is immaterial so long as the time is properly accounted for in developing the operator cycle time.

A common case of intermittent cranking is that of a milling machine operator raising the milling table with a calibrated crank to set the depth of cut for the next pass. The setting itself involves positioning to a line, discussed in a later chapter, but the intermittent cranking preceding the final setting is the result of several things: the milling table is heavy and there is friction on the vertical sliding ways; the need for care in setting the depth indicator on the rim dial requires control—overrunning of the mark would necessitate a reversal of cranking to remove the backlash in the mechanism before accurate setting is possible; and the type of control is of the same kind that prevents gathering of high momentum as mentioned earlier. However, do not confuse the movements just described with the—usually—separate hand or finger motions used to make fine or precise adjustments of cranks or wheels, especially those associated with dials or verniers.

The symbols and time values for intermittent cranking obviously differ from those for continuous cranking, and will be explained later.

Exception

A special rule concerning cranking motions has resulted from research. It states that movements of cranks, wheels, reels, etc., involving less than half a revolution are to be treated as Moves. Such motions occur, for example, during activation of indexing cranks, adjustment of machine feed wheels, and precise adjustment of a crank to a line. The latter action, besides the Move, also may involve an alignment as described in the Position chapter. These kinds of motions are recorded as Moves and are assigned time from the Move table. Thus, a Crank motion always is at least half a revolution or more.

Another exception seen in machining operations is when the operator strikes a crank "on the fly" to Move it less than half a revolution and then breaks contact with the crank in a continuous motion. This would definitely be treated as a Type III Move, and not as cranking.

To further clarify the exception to cranking motions, it will later be shown in detail that any cranking motions of from one-half a revolution to a full revolution are treated as partial cranking motions. Also treated as partial cranking motions are the completing movements following one or more full revolutions, even though they may be one-half a revolution or less; that is, a partial revolution is *any* fractional turn immediately after a *minimum* of one full revolution.

CRANK VARIABLES

The variables associated with Crank are the cranking diameter, the number of revolutions, the Type of Motion, and the resistance factor.

Cranking Diameter

Since it is logical that the time consumed increases with increasing diameters due to the longer path of hand travel, the cranking diameter is a major factor in determining the time for a given cranking motion.

Referring to the Reach and Move data, the reader will recall that most MTM times were based on the effective length of hand travel measured at the index finger knuckle. This point was chosen for convenient observation due to the essentially uniform correlation between the traversed distance and the knuckle travel. This condition is not present during cranking; the index knuckle, especially due to the whipping action described in the Definition section, follows a more complex path that is difficult to evaluate in a simple manner. Another measure point is needed.

It is simplest to measure the path diameter directly on the cranked object as shown in Fig. 5. The *radius* from the *center of rotation* to the point of grasp is normally measured on the machine part or object at the time of analysis. Twice this radius is the diameter of cranking. The point of grasp is the centerline of that part of the object held in the hand. It should be evident that the cranking diameter is much easier to obtain than would be the knuckle movement.

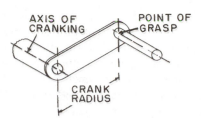

Fig. 5.   The diameter of cranking on a common crank is twice the radius from the
axis of cranking to the point of grasp.

For cranking of objects other than a common crank, the rule just given applies if certain cautions are observed.   The measured radius from the axis of rotation to the point of grasp should be perpendicular, or nearly so, to the axis of rotation.   See Fig. 6 for clarification of this point.   At the left there is shown string being wrapped around an object.   It is obvious that the desired radius for any given revolution is the taut length of the string.   In the sketch at the right, however, measurement of the string's taut length would give a radius too large; the Crank radius shown would be the correct one to use.

Wrapping string or other materials around an object normally results in a variable diameter for individual revolutions so that the diameter for all revolutions must be averaged.   First, if the object is irregular, the diameter obviously will trace an odd path.   Secondly, unless the string is allowed to slip through the fingers, the length of the string from its contact with the object to the fingers will change as the winding proceeds, until a new hold on the string is obtained.

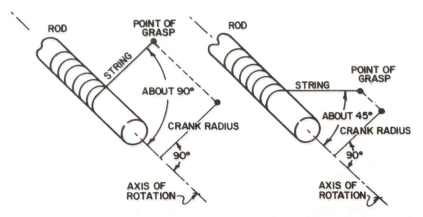

Fig. 6.   String being wrapped around an object.   (Left) The cranking radius for any given revolution is the taut length of the string.   (Right) Measurement of the string's taut length would give a radius too large; the Crank radius shown at 90° to the axis of rotation is the correct one to use.

The practicing MTM engineer will also encounter cases of cranking where one approach would be to measure or estimate the diameter of the curve described by the hand in space during the motion. This is normally difficult to do and is, therefore, not recommended except as a last resort.

If an elliptical path is observed, a sensible method would be to obtain the average diameter. This is equal to one-half of the sum of the shortest diameter and the longest diameter as shown in Fig. 7.

However, Crank data will not apply if there is need to pay reasonably close attention to exactly where the wrapping falls on the object, as when winding twine precisely and closely on the object. An example would be the reinforcing wound on fishermens' casting rods and on some older golf-club shafts.

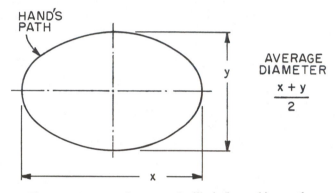

Fig. 7.   Average diameter of elliptical cranking path.

Number of Revolutions

The second cranking variable, the number of revolutions turned, must be counted or computed by the analyst at the time he writes his motion pattern. It is quite clear that the cranking time is directly related to the number of revolutions involved.

Whether the revolutions are continuous or intermittent must, of course, be noted at the time of analysis. That is, did five Cranks occur one at a time, or were five continuous rotations made? Other combinations are also possible. For example, the first three turns might be intermittent, while the last two are continuous. Close observation is certainly needed to be sure.

As explained earlier, anything less than half a revolution is treated as a Move, but all partial revolutions beyond this minimum and partial finishing revolutions should be recorded. The partial turns may again either be directly measured or estimated as conditions warrant. Three and one-half or four and one-quarter turns, for instance, would not be properly classified as four and five revolutions, respectively. The fractional portion of a turn should be included and should be shown as fractional or decimal partial turns. This is naturally more important for large cranking diameters than for small ones, because the partial turns then involve proportionately more time.

Type of Motion

The influence of the Type of Motion employed during cranking has been accounted for in the Crank time data. In fact, intelligent use of the data requires clear understanding of the effect of this variable on cranking time.

Continuous cranking includes both Type II and Type III motions. Thus, the first revolution is a Type II (in motion at the end), since no deceleration occurs at the end, while the start does involve an acceleration. The middle turns are pure Type III, since there is no start or stop during these. A Type II in motion at the start occurs for the last rotation because there is a deceleration but no acceleration in the motion.

When intermittent Cranks take place, however, each revolution is a Type I motion including a starting acceleration and a deceleration to stop. This explains why the time requirement exceeds that for continuous cranking.

Resistance Factor

The last variable affecting the time required for cranking is the amount of resistance offered by the object to the operator. This resistance is very similar to the kinds discussed in the Turn chapter. Most of the discussion there, especially that relating to torque evaluation, applies equally well here.

The effective resistance must be measured or estimated in pounds at the crank handle in a direction tangential to the path of motion at the grasping point. This may be done by spring scale, mathematical calculation, torque readings, or practical judgment depending on conditions. The weight factor data from the Move tables must then be applied to compute the resistance's effect on the time of cranking. Refer to the Turn chapter for more facts about resistant torque and lever arm measurement.

It is well to mention that high resistance is more likely to be present when intermittent cranks are performed. Since the time is longer for these also, it is likewise more important to evaluate the resistance factor than would be the case for continuous cranking.

When the resistance is variable, either a fair average value may be found for predetermined time application or else the entire cranking time could be studied by stopwatch to determine an average time which includes the resistance effects. In general, however, the Crank data do not apply to cranking against a variable resistance or when inertia effects are present, e.g., when the action is aided by a flywheel.

CRANK DATA TABLE

The MTM-1 data for the motion defined as Crank has existed since before 1952. However, these data did not appear on the MTM-1 Data Card until the 1972 revision for several reasons.

Basically, data appearing on the card have been placed there by the MTM Association for Standards and Research under the procedure described in Chapter 10. Full official international approval is indicated by data tables bearing Roman numerals, while data tables having only tentative approval by national associations are identified by Arabic numerals and called Supplementary Data. Tentative endorsement as Supplementary Data followed by application experience is a prior requirement for final approval. The Crank motion has been fully recognized as basic, and tentative approval has been given to the time data of Crank. For all practical purposes, it may be used as given below pending final action.

The cranking data originally was developed to meet the need of setting labor standards in the machining and metalworking industries where this motion occurs with greatest frequency. Because even this frequency is relatively low, and the data seldom is needed by other fields employing labor measurement, the MTM Association has no plans for further validating research at this time. So no prejudice to the cranking data is implied by its present status; long usage, indeed, has proved it has the essential qualities of good standard data.

Table 1. Crank Time Data—Light Resistance [MTM-1 Data Card Table 2— Crank (Light Resistance)—C, Supplementary MTM Data]

DIAMETER OF CRANKING (INCHES)	TMU (T) PER REVOLUTION	DIAMETER OF CRANKING (INCHES)	TMU (T) PER REVOLUTION
1	8.5	9	14.0
2	9.7	10	14.4
3	10.6	11	14.7
4	11.4	12	15.0
5	12.1	14	15.5
6	12.7	16	16.0
7	13.2	18	16.4
8	13.6	20	16.7

FORMULAS:
- A. CONTINUOUS CRANKING (Start at beginning and stop at end of cycle only)
  $$TMU = [\ (NxT) + 5.2\ ] \cdot F + C$$
- B. INTERMITTENT CRANKING (Start at beginning and stop at end of each revolution
  $$TMU = [\ (T + 5.2)\ F + C\ ] \cdot N$$

C = Static component TMU weight allowance constant from move table

When cranking motions are encountered, the time data as given in Table 1 is at present valid for standards specifications.

Concrete examples of the use of this table will be given later; but at this point, several aspects of the table may profitably be noticed.

First, the data times as listed are for light resistance only. To obtain the time effect of other resistances, it will be necessary to refer to the Move Table on the data card for appropriate weight factors. These Move factors account for the increase in cranking time under load. They were not specifically validated for cranking during the Moves-with-weight research; but, unofficially, extension of them to this situation is highly logical. Also no other factors would otherwise be available.

Remember that for any given weight factor there are a number of ways in which to measure or to calculate it. Review of this material in both the Move and Turn chapters is advisable at this point. Another important aspect of this problem is the application of the dynamic and static components of the weight factor data as explained in the chapter on Move. The static constant should be applied each and every time the cranking motion is begun from a stopped condition. The dynamic factor applies during all portions of the cranking motion in which the hand is actually moving the object being cranked.

Accordingly, the weight factors are applied in the following manner. For *continuous* cranking, the static factor is applied one time and the dynamic factor is applied once for each and every turn taken. For *intermittent* cranking, the static factor must be applied once for each and every revolution and the dynamic factor is also applied on the same basis. This application is fully clarified in the section on Practical Working Rules at the end of this chapter. Note, however, that weight during cranking is much more important from the time standpoint during intermittent cranking than it is for continuous cranking. This difference is due to the added number of times the static resistance must be overcome during the total cranking motion when intermittency brings the object being cranked to a stop.

Next look at the diameters given. These are measured as previously discussed. They range up to 12 inches by units and then by 2-inch increments up to 20 inches diameter. A plot of this data will yield a continuous curve that does not attain linearity[3]. The rate of slope change, however, is such that interpolation for the odd distances over 12 inches would seem to yield proper time results. While cranking is essentially a

---

[3] The best-fit equation to the Crank curve is: TMU (T) per revolution = $8.08\ d^{.25}$, where d = diameter in inches. This would apply directly to intermediate (Type III) continuous revolutions. For intermittent (Type I) revolutions, the equation would be: TMU (K) per revolution = $8.08\ d^{.25} + 5.2$.

one-hand motion as long as the cranking diameter is 20 inches or less, cases may arise when larger diameters are encountered. As was done for Reaches and Moves exceeding 30 inches, it would seem reasonable for Cranks exceeding 20 inches in diameter to make a straight-line extrapolation. The error will be relatively small. However, note that the length of the forearm will impose a practical upper limit near the highest data table entry.

Finally, Column T lists the times for Type III revolutions involving neither acceleration nor deceleration. Then 5.2 TMU is listed as the sum of one start and stop time, which is another way of saying that the "m" value is 2.6 TMU all through the distance range for cranking. Thus a Type I (intermittent) revolution is the T value plus 5.2 TMU.

Table 1 should be used in the following manner. For intermittent cranking, each turn's time is completely accounted for by adding 5.2 TMU to Column T; and the total time, therefore, equals the number of revolutions multiplied by this "per turn" time. In the case of continuous cranking motions, the 5.2 TMU factor is used only *once* (which will account for the start in the first revolution and the stop in the last turn) and *all* revolutions will be multiplied by the Column T time. By this procedure it is possible to list all cranking times in one, rather than two, columns and still account for the actual events of two Type II (first and last turns) revolutions with all others being Type III, if continuous cranking is done.

THE SHORTHAND OF CRANK

Crank motions are written using the following conventional symbols.

Crank

The presence of a cranking motion is recognized by the analyst when he records a capital C on his study sheet.

Cranking Diameter

Following the Symbol C for cranking will be a numeral for the inches of cranking diameter, determined as previously discussed. Decimals or fractions are to be rounded off to the nearest whole number.

Example: C9 denotes that the operator turned a crank with a 9-inch effective diameter.

Resistance Factor

After the number denoting cranking diameter, the MTM engineer will record a dash (–) which is followed by the resistance factor in pounds as obtained either by measurement or estimation.

No resistance factor need be recorded for a weight factor of 2½ pounds or less. For all other cases of higher resistance, a number to the nearest pound value will be shown to indicate the resistance factor being used. This follows the same rules that were given in the chapter on Move motions.

Example: C10–20 shows cranking, with 10-inch diameter, against a resistance of twenty pounds weight factor.

Revolutions

The numbers denoting the revolutions of the cranking will precede the letter C. The symbolism differs for continuous and intermittent cranking.

Continuous Cranking

This is symbolized simply by the number of turns taken. The numbers will be whole, decimal, or fractional depending on the actual count or visualized analysis.

Examples: 4C9 shows four continuous turns of a low resistance, 9-inch crank diameter.

8.6C12–15 describes 8.6 turns with a 12-inch crank diameter and 15 pounds weight factor.

6¼C7–10 indicates six and one-fourths revolutions against ten pounds equivalent resistance or weight factor with a 7-inch cranking diameter.

Intermittent Cranking

The only difference for this kind of cranking is that the symbol must show that the revolutions are single instead of continuous. That is indicated by showing a 1 preceding the C, and separating this from the previous numbers by a hyphen. The previous numbers will be the amount of turns taken, and are written in the same manner discussed for continuous cranking.

Examples: 3–1C16 denotes 3 intermittent Cranks of a 16-inch diameter crank, small resistance being present.
5½–1C10–25 exemplifies that a 10-inch Crank diameter was turned 5½ single revolutions, possibly single because of the resistance factor of 25 pounds.

PRACTICAL WORKING RULES

Many of the practical working rules and hints given in previous chapters will apply to cranking motions. In general, this is true for the measurements of distances and resistance and the rule of rounding off time values to the nearest tenths of TMU values is also valid.

After having identified, analyzed, and symbolized a cranking motion, the MTM engineer needs to find the time value. This may be done with the Crank data table in two ways:

The first method uses the procedure given previously in the Crank Time Data section. It is the simplest computation method. Examples follow:

Example A: Continuous Crank symbolized by 4C9–10
   One start-stop time (for part of first and last revolution) =   5.2 TMU
   Four turns at 9″ diameter are 4 × Column T or 4 (14.0) = 56.0 TMU
   Sub-total   = 61.2 TMU
   From Move table in Chap. 13, 10-pound factors give:
      1.11 (61.2) + 1 (3.9) or 67.9 + 3.9   = 71.8 TMU

Example B: Intermittent Crank symbolized by 4-1C9–10
   Four start-stop times are 4 (5.2)   = 20.8 TMU
   Four turns at 9″ diameter are 4 × Column T or 4 (14.0) = 56.0 TMU
   Sub-total   = 76.8 TMU
   From Move table, 10-pound factors give:
      1.11 (76.8) + 4 (3.9) or 85.2 + 15.6   = 100.8 TMU

The effect of intermittency on time increase is readily seen from the examples. The second approved method that may be used to compute the cranking time employs an algebraic formula on the data card to account for the same variables. In the equations:

   N = the number of revolutions
   T = the time from Column T of the Crank table
   F = the dynamic weight factor (multiplier) from the Move tables
   C = the static weight factor (additive constant) from Move data
   5.2 = the start and stop time, TMU per occurrence.

Continuous: Crank Time, leveled TMU = $[(N)(T) + 5.2]F + C$
Intermittent: Crank Time, leveled TMU = $[(T + 5.2)F + C]N$.

Checking Example A, above,

$[(4)(14.0) + 5.2]\,1.11 + 3.9$ $= (56.0 + 5.2)\,1.11 + 3.9$
                                  $= (61.2)\,1.11 + 3.9$
                                  $= 67.9 + 3.9$ = 71.8 TMU

Checking Example B, above,

$[(14.0 + 5.2)\,1.11 + 3.9]\,4$ $= (21.3 + 3.9)\,4$
                                  $= (25.2)\,4$ = 100.8 TMU

That the first method is simpler and easier is apparent, results being the same.

The above procedure must be qualified slightly when the cranking includes fractional or partial revolutions, e.g., 4.6 or 4⅓ revolutions. These instances will cause no problem for *continuous* cranking because (1) the start-stop factor is applied only once, and (2) any static weight factor—also applied only once—will not be affected by the fractional or partial portion of the total revolutions. On the other hand, any dynamic weight factor will properly be applied to the partial turns. However, for *intermittent* cranking, the non-integer portion of the total revolutions will affect *both* the number of start-stop times *and* the number of static weight factors applicable. This is simply due to the fact that both the 5.2 TMU constant and the static weight constant are all-or-nothing in nature, hence cannot be applied fractionally. To do so would *underestimate* the total cranking time for the intermittent condition, to the nearest 0.1 revolution, by 0.5 TMU to 21.4 TMU in the range of weight factors listed in the Move table.

This difficulty can be resolved in two ways. One is simply to remember, when figuring the time for intermittent Cranks which include fractional turns, that the multiplier for *both* the start-stop constant *and* the static weight factor must be increased only by full integers, never fractionally. In other words, the last (fractional) revolution requires a *full* start-stop and a *full* static constant to develop correct times. Examples should make this clear:

Example C: Continuous Crank symbolized by 4.6C9–10

One start-stop time (for part of first and last revolution) =	5.2 TMU
4.6 turns at 9″ diameter are 4.6 × Column T or 4.6 (14.0) =	64.4 TMU
Sub-total	= 69.6 TMU
From Move table in Chap. 13, 10-pound factors give:	
1.11 (69.6) + 1 (3.9) or 77.26 + 3.90	= 81.2 TMU

Example D: Intermittent Crank symbolized by 4.6–1C9–10

Five start-stop times are 5 (5.2)	= 26.0 TMU
4.6 turns at 9″ diameter are 4.6 × Column T or 4.6 (14.0) =	64.4 TMU
Sub-total	= 90.4 TMU
From Move table, 10-pound factors give:	
1.11 (90.4) + 5 (3.9) or 100.34 + 19.50	= 119.8 TMU

The other resolution is to recognize that *both* Crank formulas are special cases of a Master Crank Formula, which can be applied instead of separate formulas. This basic formula may be derived by expanding the special cases and combining them with a new symbol N' that permits full accounting for the fractional revolutions. The two expansions give:

For continuous cranking: $[(N)(T) + 5.2] F + C = FNT + (5.2 F + C)$
For intermittent cranking: $[(T + 5.2) F + C] N = FNT + N (5.2 F + C)$.

Note here that the FNT term takes care of the intermediate revolutions fully and correctly, whether or not N includes fractions, while the last term is the same for continuous or intermittent except for the N value, which is thus the only real difference and can be adjusted to correctly account for fractional revolutions. Therefore, let the $\boxed{\text{Master Crank Formula}}$ be:

$$\boxed{\text{TMU} = FNT + N' (5.2 F + C)}$$

where: $\boxed{\begin{array}{l} N' = 1 \text{ for continuous Crank} \\ N' = N, \text{ if } N \text{ is an integer} \\ N' = \text{next higher whole number than } N, \text{ if } N \text{ includes fractions.} \end{array}}$

With this single Crank formula, the possibilities effectively become:

Type of Crank	N'	F	C	Effective Formula
Continuous, Light Loads	1	1	0	$NT + \quad 5.2$
Continuous, Higher Loads	1	F	C	$FNT + \quad (5.2 F + C)$
Intermittent, Light Loads	N'	1	0	$NT + N' (5.2)$
Intermittent, Higher Loads	N'	F	C	$FNT + N' (5.2 F + C)$

Observe here that both continuous formulas are the same as on the data card, but both intermittent formulas are slightly different than the data card formula. Note also that, since N and N' *may* be different, the Master Crank Formula inherently helps the analyst to avoid the error of "shorting" the time for fractional intermittent turns.

Checking Example C, above,
$1.11 (4.6)(14.0) + 1 [5.2 (1.11) + 3.9] = 71.48 + 9.67 \qquad = \underline{\underline{81.2 \text{ TMU}}}$

Checking Example D, above,
$1.11 (4.6)(14.0) + 5 [5.2 (1.11) + 3.9] = 71.48 + 48.36 \qquad = \underline{\underline{119.8 \text{ TMU}}}$

Again, the two methods can be compared for ease of usage.

Another aspect of cranking is encountered when experienced operators reduce their effort and fatigue by using a "spinning" motion to move cross slides, raise tables, etc., by hand cranks. They take advantage of crank momentum to keep the mechanism moving for one or more revolutions. However, they sometimes keep a light hold on the crank handle or wheel rim while the spinning continues. Nevertheless, the spinning time is *process* controlled and must be determined by a stopwatch or other timing device. The MTM Crank data does not cover this "spinning" motion.

DECISION MODELS

The only Decision Model for Crank is shown as Fig. 8.

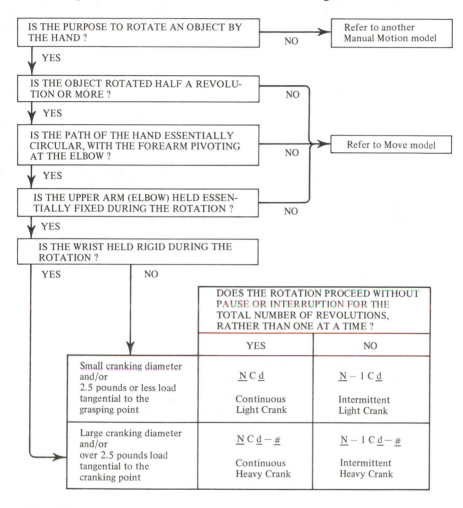

NOTE: Fill in to complete each Crank symbol, as indicated, the following:

$\underline{N}$ = Total Revolutions (may include fractions or decimals)

$\underline{d}$ = Diameter in inches

$\underline{\#}$ = Cranking Load in pounds. Apply resistance factors as found in the static and dynamic weight factor columns of the Move data table.

Fig. 8.   Decision Model for Crank.

CHAPTER 9

Apply Pressure[1]

Definition[2]

> *APPLY PRESSURE is an application of muscular force during which object resistance is overcome in a controlled manner, accompanied by essentially no motion (¼ inch = 6.4 millimeter or less).*
>
> 1. *Apply Pressure, Alone (APA) is a hesitation or lack of motion having at least a limiting Apply Force component with or without subsequent limiting Dwell and Release Force components. However, the symbol APA is not used unless all three components are present with Dwell at a minimum.*
> 2. *Apply Pressure, Bind (APB) is an APA preceded by a Regrasp, or the equivalent, indicated by a setting of the muscles.*
> 3. *Apply Pressure may be performed by any body member.*

The originators of MTM first called this element Final Tighten when it was noted after an operator had turned a screw down to its travel limit with a screwdriver. The considerable pressure exerted, characterized by a hesitation and passage of time, did not cause appreciable movement of the screw. Further evaluation showed its more general nature and application—resulting in the present name.

Understanding of Apply Pressure centers around two main concepts—the muscular force and the lack of object displacement normally identified

---

[1] The main presentation in this chapter concerns the revised international data for Apply Pressure first incorporated on the U.S.A./Canada MTM-1 card in 1972 following IMD approval. These new data superseded *both* the original data and the 1965 Supplementary Data for Apply Pressure previously carried on the card. However, because analysts may need to interpret the older data when differing symbols are found in existing motion patterns, a section for each of the superseded data versions is included at the end of this chapter. In general, updating older motion patterns to reflect the newer data will affect only the symbols and not the time values.

[2] From "Application Training Supplement No. 11, Apply Pressure" by Franklin H. Bayha, James A. Foulke, and Walton M. Hancock, MTM Association for Standards and Research, Fair Lawn, New Jersey, January 1973.

as motion. The element describes useful operator action which consumes time despite the lack of motion. To clarify these concepts, prior to a more technical description of the elemental details, several basic work measurement ideas should be mentioned.

The primary goal in assessing human performance is to determine the input a worker must contribute to achieve a given desired output. The main gauge of this input is time, so that anything *necessary* to the work which requires time must be included in the job study. However, for an operator's actions to be useful, they must accomplish some kind of required work; therefore energy must be expended on the product or service for which compensation is justified. Energy applied in this way results in two major kinds of useful work, dynamic and static.

Dynamic work is recognized by movement or displacement of the object or the worker's manual members along with varying amounts of force application. In MTM technology, dynamic work is assessed by motions involving distance, by dynamic Move weight factors, by Turn resistance levels, by dynamic factors in Crank, and by varying degrees of pressure in Position and Disengage. For each of these instances, the time effects were studied directly and incorporated into the motion data.

Static work, oppositely, is characterized by energy expenditure and force application on objects with little or no resulting movement. One instance of this in an earlier chapter is the static weight constant in Move with weight, with special study behind the time values; and the same factors apply to Crank motions. Another is "holding," which often may be necessary and unavoidable in achieving the required output, even though this action generally is discouraged in work design and special efforts are made to eliminate it. Holding often occurs while other body members perform useful motions or while process time elapses. This should be shown by a descriptive note on the analysis explaining the entry of process or holding time, and the only way to evaluate it at present is by measuring with a timing device upon each encounter. However, there are still other static work events where force is applied as a separate action and none of the previously mentioned time data are appropriate. Here the MTM element Apply Pressure is available for use in the analysis procedure, complete with appropriate associated time data determined through MTM research.

Examples illustrating the purposes for which Apply Pressure may be used are:

1. *To achieve control, or "make ready" for subsequent motions.* After the hand or other body member has *contacted* a heavy object, for example, the object cannot then be moved until control of the weight has been gained by the application of some degree of muscular

force. This example should not be confused with the static component of a weighted Move which follows *grasping* of a heavy object. In this latter case, the Grasp achieves control of the object; but the static factor accounts for part of the time required to achieve weight control, the rest being the dynamic factor which accounts for the time to maintain control during the subsequent Move.

2. *To restrain, or prevent, the motion of objects.* Apply Pressure will be used in such cases either when the period of restraint is shortly followed by another motion or when the restraint precedes the longer action known as holding. For instance, a mechanic will Apply Pressure to the top of a tire with one hand so that the wheel will not fall off the studs before he can start a holding nut with the other hand. In fact, he may use several Apply Pressures to get the wheel onto the studs before holding begins, as noted above.

3. *To overcome static friction, binding, or other resistance to motion.* This happens, for example, when a person loosens a screw-on type lid of a tightly sealed jar to empty its contents. Usage of Apply Pressure in this manner is very frequent during the MTM motions called Position in assembly actions and Disengage in cases of disassembly. Actually the Final Tighten observation which led to isolating Apply Pressure is another instance in this category. The force application in tightening assures that the last bit of movement necessary to firm placement is accomplished.

While the definition indicates the use of muscular force to Apply Pressure, it does not restrict the action to any particular muscle group. When applying force to an object held in the hand, for instance, all of the manual members up to the shoulder joint have muscle groups aiding the action. Almost any body area can also apply force to objects under suitable conditions. The shoulder may push against a bulky shipping crate, the arms or elbow could be used to hold down the flaps of an overstuffed fibreboard carton, the back can push open a heavy door while the owner's arms are loaded with carried items, the hip may be used to steady a ladder, and many other similar force applications by body areas can be cited. Whether small or great, the resistant forces offered by objects to securing of control, prevention of motion, or surmounting of resistance whichever may be the purpose of the element, all require the expenditure of energy by the muscles, and obviously this consumes time. Motion cannot ensue or be arrested until the muscular work has been expended.

The absence of noticeable object displacement during Apply Pressure also is noted in the definition. The novice to MTM is sometimes disturbed by this specific denial. It is termed an element only, because its presence during human work must be recognized—indeed its time consumption may be large in comparison to elements known as motions. When the true nature of Apply Pressure becomes clear, however, the analyst will readily recognize it due to the obvious lapse of motion during its performance. That Apply Pressure is not a motion (where "motion" implies spatial displacement of the object), but nevertheless is an important MTM element, should now be clear. It occurs *before* or *after* motions, not during them. Resistance or weight factors account for force application *during* motions such as Move.

CASES OF APPLY PRESSURE

There are two cases of Apply Pressure for normal analysis. They are defined as sub-definitions in the main definition at the start of this chapter, Case A and Case B. Each will be discussed initially as a whole. However, both cases are a combination of components, each of which has a formal definition of its own; and full clarity depends on understanding the components as well as the standard cases. In addition, practical component time data and symbols also exist, so that one can use them directly for analyzing occurrences which do not fit the defined cases.

Case A

See the definition of Apply Pressure, Alone (APA) given earlier. It can be seen that this involves the pure pressure only, since no "make ready" delay is needed prior to force application.

It is important to note that three conditions must be true in order to assign a Case A:

1. All three components (discussed below) must be present.
2. A standard force of 7.5 pounds (3.4 kilograms) must be present or assumed.
3. The Dwell time must be at a minimum.

If *all* these items are not fulfilled, then the action must be analyzed with the component data rather than with an AP symbol and a case symbol. In other words, a complete cycle of components must be present with the two variable components at standard values to assign a Case A. Technically, an Apply Pressure exists when *only* a limiting force application is present; but this would be analyzed using the symbols and time for that component alone, rather than by an APA.

Sometimes skilled operators—during highly repetitive cycles—can learn to place their hands or other body member on objects so that they can always use a Case A instead of the Case B Apply Pressure that an ordinary operator might require—thus eliminating the Regrasp included in Case B, as described below. Also, Case A occurs more frequently than Case B when the object has been held for quite some time prior to the force application. Almost all force application with the feet is done with a Case A; this often will happen when operating foot pedals, and a special MTM motion for this will be found in a later chapter (FMP).

Case B

Again, see the definition of Apply Pressure, Bind (APB) given earlier. Obviously, this is simply an APA with a prior Regrasp—a motion described fully in the next chapter—or an equivalent "shift" of the muscle group involved before the force is applied. Therefore, *all* the restrictions and conditions stated for Case A also apply for Case B, except that Case B includes the time for making the preliminary adjustment of the muscles. The sequence is first the "regrasp," "shift," "make ready," "contact adjustment," or "preliminary setting," and then the force application.

With an Apply Pressure involving the hands, the Regrasp or "shift of grip" idea is readily understood. Obviously, the action for other body members will be different. In the latter situation, the idea of an equivalent pause or slight delay accounts for the time consumed. The slightest movement or shifting of the body member prior to the force application is a clue that Case B is being seen. This action is used to orient, adjust, or attain a better location from which to apply the full force of the given body member.

Case B appears to be the most common in industrial activity. It frequently either precedes or follows a twisting action such as occurred during the screwdriver usage in the original research. Similar instances of Apply Pressure often appear before or after push or pull motions involving higher than normal resistances.

Components of Apply Pressure

As the Case A definition revealed, the Apply Pressure element has three components. Their formal[3] definitions are:

1. *APPLY FORCE (AF) is an increasing, controlled muscular force applied to an object(s) which produces essentially no resultant movement.*

2. *MINIMUM DWELL (DM) is the shortest period of time during which a force reversal reaction occurs at a relatively constant force*

---

[3] *Ibid.*

*level.   Longer periods limited to force reversal are simple Dwell; but,
if other action occurs in the interval, a holding element is being
performed.*

*3. RELEASE FORCE (RLF) is the relaxation of the muscles which
relieves applied force from an object(s).*

Some history is helpful to trace the development of these components
and show sources for the final form developed for adoption, since they are
highly technical and comprise the first basic data change to gain international approval as a replacement for original data. The very first
research report[4] of the MTM Association indicated a need for further
study of Apply Pressure, and a later report[5] contained some tentative new
data.   These justified an extensive research project with laboratory[6] and
industrial[7] phases during 1958 to 1960.   A document[8], the first of three
Application Training Supplements (Nos. 7, 9, and 11), served to facilitate
a period of field trial and was revised[9] to show the basis for the Supplementary Apply Pressure data added to the U.S.A./Canada MTM-1 card
in 1965.   Experience with that form of the new data led to its submittal to
the IMD for international consideration[10] during 1967 to 1971, gaining
approval for international data change in 1972.   The final data format[11]
was the basis for superseding both the original and Supplementary data by
the 1972 revision of the U.S.A./Canada MTM-1 card, and it is the version
presented here.   This sequence of research and validation has given firm
experimental and statistical grounding to the present Apply Pressure data
based on these components.

The new Apply Pressure research marked a departure from the former
MTM reliance upon cinefilm analysis.   Since previous projects concerned
spatial motions involving visible displacement of the body member, visual
means of measurement were appropriate.   However, motion is essentially
excluded by the definition of Apply Pressure; therefore more sophisticated
techniques were needed to measure body member force application rather

[4]R.R. 101. (Available from the U.S./Canada MTM Association.)

[5]R.R. 108. (Available from the U.S./Canada MTM Association.)

[6]Barbara Ettinger Goodman, "A Laboratory Study of Apply Pressure," MTM Association
for Standards and Research, Ann Arbor, Michigan, July 1959.

[7]R.R. 111. (Available from the U.S./Canada MTM Association.)

[8]Franklin H. Bayha and James A. Foulke, "Application Training Supplement No.
7—Position and Apply Pressure," MTM Association for Standards and Research, Ann
Arbor, Michigan, June 1964.

[9]Franklin H. Bayha, James A. Foulke, and Walton M. Hancock, "Application Training
Supplement No. 9—Apply Pressure." MTM Association for Standards and Research, Ann
Arbor, Michigan, January 1965.

[10]Franklin H. Bayha and Walton M. Hancock, "Apply Pressure Technical Description,"
International MTM Directorate, Solna, Sweden, February 1972.

[11]A.T.S. 11 (Available from the U.S./Canada MTM Association.)

than member displacement. This need was met by using electrical strain gages and other transducers which provided signals for graphical display on a multi-channel time recorder. A discussion of this type of equipment is presented in Chapter 3.

The laboratory investigation involved a representative variety of levers, push buttons, and gripping devices. It considered various modes of force application by the arm, hands, and fingers. The effects of body member location and orientation also were evaluated. The resulting statistical data were exhaustively analyzed by electronic computer equipment which produced both reliable statistical information and plotting data for curve fitting.

The basic behavior of Apply Pressure was found to include three components of a total cycle as depicted ideally in Fig. 1. This shows the

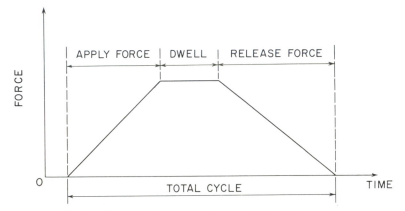

Fig. 1.   Idealized force-time diagram of Apply Pressure.

sequential application of force, some period of dwell, and the gradual removal of force from an object. Note that such a dwell period could be minimal, while the necessary feedback between the operator and the task occurs, or it can be more prolonged. It could even be extended into a holding action so that the force reduction would occur at a much later time, perhaps after another motion such as a Move or Turn had been performed. Therefore, this plot is sufficiently general to explain all modes of Apply Pressure. However, a predetermined dwell time would, of necessity, be restricted to the minimum value. Any longer time would involve process time. The components and total cycle of Apply Pressure, therefore, are now defined in terms of time, rather than motions.

With the official definitions and isolation of the significant variables in Apply Pressure resulting from the laboratory study, it was possible to structure a reliable industrial study. By using the newer measurement and analysis techniques developed in the laboratory study, the industrial research was very efficient and highly revealing. The three components were found to be valid, and additional qualifying data were generated.

Apply Force was found to involve a steady increase in force limited only by the operator's muscular capability. However, the required performance time levels off to a constant value of 4.0 TMU when 10 pounds or more of force is applied to the object. This is the maximum time value for Apply Force. Below this value, the exact equation[12] which applies is:

Apply Force = 0.893 TMU + 0.335 TMU per pound of force

The Apply Force begins after the object has been gripped manually or contacted by another body member and any preliminary contact adjustments have been completed, so that muscular force can be readily applied. The Apply Force component ends when either the force being applied no longer increases or the force is large enough to cause motion of the object. In other words, Apply Force includes only the period during which a *change in force* is occurring. Also note that the force application is *controlled*. If an object is struck with a ballistic motion, often used by experienced operators, the classification of such a blow or force application would not be an Apply Force as defined, but rather a Move.

The *Dwell* component was found to occur at an essentially constant force level. The minimum time value required is 4.2 TMU. This time for the *Minimum Dwell* component is always used when the Apply Force is immediately followed by Release Force. In all other cases, the analyst must recognize the process time involved in a clamping or prolonged holding action. In other words, Minimum Dwell is the least time required for feedback between the operator and the task situation to occur so that the muscular force application can be relaxed.

Release Force was deliberately defined as a time component because it differs from the elemental motion Release as described later, in Chapter 18. The Release (or Release Load) motion is a finger or hand element by which an object is physically separated from manual control. Release Force, however, merely shows that the pressure on the object is being removed. Physical control of the object itself has not been lost at the end of Release Force. Note also that the opposite situation occurs at the beginning of Apply Pressure. In that case, Apply Force is not the means of gaining physical control of the object but merely the application of pressure to it. To obtain physical control of the object itself requires the use of the MTM element Grasp, which will be described in Chapter 17. Another way to emphasize both distinctions is to point out that Grasp and Release are *motion* elements, whereas Apply Force and Release

---

[12] The metric equation is: Apply Force = 0.893 TMU + 0.739 TMU per kilogram of force.

Force are only *time* elements. The time value of Release Force was found to be essentially a constant of 3.0 TMU.

Another result of the industrial study was clarification of the manner in which push buttons are actuated. After the button is contacted by a normal Grasp, Apply Force—equivalent to its resistance or the counter-force of its return spring—is applied in a controlled fashion until it attains a level sufficient to cause motion. Any ensuing travel of the button, perhaps to *Move* the button from one mechanical stop or electrical contact to another, is an M—A to whatever distance is appropriate. At the end of the Move, the Dwell period begins as the button is held against the second stop or electrical contact. When Dwell is completed, a return Move occurs while the return spring balances the Dwell force; this brings the button back to its original location, but the finger is still exerting force against the spring. Then, again without motion, the Release Force begins and the pressure diminishes until it becomes zero. Finally, finger contact with the button is broken, a normal finger Release of the object. To summarize, actuating a push button requires an Apply Pressure *plus* two Case A Moves, which must be analyzed separately if the distances are significant. These Moves would be subject to any appropriate dynamic weight factor; however, accepted practice is to neglect these Moves entirely if they are only one-quarter inch or less in length (but see also the treatment of this under Practical Working Rules, later in this chapter). Also, any process time involved in holding must be added to the actuation time.

DATA CARD INFORMATION

The time data for Apply Pressure appears on the MTM-1 Data Card as shown in Table 1. Both the full cycle and component symbols and times are included at a standard force level of 7.48 pounds (3.39 kilograms) and a Minimum Dwell. As a result, the two cycle times correspond to the original two values for Apply Pressure listed on earlier data cards. Also, as given later, full cycles and Apply Force component times with other force levels—up to the time maximum for 10 pounds (4.54 kilograms)—and other Dwell periods are possible so that this new data permits maximum flexibility for methods and time variation.

It can be seen that Case A and Case B differ by 5.6 TMU, which is the time equivalent for the MTM motion called Regrasp (G2). This is the differentiation made in the material above. Further note that the data card Table III B is so numbered due to its historical relation to Table III A Turn. Apply Pressure originally was appended to Table III Turn because it was first noticed in the original film analysis as a "Final Tightening" when screws were assembled with a screwdriver. With the

Table 1. Apply Pressure Data (MTM-1 Data Card Table IIIB—Apply Pressure—AP)

FULL CYCLE			COMPONENTS		
SYMBOL	TMU	DESCRIPTION	SYMBOL	TMU	DESCRIPTION
APA	10.6	AF + DM + RLF	AF	3.4	Apply Force
			DM	4.2	Dwell, Minimum
APB	16.2	APA + G2	RLF	3.0	Release Force

new data given above, however, it was deemed wise to separate the tables; hence the new numbers.

Although not stated on the data card, the Apply Force equation given previously also is considered an integral part of the new data; for this source of component times is the main reason the new data are so flexible. Examples will be given later.

THE SHORTHAND OF APPLY PRESSURE

A complete cycle of the basic element Apply Pressure is identified by the two capital letters AP, followed by a letter A or B to show the case; however, this symbolism is correct *only* for complete cycles with Minimum Dwell and at the standardized force level. In all other instances, the analysis should show the component symbols.

The component conventions are: AF for Apply Force at the standardized force level, DM for Minimum Dwell, and RLF for Release Force. If the dwell is not minimum, the word "Dwell" is shown and the appropriate longer time value due to process or holding action is associated with it; however, it also is acceptable to merely indicate "Proc." for such non-minimum dwell. When using the exact equation for more precise determination of the Apply Force time, some variation of symbolism is accepted practice. Borrowing from the former Supplementary form of the data, one may show two numerals following the AF symbol to indicate force levels at exact pounds. For example, AF05 shows exactly five pounds Apply Force and AF10 shows the force applied for maximum time. Actually, the force can exceed 10 pounds and even be so symbolized, but the time would be for the 10-pound level. For only nominal force, usually allowed time at 0.5 pound, the symbol AF00 can be used. If the force applied is not exact to a pound level, "Actual AF" can be written in the symbol column and the force level annotated in the words of the Description column. The component symbolism, therefore, is very flexible and can easily show the unique method determined by the analyst.

OCCURRENCES WITH OTHER MOTIONS

The MTM element Apply Pressure often has appeared during research on other MTM motions either preceding or following in the motion sequences. It is helpful to discuss this briefly here, although the chapters covering the other motions have more complete information.

The MTM research on *Moves with Weight*[13] revealed some previously undefined data on force application. Before objects having other than negligible weight can actually be moved, and after the fingers have completed closing on the object, control of the mass must be gained by the application of varying amounts of pressure. The time required to apply this pressure was evaluated to develop the static and dynamic weight factors. The application of these factors therefore fully accounts for any concurrent force applications during a Move. For this reason, no separate analysis of Apply Pressure is needed when considering weight effects during object movement. Similarly, the resistance effects during Turn already have been included in the Turn time. Any included force application does not need to be separately analyzed.

When the amount of force required to assemble objects having differing degrees of fit is evaluated for the MTM motion called Position, one of the major factors in the time required is the Apply Pressure used. The purpose of the Apply Pressure in this case is mainly to overcome resistance to motion. Again, this illustrates one of the concepts in the definition for Apply Pressure.

Objects being disassembled frequently require the MTM motion known as Disengage. The need to maintain control and overcome motion resistance again justify the use of Apply Pressure in this element.

How to account for the Apply Pressure element occurring with other motions is fully clarified in the appropriate chapters. It must be said here that, in general, the rules given for Apply Pressure in this chapter are concerned with its occurrence as a distinct, separate MTM element in work performance.

PRACTICAL WORKING RULES

Apply Pressure usually appears in situations requiring the handling of heavier objects or where forces of relatively high magnitude must be employed. In such events, the analyst should look for hesitations which reveal the usage of more than normal force. He can then correctly analyze

---

[13]R.R. 108. (Available from the U.S./Canada MTM Association.)

the Apply Pressures present during work sequences. Practical study of the objects handled will generally give clues as to the proper case of Apply Pressure being employed.

Apply Pressure is sometimes used to prevent injury to the body members, avoid damage to the object handled due to dropping, and to provide other protective or preventive action. To aid in recognition of such cases, the MTM analyst can notice the physical characteristics of the objects handled. Jagged edges or burrs which might break the skin, the presence of lubricants on the surfaces which might cause accidents, objects whose nature inherently cause awkward handling, and features of objects which necessitate enough controlling pressure to prevent slipping or other accidents all flash signals to the good MTM analyst that Apply Pressures should be considered. The proper case or combination of components to allow can be found by knowledge of the variables and application rules discussed in this chapter.

Since most of the Apply Pressures occurring in industrial work are Case B, this fact can guide correct classification. Specific clues for recognizing the cases have already been given. Also, in most instances, APA and APB with their single time value will each be adequate for the kind of limiting force application ordinarily encountered. While the new capability of detailed methods-time analysis is always available with the latest Apply Pressure data, it seldom will be needed except when production or standards demands must necessarily be stringent. Perhaps its greatest value is the increased understanding of the element that promotes routine analysis with greater confidence than was possible before its development.

For application purposes, the expanded Apply Pressure data with its separate components makes possible complete analysis of the following situations:

1. *Only the Apply Force component is limiting.* Of course, no Apply Pressure will exist unless this component is present and allowable. It may be the only allowable event if the Dwell is either limited out or prolonged into a clamping or holding action *and* the subsequent force relaxation also is limited out by other motions. That is, the Apply Pressure is completed at a later than minimal point in the operation where the RLF is limited out and the Dwell period has previously been limited by the holding and/or other action which controls the allowed time. This can be true whether or not the Dwell period is minimum. When this situation exists, only the Apply Force analysis need appear in the motion pattern. If the standardized force (7.5 pounds) is assumed or actually measured, only the AF = 3.4 TMU symbol and time will be shown. For other force levels, the appropriate conventions and time values will fully reveal this analysis. An AP cycle symbol cannot be used for this situation.

2. *Only the Release Force component is limited.* This naturally will indicate the presence of other limiting motions or process time before a complete cycle of Apply Pressure is finished. The Apply Force can be standard or otherwise, and so analyzed. Likewise, the Dwell can be DM = 4.2 TMU or any longer value for actual "Dwell" which is limiting. Thus both of these components will be included in the motion pattern; but no RLF need be shown, or it should be limited out if listed. An AP cycle symbol cannot be used for this situation.

3. *Only the Dwell component is limited.* Here both the Apply Force and Release Force components must be allowed. In any event, the RLF symbol will be shown; but the Apply Force analysis may be AF or an appropriate convention and time for a different force level. Neither DM nor "Dwell" need be shown; but, if recorded, it should be limited out. Particularly for a DM occurrence, this situation would be very rare; however, for actual Dwell longer than minimum, it could occasionally occur. An AP cycle symbol cannot be used for this situation.

4. *All components are limiting, but not standard values.* This situation covers the case where either the Apply Force is at a force level other than 7.5 pounds or Dwell is not minimum, or both. Because not all three components are at standard level, all three should be shown appropriately in the motion pattern and allowed the proper time. An AP cycle symbol cannot be used for this situation.

5. *A complete cycle of Apply Pressure occurs with no binding or other cause for Regrasp (see G2 motion in Chapter 17) or its equivalent.* Here, all three components will be present and limiting, with Dwell minimum (DM) and the force at standard level (AF). This fulfills all the requirements to be shown as an APA = 10.6 TMU for Apply Pressure, Alone. Essentially, this case corresponds to the original AP2 as to descriptive content; only a change in the elemental name of the AP2 definition would make it fit this situation.

6. *A complete cycle of Apply Pressure occurs with a preceding Regrasp motion or its equivalent.* This is the last situation in 5. with an added 5.6 TMU adjustment time. It fully describes an APB = 16.2 TMU symbolizing Apply Pressure, Bind. Again, with only a name change, this situation corresponds to the original AP1. All components are present and limiting at a standard level of force and minimum Dwell. This is the *most frequent* case.

In the first four situations listed, any allowable Regrasp or the equivalent occurring prior to the limiting component(s) would be shown separately in the analysis. The "binding" phenomenon is automatically allowed only with an APB symbol and time.

One additional application rule was *suggested* in the industrial study, although it was not officially included in the new Apply Pressure data.   It concerns any separate Move time for depressing a push button.   This is a mechanically constrained movement frequently limited by internal stops to distances as small as $\frac{1}{16}$ to $\frac{1}{4}$ inch.   For this situation, the MfA time of 2.0 TMU as a minimum MTM time appears to be too large for some applications.   The suggestion of reducing the push button Move time for short actuation travel to a one-direction value of 1.0 TMU may be used as a "rule of thumb" pending future investigation.

For convenience, the exact equations are listed here for Apply Force and the two Apply Pressure cycles.   In these equations, the standardized force level is 7.48 pounds (3.39 kilograms) and the maximum allowable time is for a force level of 10 pounds (4.54 kilograms).   Also, DM = 4.2 TMU is assumed for the equations as stated.   The equations are:

Apply Force, TMU = 0.893 + 0.335 × force (pounds)
$\qquad\qquad$ = 0.893 + 0.739 × force (kilograms)

Apply Pressure, Alone, TMU = 8.093 + 0.335 × force (pounds)
$\qquad\qquad\qquad\qquad$ = 8.093 + 0.739 × force (kilograms)

Apply Pressure, Bind, TMU = 13.693 + 0.335 × force (pounds)
$\qquad\qquad\qquad\qquad$ = 13.693 + 0.739 × force (kilograms).

Finally, examples adapted from the approved training document[14] will be helpful in illustrating practical application of Apply Pressure data.

1. *Actuating Starting Buttons.*   Many types of machinery are activated by pushing the starting button with an Apply Pressure cycle including a Minimum Dwell.   Assuming standard resistance and less than $\frac{1}{4}''$ button travel, the usual analysis is: APA = 10.6 TMU.   Sometimes, the button is held in for a specific time, say 60 TMU, which will limit out the Minimum Dwell time as shown in the pattern at right.

Description	LH	TMU
Contact button	G5	0.0
Push button	AF	3.4
Hold for action	Proc.)	60.0
Minimum Dwell	DM)	—
Relax finger	RLF	3.0
Break contact	RL2	0.0
Total TMU		= 66.4

2. *Sticking Postage Stamps.*   Assume a mailing process with high repetition that justifies extra precision in assigning time for sticking the postage stamp to letters.   By a frequency study of the task on a platform scale, the average method is found to be two applications of a 2-pound force.   Without the study, the usual allowance would be: 2 APA = 2 (10.6) = 21.2 TMU.   Using the study results, the Actual AF = 0.893 + 0.335 (2) = 1.6 TMU.   Therefore, the Total TMU = 2 (1.6 + 4.2 + 3.0) = 17.6 TMU.   This allowed time is 17% less than the usual allowance would provide.

[14] A.T.S. 11. (Available from the U.S./Canada MTM Association.)

3. *Cutting Lead Wires.* The component lead wires on electronic terminal board assemblies are cut to trim by closing the pliers at the trim point, shifting the grip, and applying force to sever the wire. For the cut, a normal analysis would be: APB = 16.2 TMU. However, since a given component board may have hundreds of cuts, closer analysis is justified. If the measured cutting force is 4 pounds, the Actual AF = 0.893 + 0.335 (4) = 2.2 TMU. With this value, the analysis at right is appropriate. If the terminal board has 500 trim points, the time assigned will be 500 (16.2 − 15.0) = 600 TMU, or over 7% less than with the usual analysis.

TMU	RH Motion
5.6	G2
2.2	Actual AF
4.2	DM
3.0	RLF
15.0	TMU total

4. *Activating an Air Valve.* When activating the valve of a compressed air hose to blow chips off a part, the DM may be limited by the process time as shown at the right. Also, the RLF (and the normal Release) may be limited out by putting aside the hose. Thus, only the AF component remains to be allowed as shown in the pattern.

TMU	RH	Description
3.4	AF	Press on valve
146.8	Proc.	Blow chips, 6 sec.
—	DM	Minimum hold
13.4	M12B	Hose aside to bench
—	RLF	Relax finger
—	RL1	Finger off valve
2.0	RLl	Hand off hose
165.6 TMU	total	

These examples make apparent the versatility of the new Apply Pressure data.

The last point worthy of emphasis is that Apply Pressure consumes a relatively large amount of time as compared to many other MTM motions. Care should be taken not to miss writing it into motion patterns. This is especially true if short, highly repetitive work cycles are being analyzed. Errors in time allowed for operations can be considerable if Apply Pressures are omitted during analysis.

DECISION MODELS

Apply Pressure needs only one Decision Model, see Fig. 2.

ORIGINAL APPLY PRESSURE

This section is included to permit a comparison of the original data for Apply Pressure to the present version previously described. It is given in summary fashion to minimize space.

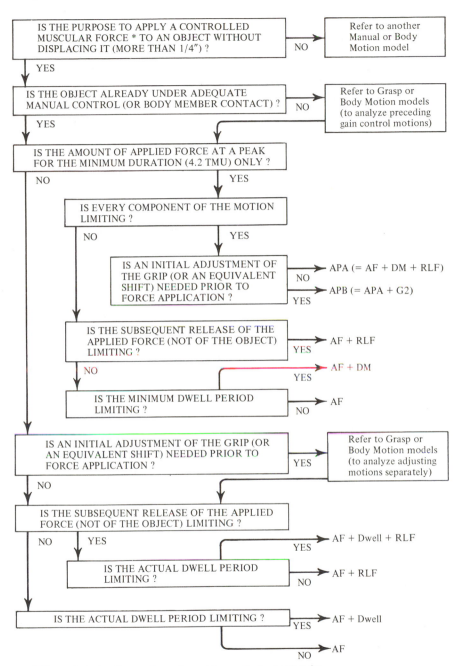

* The equivalent action can be performed by any body member, for
example, FMP (= FM + APA) is a Leg and Foot Motion.

Fig. 2.   Decision Model for Apply Pressure.

Definition[15]

> *APPLY PRESSURE is an application of muscular force to overcome object resistance, accompanied by little or no motion.*
> 1. *Apply Pressure is a hesitation or lack of motion.*
> 2. *The force required for apply pressure is greater than that required for a normal move or turn against resistance.*
> 3. *Apply pressure frequently is indicated by a setting of the muscles.*
> 4. *Apply pressure may be performed by any body member.*

Cases

Apply Pressure Case 1 exists when the body member acting requires initial orientation or adjustment to avoid loss of grip, injury, or discomfort during force application on the object.

Apply Pressure Case 2 exists when the body member acting is already fully oriented or adjusted such that full force application to the object can be made immediately.

Data Card

Former Table III of the MTM-1 Data Card prior to 1972 had the following simple data for Apply Pressure appended to the bottom:

APPLY PRESSURE CASE 1—16.2 TMU   APPLY PRESSURE CASE 2—10.6 TMU

Shorthand

AP1 showed Apply Pressure Case 1 and AP2 showed Apply Pressure Case 2.

As comparison will show, the original data contained the essential ideas; however, both the understanding afforded by the components and the more complete time analysis possible with the new data recommend it as a preferred methods-time tool. The closeness of definition and identical time for complete cyclic occurrences make it easy to update motion patterns containing the original Apply Pressure, if that is deemed desirable. Very simply, APA is the same as AP2 and APB is the same as AP1. Of course, this assumes the older analysis truly included a full cycle of the Apply Pressure element; and, if such is not really the case, the analyst can use the new component data to improve the earlier analysis which was as correct as was possible at the time it was made. The only real superiority of the newer data is the extensive research, field testing, and training development behind it which were not possible at an earlier date with the technology then available.

---

[15]From "MTM Basic Specifications" (Footnote 1, Chap. 5).

SUPPLEMENTARY APPLY PRESSURE

This section is included to permit a comparison of the interim form of Supplementary Apply Pressure data which developed into the present form. It is helpful here to explain the exact meaning of Supplementary as used in. the MTM system. Under the rules of the International MTM Directorate, which now has the authority for full approval of MTM data worldwide, national associations may perform research and develop MTM data deemed worthy of such approval. As a means of extended field testing and evaluation prior to such approval, each national association may include interim data on its own official data card as long as it is clearly labeled to be Supplementary to the international data, thus showing it lacking, as yet, the full IMD approval. Under this provision, the U.S.A./ Canada association carried the Supplementary Apply Pressure data on its MTM-1 card from 1965 to 1972, at which time the present data was substituted for both it and the original data. As can be seen by comparing this section with the earlier part of this chapter, the Supplementary period permits such experience and revision as will allow the finally adopted IMD data to be as complete and correct as possible, thereby promoting reliability and soundness in the final version.

Since the Supplementary data was derived from the same research as the newer international data, most of the earlier explanation would be identical. Therefore the summary treatment, below, mainly helps to show any differences which must be reconciled by those with extensive Supplementary practice when they proceed to apply the latest version of approved data.

Definition[16]

APPLY PRESSURE is an application of controlled muscular force to overcome object resistance, accompanied by little or no motion.

APPLY FORCE is the period of time during which no movement occurs while an increasing, controlled muscular force is being applied to an object(s).

MINIMUM DWELL is the period during which a reaction occurs for the reversal of force, where the force is held relatively constant during the reaction interval.

RELEASE FORCE is the period of time required for an operator to release muscular force.

A Total Cycle of Apply Pressure occurs each time force is applied and then released with a Minimum Dwell period intervening.

Apply Pressure, Bind is a Total Cycle with the inclusion of concurrent performance of a G2.

[16]From A.T.S. 9 (Available from the U.S./Canada MTM Association.)

Cases and Shorthand

The two cases of Supplementary Apply Pressure were: (1) simple Apply Pressure, symbolized AP; and (2) Apply Pressure, Bind, symbolized APB.

In addition, the components isolated by research were sometimes listed separately with their symbols and times: (1) AF for Apply Force; (2) DM for Dwell, Minimum; (3) Dwell for simple Dwell, not minimum; and (4) RLF for Release Force.   These were used when a complete cycle was not present.

However, the conventions for AF, AP, and APB also included two-numeral superscripts for the force level.   Here, forces up to 10 pounds were shown, e.g., AF05 for five pounds and APB 07 for seven pounds, with a special symbol of 00 for negligible force of 0.5 pounds or less.   There was no way to write a symbol for forces other than those to even pounds, therefore this condition was shown by the simple AF symbol with a note in the descriptive column for the force level involved.

Data Card

The Supplementary Apply Pressure data appeared on the MTM-1 Data Card of the U.S.A./Canada association between 1965 and 1972 as shown in Table 2.

Table 2.   Supplementary Data for Apply Pressure [Former MTM-1 Data Card Table 2 — Apply Pressure — AP (Supplementary Data)]

Apply Force (AF) =1.0+(0.3×lbs.). TMU for up to 10 lb. =4.0 TMU max. for 10 lb. and over	
Dwell, Minimum (DM) =4.2 TMU	Release Force (RLF) =3.0 TMU
AP = AF+Dwell+RLF	APB = AP+G2

No standardized force level was selected, although the Apply Force equation was simplified for application to: AF, TMU = 1.0 + 0.3 × pounds.   With Minimum Dwell, this made the values of AP08 and APB08 equivalent to the time values for the original AP2 and AP1; and these were used when it was not convenient or possible to determine the force level more precisely.   Also, the maximum stated value for AF, at 10 pounds or greater, made the maximum cycles AP10 = 11.2 TMU and APB10 = 16.8 TMU.   Note, too, that the APB differed from simple AP by Regrasp (G2) or the equivalent.

Since all of this data is now superseded by the new international data, it will serve the reader only to interpret earlier motion patterns in the record and/or to clarify for those needing updating to the new data in just what respects the data are different. The Supplementary Data served the valuable purpose of facilitating the field experience which was used to devise the final form of Apply Pressure. Comparisons of the definitions, time values, symbols conventions, and application rules make it obvious that the existing international standard is a definite improvement and benefited greatly from the Supplementary Data experience. Therefore, the analyst can use the resulting present data with the greatest confidence to meet actual analysis problems with maximum flexibility.

CHAPTER 10

Grasp

Definition[1]

GRASP[2] is the basic finger or hand element employed to secure control of an object.

1. The hand or finger(s) must obtain sufficient control of the object to be able to perform the next basic motion.

2. The object may be a single object or a group of stacked or piled objects which can be handled as though they were a single object.

The frequency with which Grasps occur is so great that examples are easy to enumerate. Besides the simplest type shown in Fig. 1, others are:

1. Obtaining hold of an MTM card to use it
2. Picking up one of a row of pencils to write an MTM analysis
3. Selecting a paper clip from a jumbled pile to fasten the analysis sheets together
4. Shifting the pencil to use an eraser on an error in the motions written
5. Transferring the completed analysis sheets to the other hand to lay them aside from the working area
6. Contacting the MTM card laying flat on the desk to pull it closer.

The performance of any of the manual motions discussed in earlier

[1]From "MTM Basis Specifications" (Footnote 1, Chap. 5).

[2]Grasp is not defined as a "motion" because it may involve one or several of the truly basic manual motions of Reach, Move, and Turn. Because *objects* are ordinarily the recipient of manual work, and control of them must first be attained to permit this work, it would be highly inconvenient if each instance of grasp required separate, detailed analysis of the motion components involved. The grouping of repetitive patterns of such components into a synthesized element such as Grasp is obviously much more practical. The Therblig Grasp is the identical element.

chapters might involve actions of any or all of the manual members up to and including the shoulder. With Grasp, however, the motions are confined almost exclusively to the fingers and hand, being generally shorter than an inch in length. Indeed, the fingers and hand are the only body members capable of grasping action. With all other members, contact is the only usual means of control.

Control of objects may also be secured by such tools as tongs, tweezers, and other mechanical devices. When this is the case, the motions used are Moves rather than Grasps. For example, closing tweezers on a very small part could require one or more M—A or M—C, depending on the precision with which the part must be gripped; the part would not be obtained with a Grasp. The Grasp definition involves the operator's fingers and hand themselves gripping the object before the MTM element Grasp is justified.

Another condition of the definition infers that Grasp is preceded and followed by other MTM motions. Since the usual purpose is to pick up or in another way to perform work on an object, a Grasp is usually preceded by a Reach and is normally followed by a Move, although other preceding and following motions are possible. This is important when recognition of Grasp is attempted because many Grasps occur during a time interval too short for the eye to actually see the action. There is no uncertainty as to Grasp being present, however, when the hand is empty one instant and appears the next instant holding an object. The purpose of Grasp is thus evident to the observer in that a controlled object must be in contact with or supported by the hand, and no other criteria is valid when deciding whether a Grasp has occurred.

GRASP VARIABLES

Grasp is a synthesized motion element. The analyst must understand this before he can clearly interpret the variables in grasping motions and the time they consume. Further help toward proper use of this MTM element is afforded by the Theory of Grasp covered in a later section of this chapter. Knowledge of both the variables and the theory will aid correct analysis of Grasp for the usual cases, as well as helping the analysis of unusual and unclassified Grasps.

Because all of the motions involved are one inch or less in length, distance variations are not considered in Grasp analysis. While the short motions of Grasping may well actually be Reaches and Moves, for which distance is a major variable, this does not mean that all the factors in the latter motions bear on the problem of identifying the type of Grasp performed. Some of the Reach and/or Move factors are pertinent, others do not apply. The established Grasp times have already taken account of the fractional (denoted f) distance involved.

Weight does not affect Grasp time because the application of pressure that control of weight involves will be made following completion of the Grasp. In other words, *control* in Grasp merely means that the fingers have closed on the object firmly enough to permit accomplishing any further motions that are to be performed. As explained in the Move chapter, the MTM procedure has placed the time to achieve the *additional* control required when grasping weighted objects exceeding 2½ pounds into the static component of the Move weight factors. The Grasp is completed as soon as the fingers have closed.

The observation time spent on motion study costs money. It would not be used economically if ordinary Grasp motions required analysis of the detailed finger movements at each occurrence. Easier, more accurate evaluation of Grasp follows when the categories and cases are defined in terms of the conditions surrounding the action rather than being defined solely on the basis of true motion content. The categories or cases of Grasp have therefore been grouped in terms of the object grasped in order to minimize the number of variables to be noted during observation and analysis of Grasp elements.

Selection Time

Selection time is caused by the necessity to choose which object to Grasp. It may be present in some types of Grasp and absent in other types. Discussion of a number of factors involved in selection will enable an MTM analyst to decide between cases and correctly account for any selection time present.

A worker uses muscular, visual, and mental control when he makes a choice. When the times required to exercise these types of control exceed that present in normal motions, the time difference is covered by selection time. The presence of such control may be noticed as a reduction in the motion speed, slight hesitation during a motion, or by minor extra motions that occur while selection is accomplished.

Part of the selection control is accomplished during the preceding Reach and the remainder will occur during the Grasp itself. Because of this, the type of Reach toward an object to be grasped depends directly on the nature or kind of Grasp to be made. The connection between any given kind of Grasp and the type of preceding Reach required will be discussed with each case of Grasp. Thus it will be seen that MTM does account for the interconnection and influence of one motion with another preceding or following.

Normal muscle control involves straightforward movement of the hand and fingers to the object being grasped with the speed usually associated with any given type of Reach. Selection time, however, may occur when the fingers must be positioned at the end of the travel before the Grasp can be made. This will slow the action.

The eyes normally are involved in almost all kinds of manual direction. When they must actually fix upon one certain object to be grasped from among many objects, however, more time will elapse. This is another kind of selection time.

A third kind of selection control, mental delay, is also present in some degree during all manual motions. The normal mental action included consists of only sufficient time to enable a rapid, simple yes or no decision. The mental delay included during selection, however, may exceed such simple decision time. When this occurs during a Grasp, the additional time must be included in the Grasp time.

Selection for grasping may include any or all of the forms of control noted and consideration need only be given to those cases where the time exceeds that normally found in motions where no selection occurs; unless the selection time is unusually prolonged, it has been included in the official data times reported for Grasp. This is because the data card times are derived from actual film studies which, obviously, included *all* the time needed to perform the particular type of Grasp in question.

When the object to be grasped lies clear of other objects, any selection time required will have been included in the preceding Reach. In other words, the fingers can readily perform any necessary motion to close on the object without hesitance. Thus single objects in isolated locations need no selection time.

Selection time is obviously needed when a single object must be grasped from a number of like or unlike objects nearby or jumbled with the object. Some of the time required will be covered in the preceding Reach, but successful grasping of the object will necessitate more finger control than for isolated objects. There is a special case of Grasp that covers this set of conditions, and all the time involved is included in the Grasp data time.

For unusual types of object selection in Grasp situations, correct times can be found by including in the motion pattern all the preceding extra MTM motions actually required by operating conditions. Selection time is no great problem so long as all of the motions or time elements actually required are correctly identified by the MTM analyst.

Nature of Object

It is evident that the time to grasp objects will vary with their size and shape. For instance, Grasping time will be less for some optimum object size than for objects either smaller or larger. Picking up a pencil will be easier than grasping either a straight pin or a common rolling pin. A readily accessible protrusion or extension of a part may simplify getting hold of it. When parts having regular geometrical forms such as cubes, balls, cylinders, and cones are handled, the mental processes are eased by much practice in grasping such shapes. When the configuration of a part will create unbalance in holding after being picked up, the worker will exercise more caution in selecting the point of Grasp to assure a firmer hold.

All such variable factors were present in the film samples on which MTM times are based, so the recognition of the proper case of Grasp based on object description will assure that the time allotted for grasping is adequate.

The effect of the nature of the object in determining Grasping time is covered in the MTM data both by case designation and size subclasses of cases. The size categories are not precise, nor are they to be literally interpreted. Rather, they are given to serve as *guides* to the relative volume or mass of objects which may be grasped in the data time given for the particular case of Grasp. The MTM analyst must use reasonable judgment in deciding the size class for the part he is analyzing. Sizes are denoted in either cubical limits or in diameter range for cylinders. Other geometrical shapes obviously are easily categorized with relation to these two basic configurations.

Object Surroundings

No one will dispute that a part lying clear of other objects will involve fewer finger motions to grasp than will a part restricted by its surroundings, or a part which must be handled in a special way. Many special kinds of restrictions imposed by object surroundings have been evaluated in the standardized cases of Grasp.

Objects contacting their surroundings on more than one surface can often be grasped with some relatively simple extra finger motions. An example of this is an MTM data card stored between two adjacent books between a pair of book ends on top of a desk (the books being close enough to barely admit the card) with one edge sticking out a bit. Sometimes a part, though hemmed in, has a protrusion which can be grasped "on the fly" or with minimum extra motion. Adjacent parts of dissimilar nature can also affect the ease with which an object can be grasped.

A need for special handling, as has been mentioned, can influence the finger motions in Grasp. Slick surfaces are harder to gain control of than usual textures. Roughness may present equivalent difficulty. Polished surfaces on a part to be grasped, or finishes sensitive to skin chemicals are other examples. Extra finger motions are needed to gain control of objects which can damage the fingers or which may be damaged by the fingers. The necessity to grip a part at a special point or within limiting areas further illustrates the special handling variable.

The procedure of evaluation given for Selection Time also holds for the variables introduced by surroundings. If the conditions fit those given for a standardized Grasp either closely or exactly, there is no problem.

When this is not so, correct motion analysis of *all* actions prior to the actual grasping will provide the proper answer to the case in question.

KINDS OF GRASP

Eleven varieties of Grasp have been standardized in the MTM procedure. There are five general types; the first and fourth types each have three cases and the third case of the first type has three subclasses for size variation.   The major division is by Type of Grasp:

Type	MTM Name	Common Description
1	Pickup Grasp	Obtain or Get Hold
2	Regrasp	Shift Hold or Realign Fingers
3	Transfer Grasp	Pass Object or Shift to Other Hand
4	Select Grasp	Search and Select or Jumbled Grasp
5	Contact Grasp	Hook or Sliding Grasp

The types, cases, and subclasses will be discussed separately.

Type I, Case A

Definition: Grasp Case 1A is used to obtain control of a single object of any size in an isolated location when the simple act of closing the fingers will suffice to complete the action.

Examples of Grasp Case 1A are numerous, because it is the most commonly encountered type of Grasp. See Fig. 1. Other instances are obtaining a desk pen from its holder or getting hold of the feed lever on a lathe.

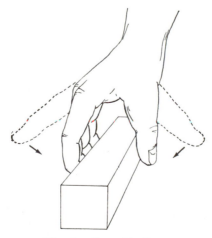

Fig. 1.   Case 1A Grasp.

The purpose of Case 1A Grasps is usually, but not necessarily, to pick up the object to Move it. This is the ordinary meaning of *obtain*. At times, however, a Case 1A Grasp may be used merely to *get hold* of the object for any reason that dictates restraint or future semi-controlled movement of the object during work cycles. The example of the feed lever of the lathe illustrates this latter purpose. It is an easy case of Grasp to recognize.

The Reach required will depend on the basic variables in the situation. As the object is situated alone, no selection time beyond that involved in the preceding Reach will be used. The nature or size of the object is relatively immaterial for this case of Grasp and, since the object is by itself, surroundings are likewise not pertinent.

All these factors indicate either a Case A or Case B Reach to be adequate. If the other hand is on the object within 3 inches of the grasping point, a Case A Reach will be used. This is because kinesthetic sense, as discussed in the Reach chapter, is operative within 3-inch limits. In other cases, where mental certainty of the object's location exists due to automaticity developed by repetition of the operation, the preceding Reach will also be Case A. In most other instances, the worker will employ a Case B Reach.

Unless all of the conditions outlined hold true, the Grasp in question will not be a Case 1A. For example, when additional elements of danger, accuracy, or other difficulty are present, some other kind of Grasp will apply. This is because the only requisite of Case 1A Grasp—simple, rapid closing of the fingers—is not met when factors such as those just noted are present.

Type I, Case B

Definition: Grasp Case 1B is used to obtain control of a single object of very small size in an isolated location or a thin object whose surface contacts closely a flat supporting surface when simple closing of the fingers is precluded by the need for accurate control or extra finger motions.

Two illustrations of Grasp Case 1B are given in Fig. 2, which shows the pickup of a celluloid ruler and a small pill. Other examples are: getting hold of the free end of the lead wire on a coil, obtaining a flat open end wrench lying on a bench top, picking up a piece of flat smooth steel from a table top, obtaining a $\frac{1}{16}$-inch diameter straight shank drill from a drill press table, and getting hold of the page of a book to flip it over.

The general purpose of Case 1B Grasps is the same as for Grasp Case 1A, the difference being that the nature of the objects and their surroundings are additional factors whose influences have been allowed for. Case 1B is an accurate Grasp for either reason. Research disclosed that both factors exert approximately equal influence on Grasp performance time.

As far as the isolated small object is concerned, no selection time is needed. The smallness of the object, however, imposes the need for accuracy in grasping. Either an accurate or controlled finger motion or two simple finger motions are involved. The term "very small" implies an object with a cross section of $\frac{1}{8}$ x $\frac{1}{8}$ inch or less, this being only a guide figure. The care needed to assure control is due to the relatively large size of the finger tip area whose receptors must sense the object and inform the brain that the object can be moved.

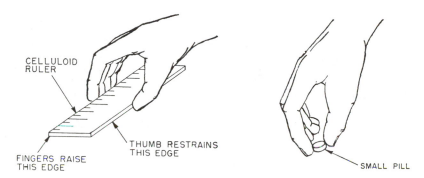

CELLULOID
RULER

THUMB RESTRAINS
THIS EDGE

FINGERS RAISE
THIS EDGE

SMALL PILL

Fig. 2. Case 1B Grasp.

Selection time may be a minor factor in separating a thin object from a flat supporting surface or other, adjacent thin objects. The extra care and accuracy needed in grasping is mostly due, however, to its shape and position in relation to the supporting surface. This generally requires the use of two simple finger motions to separate the thin object from its support by extension of the flesh of the finger tips slightly under the edge of the object, with the thumb acting as a pivot point. This action is aided if the object is flexible like the page of a book. Generally, picking up a flat object from a flat supporting surface will necessitate lifting one edge of the object to permit completing the Grasp.

One useful practice when many Case 1B Grasps must be performed is to lay a sponge rubber mat over the otherwise hard and rigid workplace surface. This may make it possible for the operator to use Case 1A Grasps to obtain the objects, which is both easier and less time consuming.

That a Case 1B Grasp will most often be preceded by a Case D Reach is logical from the definition and discussion of Case D Reach in the Reach chapter. This concept also indicates that objects presenting danger to the fingers or subject to damage will be grasped with a Case 1B Grasp, even though the same object could be obtained with a Case 1A Grasp when such factors are absent. A Case B Reach could logically be assigned as preceding this latter occurrence.

The MTM analyst should be cautious not to miss other MTM motions that may be required when Case 1B Grasp is encountered. It often happens that finger alignments, pressure application, or moving the fingers underneath the object occur in addition to the Case 1B Grasp itself. These would naturally require proper identification and extra time allowances.

Type I, Case C

Definition: Grasp Case 1C is used to obtain control of a single object approximating a cylindrical shape which contacts its surroundings with one longitudinal surface other than the one on which it rests.

The use of Grasp Case 1C in two kinds of situations is depicted in Fig. 3. Note that only one object is being obtained in both cases, although more than one object is present in the second example. Other instances seen in industry are getting hold of a straight shank drill, a welding rod, a pencil, or a stick of chalk from an orderly array of the same kind.

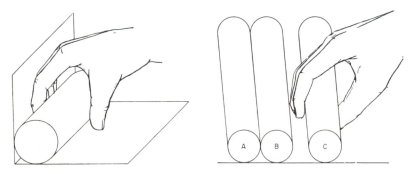

Fig. 3.   Case 1C Grasp.

While this Grasp is a Type 1 in general concept, involving the pickup of an object, its special nature has led to usage of the common name of "Interference" Grasp. Note that the objects included are rodlike in shape; and the cross section approaches cylindrical form, although cross sections of such other geometric shapes as hexagonal are not excluded, provided the length of the object is such as to satisfy the rodlike requirement. The idea is that a relatively long and slender object is being obtained from its position of rest on one long side and with a restriction to finger action on another long side.

The motion significance is that the fingers must first separate the object from its surroundings sufficiently to allow closure that will effect control. Actually, the sequence is as follows. The forefinger and/or middle fingers contact and roll the object toward the thumb, after which the leading finger or fingers shift around the object to effect closure, to complete the grasping action.   The added time for this Grasp covers these extra finger motions.   Note also that an object exceeding 1 inch in diameter will not require a Case 1C Grasp because its size relative to the size of the finger tips is large enough to permit easy closure with little difficulty.   The two lines of contact will not prevent getting the fingers on an object as large as 1 inch, so no interference exists.

To justify classification of Case 1C Grasp as such, several conditions must be met.   The shape and surroundings were noted above.   The fact that only one object is being picked up is another.   The object cannot be standing on end, but must be lying on its long side.   Jumbled conditions deny usage of this Grasp because an orderly array is not then present. In short, the MTM analyst must recognize the special nature of this Grasp and avoid using it merely because the time seems applicable when the conditions are otherwise incorrect.

The Reach preceding a Case 1C Grasp will logically be Case A, B, or D by the definition of those classes in the Reach chapter.   Cases C and E of Reach are excluded from consideration.   Decision on which of the three possible Reaches is used will depend on the nature of the object, the surrounding conditions, and the operator's usual mode of performance. Each occurrence must be analyzed by the rules for determining the proper class of Reach.   No restrictive rules can be given that will give unvarying answers for all instances of Case 1C Grasp.

Three subclasses of Grasp Case 1C are based on size variation of the nearly cylindrical cross sections involved:

1. Diameters larger than ½ inch
2. Diameters of ¼ inch to ½ inch, inclusive
3. Diameters less than ¼ inch.

Remember that these dimensions are given as relative guides for decision.   The only reason for the subclasses is to account for time variations among the actually observed instances of Interference Grasps.

Type 2

Definition: Grasp Case 2, Regrasp, is used to shift the hold or realign the fingers on an object already under manual control to improve or increase control of the object or to relocate the object in the hand for later use in the work cycle.

The frequency of Regrasp in ordinary manual motions is extremely high. A typical example is the shifting of a pen or pencil in the writer's hand before, during, and after the action of writing.   Other instances will readily occur to the reader.

Apparent inconsistencies in the original data were readily explained when Messrs. Maynard, Stegemerten, and Schwab isolated and identified this Grasp.   This was the first time it had been recognized and thus represents a contribution by MTM to the basics of motion study. Attention to time and motion together revealed its presence when independent thinking on either failed to uncover Regrasp.

Essentially, Regrasp proceeds with a rapid series of very short finger motions involving a time lapse so short that control of the object is not lost. The grip is relaxed, the fingers are quickly shifted to new holding points, and the grip is reapplied to maintain control. Actually, the fingers slide along the object in a manner which relocates the object in the hand without allowing it to drop.

Regrasp occurs with multiple frequency without intervening motions taking place in a large number of actual motion sequences. The necessity for this is evident when a major repositioning of the object in the hand is considered.

Since prior control of the object exists as a condition of performance, Regrasp will not be preceded by a Reach. The preceding motion will be a Move or another type of Grasp. Regrasp often occurs between or during other motions, a common event being either during a Case C Move or between it and an accurate placement. It also appears with high frequency as a combined Move-Regrasp when the operator employs a Regrasp to improve his grip or for orienting during object transport. The rules for combined motion discussed in the Turn chapter will apply to such cases.

The recognition and classification of Regrasp present a minimum of difficulty for the MTM analyst. This makes one wonder why it was not isolated earlier, because its utility in motion study is obvious.

Type 3

Definition: Grasp Case 3, Transfer Grasp, is used to transfer an object from one hand to the other when the shift involves brief holding of the object by the fingers of both hands.

Occurrence of this Grasp should be low in industrial operations, because its elimination by proper placement of materials and workplace layout is a constant aim of engineers abiding by the Laws of Motion Economy. However, since the complete elimination of Case 3 Grasps is difficult and sometimes impractical, it will at times properly belong in an MTM standard. An example of Transfer Grasp occurs when an eating fork is shifted from the left to the right hand after being used to hold meat to the plate for cutting. The motions included in a Case 3 Grasp are a Type 1, Case A Grasp by one hand, a reaction time, and a Release of the object by the hand first holding. The reaction time covers the pause during which mental or muscle assurance of the completion of Grasp by the receiving hand occurs. *This hesitation is the clue which distinguishes a Transfer Grasp.* If this pause is prolonged, the action is not a Case 3 Grasp—process time is involved.

When the reaction time is missing, motions other than those denoted under Case 3 Grasp have transpired. Highly skilled operators may pass an object without using reaction time by employing a Case 1A Grasp with one hand to pick the object out of the other hand, while the latter hand executes a Contact Release that involves no time allowance. A highly skilled worker may even toss the object between hands and not lose control. However, both of these situations should be analyzed for the actual motions used since assignment of a Transfer Grasp would be incorrect. Certainly, the tossing motion involves process time which should be determined by a stopwatch instead of by assigning an MTM motion.

A Case A or Case B Reach, depending on the 3-inch rule for kinesthetic sense, by the receiving hand will normally precede a Case 3 Grasp. Or a Case D Reach will be involved when the object is very small, dangerous to the hand, or easily damaged. The other hand will be either holding or moving the object just prior to the Transfer Grasp, so no Reach is involved for it.

Transfer Grasp is very easy for the MTM analyst to perceive and classify. However, its presence in a productive situation usually is undesirable. The analyst should be alert to the possibility of eliminating it by methods improvement based on relocation of the parts in the operation. When such deletion is not possible or practical, however, he should allow the observed Transfer Grasp.

Type 4

Definition: Grasp Case 4, Select Grasp, is used to obtain control of a single object from a jumbled pile of objects when the act of search and select must precede closing of the fingers.

This Grasp has become known as the "Totepan" Grasp because of typical examples like these:

1. Selecting one screw from a totepan full of screws
2. Picking out a nut from a pan full of screws, nuts, and washers.

Selection time, as previously discussed, is obviously a large factor in Case 4 Grasps. Because the fingers are definitely aiming for one particular piece, the same amount of selection will be needed whether the surrounding objects are similar or dissimilar. This time for aiming the fingers will always exceed the relatively rapid mental action additionally needed due to dissimilarity, as compared to selection of similar objects, so that no appreciable or discernible difference in time will inhere in the two events.

The definition requires that jumbling—lack of orderly array—be present to justify classification of a Case 4 Grasp. Parts neatly stacked, layered, or placed systematically will require less Grasp time or at least some other kind of Grasp.

On the other hand, a Case 4 Grasp implies that the object being aimed at can be easily separated from the others around it. Interlocking, tangling, or twisting together of parts presents a different problem. To obtain one from such parts will likely require Reaches, Moves, etc., other than those normally involved in a Case 4 Grasp. One part may possibly be grasped by Case 4 action, but subsequent additional Moves will probably be needed to separate it from the other parts or to otherwise surmount the adverse condition of the parts. Occurrences of Grasp for such situations require detailed motion analysis rather than cursory assignment of a Case 4 Grasp and its time value, even though the derived time value might equal the time for a Totepan Grasp.

The Reach preceding a Case 4 Grasp must, by definition and control involved, be a Case C Reach. No other class of Reach will logically connect the motions. Case D Reach might be thought proper when the possibility of danger or damage is present; but the time would not differ and the selection time idea in Reach Case C will overrule.

The relative size of object grasped justifies three subclasses of Case 4 Grasp:

1. Objects larger than 1″ x 1″ x 1″
2. Objects from ¼″ x ¼″ x ⅛″ to 1″ x 1″ x 1″, incl.
3. Objects smaller than ¼″ x ¼″ x ⅛″.

As a container of objects being grasped approaches the empty state, Case 4 Grasp would obviously no longer apply. However, the analyst should remember that MTM work patterns are always written for the normal or average situation encountered. If the container is usually full, he would specify a Case 4 Grasp. If it normally contains mostly isolated objects, he should fairly judge another kind of Grasp.

Type 5

Definition: Grasp Case 5, Contact Grasp, is used when enough control of a single object is gained merely by purposeful touch of the object surface by the finger(s) and/or hand. It is actually not a motion but rather a descriptive term.

Most commonly, this type of Grasp is called the Contact, Sliding, or Hook Grasp:

1. Contact because this is the aim of "purposeful touch";
2. Sliding because this is often the action on the object after the Grasp;
3. Hook because the hand and fingers often assume this shape in such Grasps.

Instances of Case 5 Grasp occur when a worker contacts the hinged cover of a box to swing it shut, lays his fingers on a part to slide it out of the way, or hooks his hand on the lip of a totepan to draw it toward him. Other typical examples are the *contacts* prior to pressing a typewriter or adding machine key, pressing an electrical contact button, pushing a door shut, etc.

Logical analysis makes it evident that Grasp Case 5 requires no time. One instant the manual members are reaching; immediately upon contact, they are either moving or constraining the object. The Contact Grasp, however, should be recorded in the MTM pattern to show that control of the object has been gained. In this way, the methods description is completed, even though no time is involved.

The use of Contact Grasp has no effect on the preceding Reach. The case of Reach used must be noted solely by examination of the part and analysis of the Reach by its own rules. It is obvious from this that the only type of Reach excluded is Case E; any other type might be required.

The absence of any perceptible pause between Reach and Move is a clue to the performance of a Case 5 Grasp. Pushing or pulling of objects is frequently the end result of a Contact Grasp and therefore a good clue to its occurrence. Another clue is when actions are taken on objects which remain on a surface or are mechanically constrained, such as a lathe feed lever. These situations usually involve Case 5 Grasps because the hand, after contacting, must supply only slight extra control—the object is basically controlled by the surface or mechanism.

Note that contact alone will not ordinarily permit the object to be lifted nor will an object's weight usually be controlled without MTM motions in addition to the Contact Grasp. If there is space to reach the hand under the object and lift it with a flattened palm, however, it is possible to do so with a Contact Grasp only.

Since Contact Grasp requires no perceptible performance time, the workplace and operating facilities should be designed to utilize this form of Grasp whenever possible.

General Comment

At this point the reader would benefit substantially from a quick review of the Kinds of Grasp. An efficient way is to study the Decision Models found at the end of this chapter.

THE SHORTHAND OF GRASP

Grasp

The MTM engineer will show a capital letter G on his study to denote Grasp.

Type of Grasp

An arabic numeral shows the type of Grasp employed. Thus G1, G2, G3, G4, and G5 will show the complete gamut of types.

Case of Grasp

The case, as per previous discussion, is noted by a capital letter after the type number. The possibilities are: G1A, G1B, G1C, G4A, G4B, and G4C.

Size Subclass

A number will show the size subclass for a Type 1, Case C Grasp. The following listing thus completes all combinations of symbols: G1C1, G1C2, and G1C3.

DATA CARD INFORMATION

Table IV of the official MTM-1 card lists the time values for Grasp. These are shown here in Table 1.

Table 1.   Grasp Data (MTM-1 Data Card Table IV—Grasp—G)

TYPE OF GRASP	Case	Time TMU	DESCRIPTION	
PICK-UP	1A	2.0	Any size object by itself, easily grasped	
	1B	3.5	Object very small or lying close against a flat surface	
	1C1	7.3	Diameter larger than 1/2″	Interference with Grasp
	1C2	8.7	Diameter 1/4″ to 1/2″	on bottom and one side of
	1C3	10.8	Diameter less than 1/4″	nearly cylindrical object.
REGRASP	2	5.6	Change grasp without relinquishing control	
TRANSFER	3	5.6	Control transferred from one hand to the other.	
SELECT	4A	7.3	Larger than 1″ x 1″ x 1″	Object jumbled with other
	4B	9.1	1/4″ x 1/4″ x 1/8″ to 1″ x 1″ x 1″	objects so that search
	4C	12.9	Smaller than 1/4″ x 1/4″ x 1/8″	and select occur.
CONTACT	5	0	Contact, Sliding, or Hook Grasp.	

Examination of MTM-1 data cards prior to the 1972 revision reveals that some of the time values formerly were in light-face type, others in heavy-face type. The heavy figures were obtained directly from actual research data. The tentative light figures were either extrapolated from research data or synthesized by the Theory of Grasp described below; these values were for G1C1, G1C3, and all G4 performance times. All of the values, however, have yielded good results in thousands of MTM analyses and may be used confidently.

The MTM Association recognizes the need for further data and research on Grasp. Such research may either validate the tentative values or provide new information to improve the Grasp data. For example, the original MTM-1 data card showed a performance time of 1.7 TMU for G1A based on research done with a film speed taking just that amount of per-frame time. In the research on *Short Reaches and Moves*[3], using a film speed four times as fast, the newer and more accurate value of 2.0 TMU was found for G1A. After validation, this value was made official on all MTM-1 cards since the March, 1955 revision.

THEORY OF GRASP

It was mentioned at the outset that Grasp is by nature and definition a synthesized MTM element composed of very short basic finger and hand motions. Theoretically, all grasping is done by some combination of reaching or moving with the fingers, and the more complicated Grasps may also require aligning of the fingers. An attempt to explain and isolate these more basic motions resulted in what is known as the Theory of Grasp.

There are three purposes or uses for the Theory of Grasp:

1. To promote the ability of the MTM analyst to understand and visualize the standardized cases of Grasp now included on the data card.
2. To synthetically arrive at usable times for unusual kinds of Grasp not now included in the MTM data. This should be done in a cautious manner, however.
3. The data card gives time values for grasping only single objects, as reference to the definitions for each case of Grasp will show. In actual practice, multiple Grasps or Grasps of a handful of pieces are a frequent event. The Theory of Grasp enables the MTM engineer to evaluate such situations because it both increases his background data and shows the actual method of synthesis he may profitably follow.

The Theory of Grasp is presented in tabular array in Table 2. Both the basic motions and time values are given, along with comparison to research[4] values and the data card for each kind of Grasp. The reader should study this information carefully. It is self-explanatory, except for the following specific comments.

The listing for motion significance at the bottom of Table 2 calls attention to the theoretical basic motion components used to synthesize each case of Grasp. It includes three kinds of fractional finger reaches, one fractional

[3]R.R. 106. (Available from the U.S./Canada MTM Association.)
[4]Ibid.

Table 2. Theory of Grasp and Release

Case of Motion	Motion—Time Synthesis f = fractional inches or less than 1″	TMU Value		
		Theory	Experiment*	Data Card
G1A	RfA + G5 = 2.0 + 0.0	2.0	1.972	2.0
G1B	RfA + G5 + MfB = 2.0 + 0.0 + 2.0	4.0	}4.351	3.5
	RfD + G5 + MfB = 2.0 + 0.0 + 2.0	4.0		
G1C1	G5 + MfB + RL2 + RfA + G5 + MfB = 0.0 + 2.0 + 0.0 + 2.0 + 0.0 + 2.0	6.0		7.3
	G5 + MfB + G2 = 0.0 + 2.0 + (4.0 or 6.0)	6.0 or 8.0	}7.85	
G1C2	RfD + G5 + MfB + RL2 + RfA + G5 + MfB = 2.0 + 0.0 + 2.0 + 0.0 + 2.0 + 0.0 + 2.0	8.0		8.7
G1C3	P1SE + G5 + MfB + RL2 + RfA + G5 + MfB = 5.6 + 0.0 + 2.0 + 0.0 + 2.0 + 0.0 + 2.0	11.6		10.8
G2	RL2 + RfA + G5 + RL2 + RfA + G5 + MfB = 0.0 + 2.0 + 0.0 + 0.0 + 2.0 + 0.0 + 2.0	6.0		5.6
	RL2 + RfA + G5 + MfB + RL2 + RfA + G5 = 0.0 + 2.0 + 0.0 + 2.0 + 0.0 + 2.0 + 0.0	6.0	}5.899	
	MfB + RL2 + RfA + G5 + RL2 + RfA + G5 = 2.0 + 0.0 + 2.0 + 0.0 + 0.0 + 2.0 + 0.0	6.0		
	Occasionally RL2 + RfA + G5 + MfB = 0.0 + 2.0 + 0.0 + 2.0	4.0		
G3	RfA + G5 + Reaction time + RL2 + RfE = 2.0 + 0.0 + 1.6 + 0.0 + 2.0	5.6	5.257	5.6
G4A	P1SE + G1A = 5.6 + 2.0	7.6		7.3
G4B	P1SE + G1B = 5.6 + 4.0	9.6	}9.204	9.1
G4C	P1SE + G1C2 = 5.6 + 8.0	13.6		12.9
	P1SE + G1A + G2 = 5.6 + 2.0 + (6.0 or 4.0)	13.6 or 11.6		
G5	Contact only—logically no time	0.0	0.0	0.0
RL1	RL2 + RfE = 0.0 + 2.0	2.0	2.041	2.0
RL2	Break contact only—logically no time	0.0	0.0	0.0

Motion Significance:

RfA	finger reach with automaticity
RfD	finger reach with care (smallness, accuracy)
RfE	finger reach to clear
MfB	grip or move of object to hold
P1SE	placement of fingers (including selection time)

*From R. R. 106. (Available from the U.S./Canada MTM Association.)

move of the object being grasped, and a case of Position (see Chapter 19) which covers aligning of the fingers to the object being grasped. Also note the listing of "Reaction time" in the synthesis for G3 Transfer Grasp. This phenomenon was mentioned in the discussion of this element as being the hesitation time which covers the pause involved. It is the reason why 5.6 TMU is required instead of the 4.0 TMU otherwise taken by the RfA and RfE components of G3. Because Release (see Chapter 18) motions are like those in the Theory of Grasp, the table also shows the Theory of Release.

It might be noted that there seems to be a lack of proper correlation among some theoretical, experimental, and data-card times. This should not disturb the reader, because the data-card times have yielded acceptable time standards in thousands of applications to date. Until data-card revisions are forthcoming, the analyst should rely on the present data and application rules.

PRACTICAL WORKING RULES

All of the major types of Grasp occurring with high frequency have been included on the MTM card in standardized form. That many other infrequent varieties of Grasp may well be possible is rather obvious, and for these a few practical hints will help the engineer to make reasonable analyses.

For the grasping of objects that are very large in relation to the hand or fingers, an analyst must decide what motions are most natural and likely. If the object is too large to permit the hand to close around it, then only G5 is possible and other MTM motions will be needed in addition. It is very likely that G1A is the logical choice for sizes between that extreme and the range of smaller sizes covered by the data card. For example, recall the 1-inch diameter limit for a G1C suggested in the discussion of that motion.

The presence of a readily accessible protrusion or extension on an object, whether jumbled or not, will tend to indicate use of a G1A. This is a case where the ordinary rules for grasping do not apply. However, the situation is not exactly the same as an ordinary G1A with a preceding A or B Reach. It is probable, when jumbling is present, that the Reach will be a Case C even though the Grasp is G1A. In other words, the controlled Reach will locate the fingers well enough to permit them to close easily on the prominent feature of the object without further selection time being needed during the Grasp itself.

The data on the MTM-1 card applies directly to the use of bare hands. Then what motions should be written when gloves are worn?   The answer[5] will depend mostly on the texture, dryness, and fit of the gloves.   In any event this must be judged in the light of whether the hand and fingers can act in a normal fashion.   If they can, the regular MTM Grasps will apply.   If not, other motion analysis or a stopwatch should be employed.

Another assumption in developing the data card values was that the object being grasped was visible to the operator.   Special rules have been developed and successfully applied in the field for blind Grasps[6]; but these cannot be fully sanctioned until validated by further research. One is that blind Grasps following a Case A Reach must be classified by the nature of the object grasped; they will tend to be some kind of G1 or a G5.   Another rule used in the field says that blind Grasps preceded by B, C, or D Reaches appear, in researches to date, to require some case of G4.   These rules will help the analyst until a final answer has been obtained.

Finally, every case of observed Grasp that does not reasonably fit the standard categories of MTM should not be given a standard designation. Neither should it be assigned the data card times unless they happen to be equivalent.   When possible, the Theory of Grasp should be used to synthesize the true condition and the time assigned according to the motions determined necessary.   When this approach is impractical or fails, the observer should rely on such other means of work measurement as time study rather than erroneously apply the MTM data.

---

[5] On this topic, but not verified in industrial study, was: Walton M. Hancock and William R. Tiffan, "A Laboratory Study Concerning the Effect of Gloves on Performance Times," *Journal of Methods-Time Measurement,* Vol. XIV, No. 2, 1969, pp. 14–28.

[6] Also reporting only laboratory trials was: Stanley M. Block, "Recent Research on Blind Grasp," *Journal of Methods-Time Measurement,* Vol. V, No. 4, Nov.–Dec. 1958.

DECISION MODELS

The Decision Models for Grasp analysis follow as Figs. 4A and 4B.

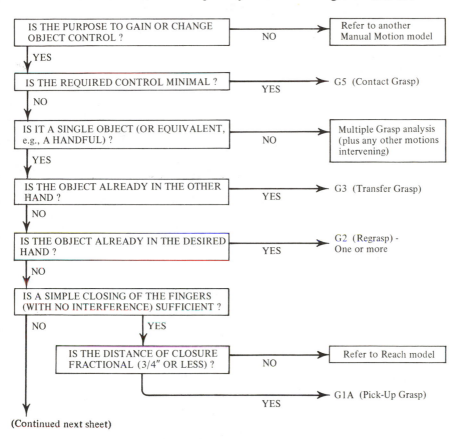

Fig. 4A. Decision Model for Grasp.

(Continued from last sheet)

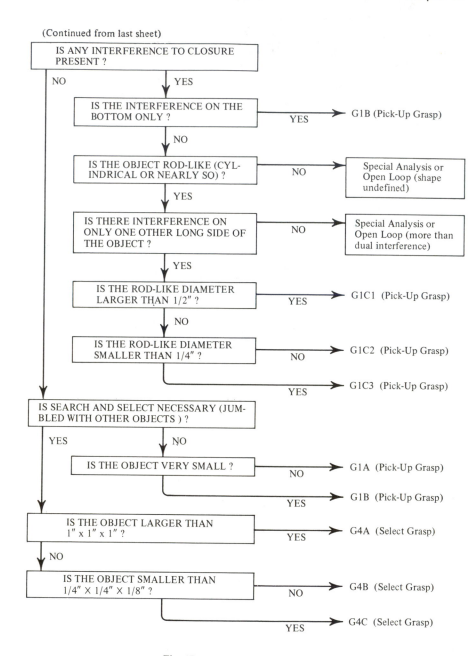

Fig. 4B.   Decision Model for Grasp.

CHAPTER 11

Release

Definition[1]

RELEASE[2] *is the basic finger or hand motion employed to relinquish control of an object.*

1. Release is performed only by the fingers or the hand.

The purpose of Release is twofold. First, control of the object is no longer desired by the worker and it is thus, in common words, "let go." Second, Release frees the hand and fingers for use in other motions. The latter reason is a distinction which aids recognition of the necessity for Release in motion patterns where its occurrence might otherwise be missed. This is especially true when the motions are being visualized rather than observed directly.

Being one of the simplest possible MTM motions, Release consumes the minimum amount of motion time. Since the eyes cannot actually see the motion, recourse in its identification must be based on logic or the observable effect it produces on the object. This is similar to the situation discussed in Grasp; and in a sense Release is really the opposite of Grasp. When an object is seen to be controlled manually one instant and completely free of control the next instant, it is obvious that Release was performed. This is the recognition clue.

The performance of Release is restricted by the definition to direct action by the fingers or hand with no intervening devices between them and the object concerned. When various hand tools such as tweezers, hand clamps, and the like have been used to hold or control the object, it is then freed

---

[1]From "MTM Basic Specifications" (Footnote 1, Chap. 5).

[2]The well-known Therblig called Release Load is the identical motion as defined. For brevity in use, the term Load is usually omitted in MTM practice. Recognition of it is implied, however, in the second letter of the MTM symbol.

by Reaches or Moves which cause or allow the device to open. For example, letting go of a part held in the jaws of spring-loaded wiring pliers could involve only a Case B Move by the fingers, which hold the plier handles by contact pressure against the spring resistance. If the pliers are allowed to open fully, however, the Move would be a Case A to the maximum travel of the plier handles. If a pair of common pliers are being used, the motion analysis is more complicated because at least one of the plier handles must be "hooked" between the fingers to avoid dropping the pliers when the part is released. In this case, either the analysis above is possible or else the part could be released by performing an RfE with the thumb. It is also possible that a Move-Turn of the pliers prior to or combined with the opening motion would be needed to allow the part to fall free of the jaws of the pliers. If this is true, the actual releasing motion could be limited out in the analysis.

KINDS OF RELEASE

Because there are only two logical ways in which the fingers or hand can end control of objects, only two cases of Release are needed.

Case 1

Definition: Release Case 1 consists of opening the fingers as a separate, distinct motion.

Most MTM analysts refer to Case 1 Release as a "Normal" Release because it is a simple, easy, natural motion. Reaching of the fingers out of the way of the object is the only action involved. Research has definitely shown this to be a short Case E Reach, generally an inch or less in length. Refer to Fig. 1.

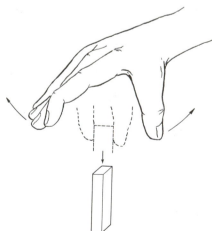

Fig. 1. Case 1 Release.

The value of isolating Release as an independent motion is in the descriptive worth of showing in motion sequences that an object is no longer controlled or restrained by the worker's hand.   For purposes of merely accounting for the time taken by the motion, recording of an RfE would otherwise suffice.   But recall that a good work measurement analysis shows *both* what the worker is doing and what happens to the workpieces. Writing a Release in the motion pattern achieves both aims, while an RfE would only show the operator's action.

The definition of Case 1 Release recognizes that the fingers may be opened while other MTM motions are performed.   For example, in several of the kinds of Grasp there are internal events when the fingers open momentarily but the aim in that case is to *gain* control of the object. When, however, the same motion occurs in a separate and distinct manner with the aim of losing object control, the justification of Release as an independent motion category is readily evident.

The result of Case 1 Release on the object need not be identical for all occurrences.   The object may be dropped, freed from restraining action, or merely left to rest on a supporting surface.   The manual action is the same in each of these possibilities.

Case 2

Definition: Release Case 2 consists of breaking contact between the object(s) and the fingers or hand, when no discernible motion occurs.

The sole purpose of Contact Release, the common name for a Case 2 Release, is to provide an MTM element describing the fact that the object is no longer under manual control.   Since no motion is involved, lack of this element in a motion pattern might cause the reader of it to wonder whether the object had been freed or was still under control.

Logically, since no motion occurs during a Release Case 2, this element will consume zero time.   One instant the hand is in contact with an object, the next instant it is reaching away with no perceptible pause between. This element is thus the opposite of a G5, or Contact Grasp.   The MTM analyst must discern the presence of Contact Release to account for operation method only.   He can do this by being alert to relinquishing actions during work sequences.

THE SHORTHAND OF RELEASE

The conventional symbols for Release are as simple as the motion element.

Release

The basic element Release is shown by the two capital letters RL, which are an abbreviation of Release Load.

Case

An arabic numeral denotes the case of release performed thus:

RL1 shows performance of a Case 1 or Normal Release.

RL2 shows that a Case 2 Release was done, a Contact Release.

DATA CARD INFORMATION

Data for Release in Table 1 appears in Table VI of the MTM-1 card.

Table 1.   Time Data for Release (MTM-1 Data Card Table VI—Release—RL)

Case	Time TMU	DESCRIPTION
1	2.0	Normal release performed by opening fingers as independent motion.
2	0	Contact Release

THEORY OF RELEASE

The reader probably noted in the Grasp chapter, under the theory section, that the Theory of Release was combined with the Theory of Grasp for convenience.   The only purpose served by a Theory of Release is to show that RL1 involves an RfE.   The time for an RfE, and thus the time for RL1, was fully validated in the research on *Short Reaches and Moves*.[3] The films on which this data was based were taken at the high speed of 4,000 frames per minute (0.42 TMU per frame), therefore, the value for RL1 should need no further change.

PRACTICAL WORKING RULES

Recognition of the performance of Release during work cycles and correct decision as to the case observed is aided when the analyst's attention is focused on the type of control being relinquished.   One key to this goal is the type of Grasp with which the object being released was originally obtained.

In most instances where an object was originally subjected to a Pickup Grasp, it will normally be abandoned with an RL1.   Another possibility

[3]R.R. 106. (Available from the U.S./Canada MTM Association.)

is when the use of a G5 to obtain is followed during the ensuing motion sequence by a G2 to gain further control or enclose the fingers around the object. This will almost always cause the Release to become an RL1.

The analysis is reversed for RL2. Normally an object obtained with a G5 will be abandoned by an RL2. Next assume the object was lifted by a pickup type of Grasp. If either a Regrasp (for example after the object is being guided along a surface) or an R—E of the fingers from the object occurs during the operating cycle, the use of RL2 to merely finalize loss of contact will logically be the correct Release classification.

An event where an R—E might be used *instead* of Release was discussed under that case of Reach in the Reach chapter. This is where a handful of small parts are dropped by opening and spreading the fingers, which would require that the fingertip travel be measured to assign an R—E or a sequence of R—E motions.

When sticky substances are present on the surfaces of objects to be released, some delay time may occur during the Release. In such cases, the time of releasing must be found either by analyzing any other MTM motions involved, by means of a stopwatch, or by film analysis. The MTM Release motions do not include any difficulty in letting go of the object.

Releases also occur at times in combined motions, as when objects are tossed aside or dropped "on the fly." The pattern at right fully shows what happens to the motions and time in such cases. The dash (—) in the TMU column for RL1 shows the time is limited out.

M12Bm	10.0 TMU
RL1	—
mR5E	4.6 TMU
Total Time	14.6 TMU

Release also occurs with great frequency during many such common motion sequences as the Reach, Grasp, Move or Turn, and Release as would be employed to rotate a nut on a bolt. The MTM analyst should be alert for such short motion patterns involving Releases.

Because the performance of Release is so repetitive in motion patterns, the analyst must use caution so as not to miss it. The time for a single Release is usually not sufficient to invalidate the total time for a motion pattern of normal length; but, if a number of Releases are missed, especially in an operation of short duration, the cycle time can be affected adversely.

DECISION MODELS

The Decision Model for Release is given as Fig. 2.

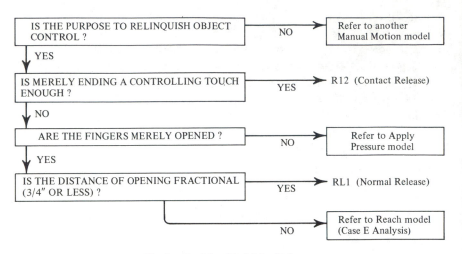

Fig. 2.   Decision Model for Release.

CHAPTER 12

Position

Position[1] is one of the most important single motion categories in the MTM system. Conceptually, it accounts for the highest precision and/or accuracy of placement at the end portion of object movement. It occurs with high frequency in many types of work; and, in addition, the time requirement is high when compared to many other MTM motions. Care in analysis and a clear understanding of the concept of position, therefore, are vital to the determination of correct times for motions and motion patterns. The validity of many otherwise correct MTM analyses can be reduced significantly by wrong identification and classification of the Position elements involved.

For reasons that will soon be apparent, the formal definitions of Position have been deferred in this chapter, unlike the practice in previous chapters where the formal definitions were given first. To start the discussion, only a general statement or concept is presented, together with a common core explanation that is followed by two subchapters where formal definitions of Position will be found. Despite this special textual approach, the numbering of footnotes, figures, and tables in this chapter will be continuous for the chapter as a whole to facilitate reference. This unifying introduction will also present certain ideas that apply equally to both subchapters.

In the third edition of this text the authors reported on three sets of Position data. These were International Position, Supplementary Position, and Alternate Position. International Position is fully approved by the IMD and by *all* CNA organizations. Supplementary Position is approved for use *only by the U.S.A./Canada CNA*. Both of these are described in detail in this text.

---

[1]As was true for Grasp, Position is a synthetic combination of the basic manual motions; it is, therefore, defined as an "element" to distinguish this fact. No Therblig matches this concept exactly, although combinations of Use, Pre-Position, Position, and Assembly might connote the same action. The utility of Position (MTM variety) in expediting motion analysis will be appreciated by gaining an understanding of this chapter.

Alternate Position, eliminated from this edition of *Engineered Work Measurement* for reasons that are later explained, was at the time of the writing of the third edition identified by the IMD as being acceptable as optional Position data by any CNA organization believing its use to be appropriate.

The U.S.A./Canada CNA chose not to accept Alternate Position as optional data even though they had performed the additional study that produced it. The changes in the Standard Time resulting from its use from that produced by using Supplementary Position were not deemed to be significant. Related to the decision not to approve Alternate Position for use by the members of the CNA was that they had previously adopted the set of data identified as Supplementary Position, a choice allowable to them since they did the research on Supplementary Position (as well as doing the research on Alternate Position).

International Position is described in the first subchapter of Position, and Supplementary Position is fully covered in the second subchapter.

The Alternate Position research endeavored to extend the Supplementary Position research. It did advance knowledge slightly, but its utility apparently has been deemed to be marginal. Not only was the use of the Alternate Position data rejected by the U.S./Canada Association, but an effort by the author during 1985 to ascertain the degree of usage of Alternate Position by the IMD CNA organizations indicated that virtually no significant usage was being made of Alternate Position. It appeared that the overseas CNAs are really only using International Position.

Alternate Position, for those who want more information, is based on the same research[2] as was Supplementary Position, but was refined and enlarged in accordance with the feedback the latter generated from field application. Certain rules and time values were altered slightly after reexamination of the research data and a closer interpretation of it, but the over-all time change was less than 0.2 per cent increase for the complete data set. This revision more strictly structured the data and its design as a prelude to its possible extension to the tighter fit part of the range (not included in the research). One important specific, the selected point values for certain component times were found to be less advantageous than adoption of the exact numbers and formulas developed in the research.

As to the sequence of events leading up to the current situation, one must take into account the actions of the International MTM Directorate (IMD) which in 1966 assigned evaluation of the Supplementary Position data to its Basic Research Committee. An initial draft report dated May 17, 1968 was the basis for technical replies by the Cooperating National Associations for consideration at the 1969 General Assembly. A revised draft using

---

[2]R.R. 109 and R.R. 110. (Available form the U.S./Canada MTM Association.)

these critiques was further considered in 1970, when it was decided that, due to the lack of data at the tighter fit end of the application range, the International Position data could not be superseded. In addition, many suggestions led to further development justifying the new name of Alternate Position and the final document was published in May 1972. This background is necessary to emphasize the exact official status of all three sets of position data as follows:

1. International Position (sometimes erroneously called "original" Position) is the *only* fully approved Position data, and recognition is binding on *all* member associations of the IMD.
2. Supplementary Position is sanctioned *only* by the U.S.A./Canada MTM Association. Initially this was done to permit field application trials, which were found successful enough to justify its continued usage by that national group until any final resolution of the whole Position data question.
3. Alternate Position is not recognized or sanctioned by the U.S.A./Canada MTM Association and therefore should not be used in these countries. In fact, as previously stated, it does not seem to have been adopted and or used by overseas CNAs.

Because the inclusion of full information on Alternate Position in the third edition of this text caused some applicators in the U.S. and Canada (where the text is mainly used) to assume that it or else all three sets of data could be used, Alternate Position is deleted from this fourth edition at the suggestion of the cognizant CNA. Also, as stated earlier, no one seems to be using it on a regular basis. Any reader interested in the details of Alternate Position can refer to a library copy of the third edition of this text or else they can secure copies of the underlying research documents from their cognizant CNA organizations.

INITIAL CONCEPT

Position, in the general descriptive sense, is a manual action used to place (align, orient, and engage) two objects or the equivalent in an accurate fashion, after they have been brought closely together with Move motions. This is *not a definition,* but will serve to begin the presentation.

The concept of Position was invented by the originators of MTM to overcome difficulty in the recognition, usage, and timing of the physical actions included when they tried to adhere to earlier micromotion categories. The Therbligs separated the align and orient motions as Pre-

Position and Position (not the same as the MTM variety); the blending of these into engagement resisted the distinguishing of the end point, even on cinefilms.   Likewise, the Assemble Therblig covering the engaging action included such short distances as to make difficult the analysis and timing of minor depths of insertion.   Visual and film analysis were notably easier when Position was defined to include a limited amount of insertion, the only necessity being the treatment of deeper engagements as separate elemental motions.

Since a major amount of industrial and business activity concerns assembly routines, examples of Position come readily to mind:

1. A machinist centers a punch to a cross mark on a template to punch mark it preparatory to drilling a guide hole.
2. A test engineer brings the adjustment knob of a variable resistor to a close setting.
3. A clerk installs an index tab into the metal slot of a file divider.
4. A saleslady fits her coded key into the lock of a cash register.
5. A factory hand locates a bolt into a hole.

Note, in all of these instances, that the object being placed was moved to within an inch or so of the final location just prior to the act described. Also, until the act of Position was completed, no further displacement of the object exceeding about an inch would normally occur.   Possibly movements exceeding an inch after positioning could occur in Examples 3 and 5.   Insertion deeper than an inch, if required, would be done by a separate Move following the positioning motion.   Several important application rules derived from the set of facts just stated will be discussed in detail later.

A very important fact regarding the Move preceding a Position emerges from considering the necessary spatial relationship of the parts at the start of *any* positioning element.   Referring back to the Move chapter, a Case B Move will locate an object within an inch or two of a given point, whereas a Case C Move will bring it to about half an inch of the final location.   In the examples of Position cited above, the engaging object was moved to within an inch or less of the engaged object prior to the act of Position.   Position cannot be accomplished unless the object has first been moved to within half an inch to an inch of the engaged object, therefore the following MTM rule is important enough to merit strong emphasis: *All Moves immediately preceding Positions must be Case C Moves*.   Only rare exceptions to this rule will be found.   The MTM analyst must never forget this rule if correct analysis of the Moves preceding Position is to be made.

It should also be noted that if a part can be located adequately with either a Case B or Case C Move alone, no Position is needed.

GENERAL DISCUSSION

Note that Position is a *manual* element; that is, it can be performed only by the fingers and/or hand-arm muscle combination. The reason that other body members cannot perform Position is that it is a set of very basic motions involving moderate to extreme control beyond that normally expected of other body members. Accurate location of the foot or with the foot, for example, cannot be made with a degree of precision comparable to that possible with manual members. The leg muscles can only rarely situate the foot closer than $\frac{1}{4}$ to $\frac{1}{2}$ inch of a certain location. With hand control of objects, however, it is possible to place objects to within thousandths of an inch.

For a Position to occur, more than one object necessarily must be involved. Obviously, at least one of the objects must be mobile, although it need not be completely free of restraint in space. The other object(s) may either be stationary or mobile. The purpose of performing positioning motions is to bring the objects concerned into some definite physical relationship *relative to each other.* The nature of contact desired, final configuration, or spatial proximity necessary will dictate the kind and number of small motions required to achieve positioning—remembering that *all* positioning motions (except for insertions exceeding an inch) usually occur within a sphere of space *two inches* in diameter at most. These motions result in one or more of the physical effects known as *aligning, orienting,* and/or *engaging*; detailed consideration of these three actions will follow later. The motion analysis must account for the individual and combined effects of these three factors, as discussed under Position variables.

It is possible for more than two objects to be involved in a Position. For example, when a waitress inserts a fresh handful of toothpicks into an empty toothpick pot, the handful comprises a single piece of wood or plastic as long as individual toothpicks do not separate during handling. If the pot were already partially filled, causing separation of the handful, additional motions besides the standard Position would be needed to gather them together again. Another example, involving another type of material, is the positioning of a prepunched stack of notepaper into a three-ring notebook. Still another instance is seen when a stack of punched computer cards is placed into the holder of a card feeder mechanism.

Earlier it was indicated that any major distance displacement of objects will require ordinary Moves (exceeding an inch) by the hands of the operator. Reach, Move, Turn, Crank, and Apply Pressure are considered to be the truly basic motions; all other hand motions are synthesized combinations of these more basic motions.   This idea was clarified in the Grasp chapter, particularly in the explanation of the Theory of Grasp.

Such minor motions as Moves (an inch or less in length), together with whatever Turns and Apply Pressures might be needed, have been grouped for utility into the category known as Position.   It would be impractical, from an analysis time economy standpoint, to make the detailed analysis of these short motions on each occasion during which their combined effect in positioning objects fitted a recognizable, recurring pattern. Moreover, one cannot see many of the short motions with the unaided eye, a camera and film analysis would be needed.   This pattern, having been grouped conveniently into measureable categories, can more justifiably be applied directly from predetermined time tables as a category in its own right.

In common time study language, Position represents a set of standard data for grouped minor basic motions.   The development of such data is explained in a later chapter.

The reader must be cautioned that the analysis of Position departs somewhat from one MTM rule given earlier.   It has been stated that the MTM analyst is concerned with the actions of the *operator,* not their effect on the workpiece.   While this still holds for Position, correct analysis in this case demands attention be directed to the objects as well.   This is true because the factors used to classify the given occurrence of Position into the data card categories have been based on physical relationships between the parts being positioned.   The variables of classification are best determined by the analyst's examination of the parts relative to each other.   However, attention to the operator's motions must not be forgotten, since such observation will disclose instances where *required* motion sequences actually used do not coincide with standard Position sequences.   Such discoveries may not occur unless the MTM engineer is alert to the worker's performance.

It has been stated several times that MTM is a practical procedure. The deviation from the regular MTM work procedure during Position analysis and the substitution of a second set of important guides—an analysis of the variables of the parts—is another instance of practicality.   It is a sound approach to accounting for the alignment, orientation, and engagement which the operator performs on the parts.   Different configurations *require* varying motions that may be difficult to recognize readily, especially if no prior clues exist as to what may be expected.   Prior study of the more easily measured and judged parameters of configuration in lieu of merely determining work motions obviously is expedient and practical.

One predetermined time system, DMT, is founded and developed on this very concept used in establishing and applying MTM Position data—that performance time is related to the physical configuration of objects handled. The MTM approach has been completely open-minded and this attitude continues to permeate the research activity and continuing development of the MTM Association.

POSITION DATA SETS

The data found in subchapter 12A describes International Position, which resulted from the original MTM research by its developers. The Position data in subchapter 12B, identified as Supplementary Position, resulted from the first additional MTM research on Position. It was performed under the auspices of the U.S.A./Canada MTM Association.

Understanding the involved situation with regard to Position, some direct explanation and historical perspective is needed so that particular readers can properly understand (and to a degree select) and apply Position data in MTM analysis. Both the background status and technical features of the sets of data presented in this chapter provide reasons for their concurrent existence.

International Position

This name properly indicates its status as the *only* set of Position data fully recognized by all national MTM associations. When Position is mentioned without qualification among MTM users, this set of data usually is understood.

Before the additional Position data set(s) existed, this set was called simply "Position." It is the data format derived from the research of the MTM originators which appeared on the first data card issued in 1951 by the MTM Association for Standards and Research. It has remained unchanged on all data cards issued by every MTM agency since then, having been adopted as the international standard shortly after formation of the International MTM Directorate. This fact explains its present name, since it is the only Position data recognized on the official IMD data card—which is binding on all Cooperating National Associations. As explained in the Apply Pressure chapter, national groups may show other motion-time data on their data cards only when clearly marked to differentiate it from such fully-approved IMD data.

The purpose in mentioning the general technical features of each set of Position data is to familiarize the reader with their similarity and differences in advance of detailed study. In this way, each set can be placed in perspective as it is learned and applied. This is important because, if valid, all Position elements will fall within a certain time continuum; but, for practical application, discrete point values must be selected and explicitly defined within this time span. The International Position data achieves this division of the continuum in a different way than is used for the other two sets of Position data. However, all sets deal directly with the Align, Orient, and Engage actions of which Position is composed. For valid results, therefore, one must apply each set in accordance with its own design criteria and application rules—whether or not the sets have many similarities of action, criteria, and rules.

Noteworthy features of the International Position data are the facts that it is *behavioral* and *discrete* as to time values based upon the Theory of Position. To elaborate the first point, the critical question for motion classification is whether the means of determining the subdivisions involve the physical characteristics of objects (e.g., linear sizes or shapes) or depend instead on perceptive observation of the operator's behavior. The objective-type measurements are more readily reproduced and interpreted, whereas the behavioral-type analysis will be subject to observational errors mentioned earlier in this text. Such errors can be minimized by reference to the objects after the main operator actions are determined. This modified behavioral approach is the one used in the International Position data. It requires direct method observation or visualization coupled with careful judgment of the objects to select the proper category of Position according to a synthetic Theory of Position. The latter was derived from consideration of the underlying operator motions to approximate the time requirements found in the original research, so that an organized choice of the discrete time values can be made with minimal hesitation. The data card values are those developed in the theory. For this latter reason, the U.S.A./Canada data cards issued prior to 1972 showed all International Position data values in light-face type to indicate that they were not direct research observations.

All of the technical factors will be made clear in the sub-chapters; but, naturally, many are common to all the data sets. One of these is the treatment of object geometry as it influences the motion(s) required to effect the Orient action during Position, in fact this parameter will be the basis for the only time plots which can adequately show all three data sets. One common factor applied differently between the sets is coverage of the variable motions caused by object handling. These can be integrated into the Position data, as in International Position, or they can be separately considered as external variables, as in the other two data sets. A key variable—actually the differing treatment being the reason all the data sets must coexist pending research resolution—is the way the parts being positioned fit together. The International Position uses *behavioral* decision rules for classification and the *full range* of fits is covered, from the defined limits of the preceding Move down to and including fits so tight they require heavy pressure to ensure engagement. Both of the other data sets use *objective* decision rules but only cover the *partial range* down to a total object clearance of .010 inch, which is a relatively loose sliding fit requiring much less engagement pressure than the tightest category of International Position. Until tighter fits are covered by the newer data, the International position must remain available at least to permit measurement of that part of the range. This is the principal reason it was not replaced by the newer data, even though the latter has many advantages and may, perhaps, apply in almost all of the Positions actually found in the real working world.

Supplementary Position

New research described in the sub-chapter on Supplementary Position was the basis for its development, as well as the subsequent Alternate Position development. As anticipated by the Theory of Position in the earlier data, this research established definitive descriptions and time values for the action components of Position. It also demonstrated the independence and additivity of these components, an important basic consideration in designing predetermined time systems. This component data is separately available in the new Position sets in both tabular and equation formats. However, for efficient application, the most frequent combinations of components have been compiled into tables of unit values for the whole set of motions within a given category of Position; these are the tables found on data cards for the Position sets. As mentioned earlier, the newer data is *objective* rather than behavioral and its main limitation is its lack of coverage at the tighter object fits. Further research to cover this part of the dimensional range would be necessary for the newer data to completely replace the International Position data.

The name "Supplementary" indicates that this data has the limited approval of the U.S.A./Canada association that issued it, a permissible practice (see Apply Pressure Chapter) for the extensive field testing of new data. It was added in appendix form to the 1965 data card for that purpose and has been carried since because of its great success; indeed, many MTM users have found that it covers almost all the Positions they encounter with recourse to the International Position data seldomly needed. When the Supplementary data was submitted to the International MTM Directorate, it was not adopted for two reasons. First, as already explained, the full range of fits is not covered. Second, the IMD upon reconsidering both the research and practice results made further refinements in the data to effectively develop another set which was named "Alternate Position" data. Most of the latter is identical to the Supplementary, but added features and minor time changes necessitated its recognition as a different set of data.

At this point, further description of the Supplementary Position data is deferred to its sub-chapter; since the technical terms and data needed to fully clarify the characteristics are not available to the reader at this stage.

12A—INTERNATIONAL POSITION

Definition[3]

POSITION is the basic finger or hand element employed to align, orient, and engage one object with another to attain a specific relationship.

 1. An accurate and predetermined relationship between the objects must be attained.

 2. The relationship may be a nesting or mating of the objects, or may be a visual locating of one object to another.

 3. Normally, only objects can be positioned; occasionally, the finger or hand may be used as a tool and considered as an object in positioning.

 4. Align is to line up the two parts so that they have a common axis.

 5. Orient is to rotate the part about the common axis of engagement so that it can be mated with the other part.

 6. Engage is to enter one part into the other part.

The nature of Position as defined has already been described with examples at the start of the main section of Chapter 12. This definition, however, applies only to International Position; but the concepts are equally valid for other Position data. The illustrations of this sub-chapter especially apply to the other part of this chapter, 12B.

POSITION VARIABLES

The principal motions involved in Position are Align, Orient, and Engage. Each of these includes many variables that have been evaluated and covered by practical rules of the MTM procedure. For simplicity in the following elaboration, attention is restricted to two objects, although more than two can be involved in a given Position, as previously explained. One is called the engaging part, the other is termed the engaged part. It is assumed that the engaging part is the one held in the operator's hand, that is, the part being moved during the Position. The engaged part may be stationary, manipulated in the hand, or move in synchronization with the engaging part. The intent is to match the engaging part to the engaged; in most cases this also involves insertion.

[3]From "MTM Basic Specification" (Footnote 1, Chapter 5).

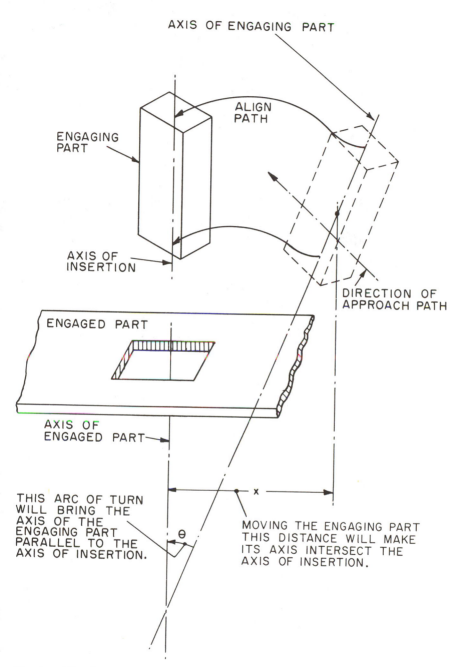

AXIS OF ENGAGING PART

ALIGN PATH

ENGAGING PART

AXIS OF INSERTION

DIRECTION OF APPROACH PATH

ENGAGED PART

AXIS OF ENGAGED PART

THIS ARC OF TURN WILL BRING THE AXIS OF THE ENGAGING PART PARALLEL TO THE AXIS OF INSERTION.

θ

x

MOVING THE ENGAGING PART THIS DISTANCE WILL MAKE ITS AXIS INTERSECT THE AXIS OF INSERTION.

Fig. 1.   Align in Position.   The general Align motion shown consists of a Move of distance X and a Turn of angle θ during the same time interval.

In one variety of Position to be discussed later, no insertion occurs. This presents no difficulty, however, since it merely indicates the absence of one of the possible actions during Position. The concept of engaging and engaged parts will not suffer when applied to this type of Position, rather the engagement in this case may be thought of as one of the following: contact only, potential contact, or potential insertion. This type of Position is called Align Only. It may actually include both Align and Orient as discussed later, although usually the Align is the sole action.

Each of the three major actions will now be discussed separately.

Align

Definition: Align consists of all basic motions required to make coincident the insertion axes of the engaging and engaged parts during a Position.

The illustration of aligning in Fig. 1 makes use of three different axes. These axes are imaginary lines that facilitate description of aligning or other positioning motions. Examination of these axes and the motions needed to make two of them coincident will help in understanding the nature of alignment. In order to avoid confusion, it is necessary first to state explicitly what each axis represents.

The axis of the engaging part is the line along which the engaging part will later travel if it is inserted into the engaged part. The axis of the engaged part is the line or direction in which any successful insertion of another part must proceed relative to the configuration of the engaged part. Finally, the axis of insertion is the coincident line formed by the axes of the engaging and engaged parts during and after engagement. Note also in Fig. 1 that the general direction of the path by which the engaging part approaches the engaged part during the Move preceding the position may bear almost any arbitrary geometric relationship to these defined axes.

Clarity on the nature of alignment is gained by examining the various possible spatial relations of the axis of the engaging part to the other axes at the completion of the part's approach, just prior to the Align motions of Position. Obviously, when it coincides with the axis of insertion (1) no Align is needed. If it intersects the axis of insertion at an angle, (2) a pure Turn of the engaging part will usually be used to finish Align. If it is parallel to the axis of insertion, (3) only a short lateral Move is required to complete the Align. In most cases, none of these three simpler conditions applies. Therefore, the most general relationship of the axis of the engaging part to the other axes dictates the need for (4) *both* a short Move *and* a pure Turn of the engaging part relative to the engaged part to accomplish alignment.

Motionwise, the MTM procedure must account for the short Move and/or pure Turn as may be required for alignment. These motions are very short, indeed they cannot normally be detected with the naked eye. However, the film analysis showed that, considering the usual frequency of each of the four possibilities noted above, an average value of 5.6 TMU will suffice for the time needed to Align two objects *and* perform insertion up to one inch. This has been included in the Theory of Position.

As a reminder at this point, the reader should review the main section of this chapter regarding the analysis of Moves preceding Position. With the discussion of Align just given, the logic of the Move analysis should become clearer.

Orient

Definition: Orient consists of all basic motions required to geometrically match the cross section of the engaging part (or its engaging projection) with that of the engaged part relative to the axis of insertion as determined by their projections on the plane of initial engagement.

It is called to the reader's attention that the hole can be in either the engaged or engaging part. The Position data is equally applicable to either situation.

Reference to Fig. 2 will reveal that orienting accounts only for the additional motions needed for matching after alignment but prior to insertion. Actually, much of the total orientation of the engaging part may be partially accomplished during both alignment and the Move preceding a normal Position.

The axis of insertion provides a convenient reference line for evaluating the geometric relationship between the shapes of the parts. In addition to this line, the plane on which the projected cross-sections of the engaging part and the engaged part are to be evaluated must be selected. While an infinite number of planes perpendicular to the axis of insertion would show the projected cross-sections, one particular transverse plane is most convenient for gaging the geometric relations. This is the plane of initial engagement, which is defined as the plane perpendicular to the axis of insertion at which any lateral motions of the engaging and engaged parts would cause point or line contact between the parts. If, as is shown in Fig. 2, the cross-sections of the parts are projected on the plane of initial engagement, it is then possible to compare their geometric relationship with ease. Of course, this is usually a mental process on the part of the MTM analyst, although it may often be judged correctly by manipulation of the parts being positioned.

The stated purpose of orienting is to match the cross-sections of the parts. With the means of observing or visualizing the cross-sections given above, comparison of the square, rectangular, circular, triangular, or other shape of the engaging and engaged cross-sections will readily reveal the amount

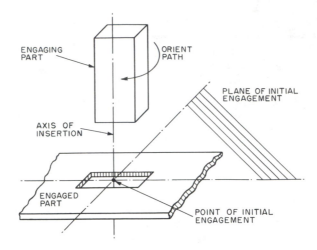

CROSS SECTIONS
(ON PLANE OF INITIAL ENGAGEMENT)

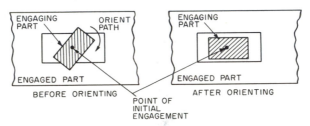

Fig. 2.   Orient in Position.

of rotation about the axis of insertion which the operator must perform to permit insertion.

In terms of motions, orienting is accomplished with finger turns or equivalent manual actions of the worker which provide the same rotation. The number of such turns or their angular magnitude will vary with the relative cross-sections of the parts. In the Theory of Position, three general classes have been categorized, although it is recognized that the possibilities are numerous. Practical considerations, however, would indicate the sufficiency of limiting categories to a few easily determined classes, the details of which are later discussed.

Engage

Definition: Engage consists of all motions required to effect up to 1 inch of insertion of the engaging part into the engaged part following completion of aligning and orienting.

The act of engagement shown in Fig. 3 is the easiest Position component to understand, although it involves more motions and more judgment to evaluate correctly. Basically, the motions of Move (1 inch or less), Apply Pressure to overcome any resistance to insertion, and Regrasps to maintain control of the objects are all potentially present during engagement.

Referring again to the examples at the start of this chapter, it will be noted that Position time does not include any Move greater than 1 inch. Another MTM rule for Position results from this manner of limiting the scope of Position motions: *All insertion beyond 1 inch from the plane of initial engagement must be analyzed and assigned additional time.* The

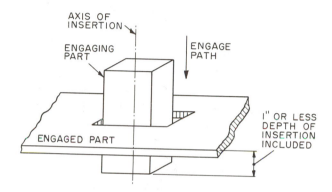

Fig. 3.   Engage in Position.

analysis of this additional insertion usually shows that it is some case of Move; time assignment should follow the regular rules for that motion. When other motions are appropriate, the additional time will be determined by the rules for such motions.

The next factor in engagement, overcoming resistance to insertion, relates to the amount of force the operator must apply to effect insertion. Apply Pressure, or actions equivalent to it, has been isolated in MTM as the motion which accounts for this force. In order to judge the amount of force, or the number and case of Apply Pressures needed to meet the many possible conditions of insertion resistance, Classes of Fit of the engaging part relative to the engaged part have been categorized in the Theory of Position. Again, while many fit relationships are possible, practicality suggests a limited number of classes.

Finally, control of the objects must be maintained if successful insertion is to be accomplished. The MTM motion, or its equivalent, needed to improve or maintain hold on the objects is Regrasp. The Theory of Position has accounted for the number of Regrasps needed in two ways. Part of them (what can usually be expected) have been covered during the Engage time assigned to various Classes of Fit; an additional Regrasp is then assigned if the object is of more than usual difficulty from the Handling standpoint. Note that an APB, consisting of the equivalent of an APA and a G2, also includes part of the Regrasps that might be included in the Engage values for the theory. The total number of G2 needed during a given Position is also a matter of frequency and observation of a large number of Positions on motion picture films, so the MTM analyst again relies on research results rather than attempting to evaluate this for himself.

Effect of Weight

The present MTM data for Position does not make direct accounting for the weight of the objects being handled. Obviously, the weight will have some effect on the time to Position an object. The MTM procedure up to now, however, has been based on the idea that the presence of significant weight would make necessary observably different motions in order to Position a significantly weighted object. Logical deduction would indicate, however, that any possible effect would not involve the static weight component isolated in the research on weighted Moves. This is because the object weight obviously is under close control at all times during a successful Position. If control is lost, new positioning motions will be needed, and not a revision of the normal Position in question.

That the dynamic component of weight might affect Position time is more reasonable, although the numerical percentages might not be the same as for spatial Moves. As discussion of the preceding variables has shown, the Position times are an admixture of the MTM motions of Move, Turn, Regrasp, and Apply Pressure. At all times during the action, the full weight of the object being located is in the hand. Weight with Moves has been fully explained, and the use of three categories to account for weight in Turn is also known to the reader. How weight affects regrasping is less well-defined, but the G2 for difficulty in handling is an example of MTM thinking, since weight increases handling difficulty. And Apply Pressure, with its explanation in terms of force, evidently varies in magnitude to some extent by the weight handled. Therefore, unless weight effects are very significant, their influence in normal Positions is already indirectly included in the Theory of Position upon which the present time values are based.

No further adjustment for weight is needed—except when the weight causes motions other than those found in a normal Position. If the weight is substantial and/or the motions are different, a separate MTM analysis or even a time study is needed.

KINDS OF POSITION

The present data includes eighteen different kinds of Position. Each kind is a combination of three Classes of Fit, three Cases of Symmetry, and two Handling categories.   The manner in which the fit, symmetry, and handling account for the Position variables of Align, Orient, and Engage will be the next topic.   They are the factors in Position which are judged directly in practical application of MTM time values to positioning motions.

Class of Fit

Definition: The fit of engaging and engaged parts is a measure of the clearance between them and/or the pressure required for insertion.

The fit between parts that can be assembled by manual action will vary from practically no clearance to very generous spaces of the order normally associated with Moves only; it is rather obvious that the time for insertion will vary accordingly.   Actually, there is a continuous relation between fit allowance and performance time.   As practicality dictates, however, the MTM procedure has assigned arbitrary limits to restrict to three the number of Classes of Fit which must be distinguished.

Since the Class of Fit is directly related to Engage time, Engage as a Position variable is evaluated by determining the Class of Fit.   The point of initial engagement is the only point from which measurements of fit can be made correctly.   This point in a Position action occurs when the engaging part first contacts the plane of initial engagement as defined earlier.   In other words, engagement has begun when any lateral motion of the parts will cause line or point contact between them.   It is incorrect to judge the Class of Fit at any other plane transverse to the axis of insertion.   The fit may be judged by either the amount of *total* lateral clearance or the force of insertion needed, provided the point of initial engagement is used to determine these.

The total lateral clearance for a $\frac{1}{4}$-inch rod being positioned into a $\frac{5}{16}$-inch hole, for example, would be $\frac{1}{16}$ inch.   Finding the clearance is more difficult with irregular or odd shaped parts, but the same principle is used as for common geometric cross-sections.   When the clearance cannot easily be judged, recourse to the pressure required to engage must be employed.

A commonly used means for determining the amount of pressure required for insertion is known as the *gravity* test.   In this test the engaging part is held so that the axis of insertion is vertical and the part is close to the engaged one, following which it is released and the resulting effect observed.   If the parts engage readily, no pressure will be needed to Position them.   If a slight nudge or push of the fingers is needed to complete insertion, light pressure will accomplish the engagement.   If the parts will not engage unless the tester must grip them and apply appreciable force, heavy pressure is indicated during performance of a Position with the parts.

These engaging pressures and the clearances associated with them have been assigned Class of Fit designations as follows:

Class of Fit	Clearance	Pressure
1. Loose	Appreciable, but not in excess of ½ inch total	None
2. Close	Visible, parts are snug but will slide with only slight resistance or friction	Slight
3. Exact	None visible, very slight tolerance, but parts can still be engaged manually	Heavy

A rule of thumb states that when the fit is so loose as to raise a question whether a Position is proper, Class 1 is involved; when difficulty in insertion is evident, Class 3 is indicated; all other conditions are Class 2. Class 2 fit occurs with highest frequency, Class 1 next, and Class 3 is rare in industrial operations, since mechanical devices are normally used when heavy pressure is needed to engage parts. Admittedly, judgment is involved in any case—but this judgment is objective and does not subjectively evaluate the operator. Practice will enable the MTM analyst to correctly judge the Class of Fit.

Case of Symmetry

Definition: The Case of Symmetry describes the geometric properties of the engaging and engaged parts as they affect the amount of orientation required prior to insertion.

Just as the Class of Fit gages the engagement of parts, the Case of Symmetry accounts for the Orient motions needed by examining the relationship between the cross sections of the parts. The cases have been set up in terms of the number of ways in which the relative configurations of the parts will permit insertion to occur. Symmetry, at least for the simpler shapes, is thus made easy to determine; difficulty in assigning the Case of Symmetry will be met only when borderline shapes are present or when the analyst attempts to assign the case without remembering that alignment is theoretically assumed to be completed before orienting begins.

The basis on which symmetry is judged is the amount of orientation needed during the Position motions themselves; any pre-orienting of the object during the Move preceding Position is discounted. Many operators either pick up parts so that little or no orientation is needed, or they employ Turns and/or Regrasps during the preceding Move to reduce the orienting motions. In fact, it is the aim of good motion analysts in reducing Position time during assembly to so design the workplace and material delivery as to reduce or eliminate the need for orienting of parts during Positions. This would permit the legitimate assignment of a lower Case of Symmetry.   Recall, however, that the MTM standard is set for an average operator working with a given set of conditions. High levels of skill and/or effort will also disclose observed Positions that are of lower case than would be true of average performances, but allowing these would not set a standard for the average operator.

If the amount of legitimately allowed pre-orienting is correctly analyzed prior to classification of the Position, there should be little difficulty in determining the Case of Symmetry involved.

The three Cases of Symmetry used in MTM (based on *no* pre-orientation) can be described as follows:

Case of Symmetry	*Description*
Symmetrical	The parts can be engaged in an infinite number of ways about the axis of insertion
Semi-Symmetrical	The parts can be engaged in two or several ways about the axis of insertion
Non-Symmetrical	The parts can be engaged in only one way about the axis of insertion

Figure 4 shows obvious instances of all three Cases of Symmetry.   A *suggestive* limit to the meaning of "several" in the Semi-Symmetrical category is up to ten ways.   However, the number cannot be infinite.

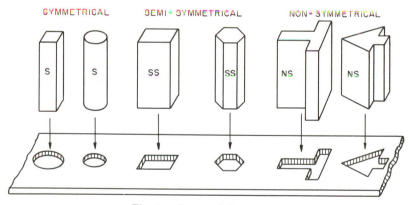

Fig. 4.   Cases of Symmetry.

Note that the geometry of both parts must be considered in deciding symmetry.   For example, it is incorrect to state that positioning of a hexagonal shaft is Non-Symmetrical.   This would be true only if the hexagon were not of equal sides and the hole was of like shape.   If the hole and shaft were both regular equal-sided hexagons, the insertion would be Semi-Symmetrical.   If the hole was circular and of sufficient diameter to permit insertion of the hexagon shaft, regardless of the regularity of the hexagon sides, the Position would be Symmetrical. So for the one hexagon shape, if no further data are known, the Case of Symmetry could not be decided; the shape and size of the hole must also be known.

When an object being positioned is mechanically constrained, an exception to classification by shapes may be taken. The use of guides, swivels, or fixed arms may enable an operator to perform Symmetrical or Semi-Symmetrical Positions even though the parts configurations would otherwise indicate Non-Symmetrical insertion. No fixed rule for such cases can be given; each must be examined for the orientation required of the worker.

Ease of Handling

Definition: An engaging part is considered Easy-to-Handle if no change of grip during Position is necessary, desirable, or made for convenience.

Justification for this factor in Position has already been explained in the discussion of Engage. What must be added is reasons to which handling difficulty justifying the extra Regrasp can be attributed. With this data, visualized analysis becomes possible; observations would reveal directly the extra G2 for handling difficulty.

The decision as to whether the part is Easy-to-Handle or Difficult-to-Handle must be based on considerations of the rigidity of the part, its size relative to its weight, and the manner of grip the worker has upon it. These factors determine how well the part can be controlled during Position. It is also evident that, since control requirements are greater when the fit is Close, the difficulty of handling will tend to increase with close tolerance Positions. No definitive rules have been set in MTM for these factors, but a few examples are suggestive.

A rigid part is stiff and can be easily controlled unless it is held at some distance, depending on the fit, away from the point of engagement. Flexible items such as string, thread, stranded wire, and some types of springs are nard to handle unless held very closely, even when the fit is generous. Maynard, Stegemerten, and Schwab suggest that a $\frac{1}{16}$-inch diameter rod would be difficult to insert into a closely fitting hole unless the point of grasp were within $\frac{1}{2}$ inch of the point of engagement. On the other hand, they state that it would be easy to handle a 3-inch diameter rod of light plastic while positioning it into a Loose fit though the holding point was 4 to 5 inches from the initial insertion.

SHORTHAND OF POSITION

Position

Recording of a capital P will show that Position has occurred.

Class of Fit

The Class of Fit is indicated by the arabic numerals 1, 2, or 3 following the symbol for Position.

Example: P1 shows the presence of a Position with a Loose fit.
P3 shows the presence of a Position having an Exact fit.

Case of Symmetry

Capital letters following the symbol for Class of Fit identify which of the three Cases of Symmetry is involved in a Position.   The letters are S for Symmetrical, SS for Semi-Symmetrical, and NS for Non-Symmetrical.

Example: P1NS describes a loose-fit Position with a Non-Symmetrical relationship between the engaging and engaged parts.
P2SS shows the insertion of a close-fitting part into a hole that it could enter in two or more, but not an infinite number of ways.   This is the Semi-Symmetrical situation.

Ease of Handling

The capital letters E and D, respectively, symbolize Easy and Difficult cases of parts handling during Position.

Example: P1SE fully symbolizes the Symmetrical positioning of an Easy-to-Handle part into a Loose fit with the engaged part.
P2NSD are the symbols needed to indicate the close-fitting insertion of a Difficult-to-Handle part into a destination of a configuration permitting only one manner of orientation for positioning, the Non-Symmetrical case.

THEORY OF POSITION

Up to this point, the *variables* in Position have been discussed to show what manual actions are involved and the *factors* by which these variables are evaluated, timewise, were explained.   The Theory of Position inter-relates the variables and factors in such a way as to obtain the present time data shown on the MTM card.   As indicated earlier, the Theory was developed to explain the original observed experimental data.   The combination of theory and experiment resulted in such practical data that it became the MTM standard.   Further experimental checks have tended to confirm these time values, and usage has confirmed that they produce adequate answers.

The Theory of Position serves two useful purposes: (1) It helps the

KEY TO MOTION VARIABLES IN POSITION

Variable		Factor Evaluated	Motions Assigned by Theory	Time, TMU
Symbol	Meaning			
C	Constant	Align + Insertion	Constant for all classes (up to 1″ of depth included)	5.6
T	Orient (Case of Symmetry)	S	T0 (Symmetrical)	0.0
		SS	T45 (Semi-Symmetrical)	3.5
		NS	T75 (Non-Symmetrical)	4.8
F	Engage (Class of Fit)	1	None	0.0
		2	APA	10.6
		3	$\overbrace{\text{APA} + \text{APA} + \text{G2} + \text{APA}}^{\text{APB}}$ 10.6 + 10.6 + 16.2	37.4
H	Handling (Ease of)	E	None (Easy to Handle)	0.0
		D	G2 (Difficult to Handle)	5.6

MOTION–TIME SYNTHESIS OF THE DIFFERENT KINDS OF POSITION

Factors			Variables		Total Time, TMU		
Class of Fit	Case of Symmetry	Ease of Handling	Motions (Refer to Key)	Time, TMU	Theory	Exp.*	Data Card
1 (Loose)	S	E	C	5.6	5.6	5.6	5.6
		D	C + H	5.6 + 5.6	11.2	9.0	11.2
	SS	E	C + T	5.6 + 3.5	9.1	6.6	9.1
		D	C + T + H	5.6 + 3.5 + 5.6	14.7	13.7	14.7
	NS	E	C + T	5.6 + 4.8	10.4	?	10.4
		D	C + T + H	5.6 + 4.8 + 5.6	16.0	16.3	16.0
2 (Close)	S	E	C + F	5.6 + 10.6	16.2	?	16.2
		D	C + F + H	5.6 + 10.6 + 5.6	21.8	19.6	21.8
	SS	E	C + T + F	5.6 + 3.5 + 10.6	19.7	14.1	19.7
		D	C + T + F + H	5.6 + 3.5 + 10.6 + 5.6	25.3	25.5	25.3
	NS	E	C + T + F	5.6 + 4.8 + 10.6	21.0	21.9	21.0
		D	C + T + F + H	5.6 + 4.8 + 10.6 + 5.6	26.6	27.2	26.6
3 (Exact)	S	E	C + F	5.6 + 37.4	43.0	39.4	43.0
		D	C + F + H	5.6 + 37.4 + 5.6	48.6	?	48.6
	SS	E	C + T + F	5.6 + 3.5 + 37.4	46.5	43.9	46.5
		D	C + T + F + H	5.6 + 3.5 + 37.4 + 5.6	52.1	?	52.1
	NS	E	C + T + F	5.6 + 4.8 + 37.4	47.8	53.1	47.8
		D	C + T + F + H	5.6 + 4.8 + 37.4 + 5.6	53.4	?	53.4

* Data per original research; ? indicates no value reported.

Fig. 5. Theory of Position.

analyst to understand precisely the motions being evaluated in Positions; and (2) it enables one to examine and develop standards for unusual types of positioning which do not fit the standard categories so clearly defined. In this respect, it is much like the Theory of Grasp previously discussed. Full development of the Theory in self-explanatory form is given in Fig. 5.

Align is always the minimum motion present during Position; and, together with one inch of Move for insertion, it explains the total time effect in P1SE. Orientation is accounted in terms of the degree of Turn needed for the three Cases of Symmetry—none for Symmetrical, 45 degrees for Semi-Symmetrical, and 75 degrees for Non-Symmetrical. These angular values were determined from a series of averaged observations. The reason they are not greater, as might be suspected, is the normal occurrence of pre-orienting in most cases during the Move preceding Position. Engagement, in terms of Class of Fit, is explained by the amount of force or pressure indicated by varying amounts of APA and G2 employed by the worker. The one-inch Move time for Engagement is combined with Align in the P1SE time value. Finally, difficulty in Handling causes the usage of an extra G2 as previously explained. More detailed accounting of the data backing up the Theory can be found in the text by Maynard, Stegemerten, and Schwab. The preceding should, however, satisfy the average reader.

DATA CARD INFORMATION

The data card issued initially by the originators used the experimental values, but all data cards issued by the MTM Association have used the

Table 1.   Time Data for International Position (MTM-1 Data Card Table V— Position*—P)

CLASS OF FIT		Symmetry	Easy To Handle	Difficult To Handle
1—Loose	No pressure required	S	5.6	11.2
		SS	9.1	14.7
		NS	10.4	16.0
2—Close	Light pressure required	S	16.2	21.8
		SS	19.7	25.3
		NS	21.0	26.6
3—Exact	Heavy pressure required.	S	43.0	48.6
		SS	46.5	52.1
		NS	47.8	53.4
SUPPLEMENTARY RULE FOR SURFACE ALIGNMENT				
P1SE per alignment: >1/16 ⩽1/4"		P2SE per alignment: ⩽1/16"		

*Distance moved to engage—1" or less.

theoretical values.   Data cards before the 1972 revision showed this by light-faced type, but the current card does not use this distinction.   Note that the synthetic (theoretical) values are generally on the safe side.   This was done to give benefit to the average operator for whom the MTM technique is intended.   Data in Table V of the official MTM-1 data card is shown in Table 1.   Note that it is arranged so that it reads directly from the shorthand symbols for a given Position.

Also note that the 1972 revision includes Supplementary Data for Surface Alignment, which is explained below.

PRACTICAL WORKING RULES

A number of special types of events connected with Position motions deserve further explanation to aid the MTM engineer in making correct analyses.

Positioning to a Line or Point

Reference was made earlier to a type of Position with no insertion, previously known (misleadingly) as "Align Only," but identified on the 1972 data card as "Surface Alignment."   Actually, this type of action usually includes both Align and Orient motions, but the *Engagement* is restricted to contact of the object with the line or point in question.   The engaging part may be said to be engaged as soon as it is properly located to or has contacted the line or point desired, this usually being on a stationary object.

Examples of this type of action are:

1. Locating a pencil point or centerpunch to a line, point, or line intersection scribed on a surface (Align only)
2. Immersing an object into a liquid to a mark or line on the surface of the part (Align only or Align including orienting to final limit)
3. Aligning a ruler to a point, points, or line for measurement or for drawing of a line on an object (as explained below).

When the rule of thumb for "Align Only" was adopted, it was understood to include the minor distance of approach from the spatial end of the preceding Move down to the surface; the application still follows this rule.   However, later research shows that the rule is strictly correct only when the full action takes place entirely on the surface, which the new designator of "Surface Alignment" recognizes.   This time difference is small, but officially the *original* rule given below still holds for analyzing such Positions.

Since the alignment involved can be approximate, close, or exact, a special rule of thumb has been developed to standardize the practice for such actions.   This rule has four parts:

A. For alignments over ½″   M__B only
B. For alignments within ½″ to ¼″   M__C only
C. For alignments between ¼″ to ¹⁄₁₆″   M__C + P1SE or P1SD
D. For alignments within ¹⁄₁₆″ or less   M__C + P2SE or P2SD

Notice that all such alignments are considered to be Symmetrical.   The fit can vary and the object may be Easy- or Difficult-to-Handle.   The fractional inches should be understood as the *total* clearance, which means the permissible entry area around the target into which the object must be placed.   Now, if the target can be approached from only *one* direction, then the dimensional variation is measured as a total from one side of the target.   However, if the target can be approached from *any* direction, the dimensional deviation permitted from one side of the target will be half of the value listed, so that the total clearance will remain as listed.

Judgment is required to use this rule of thumb.   For instance, roughly locating a ruler within ¼″ of a point would be adequately covered by a P1SE.   For more exact location, this could change to a P2SE.   These analyses would even hold for several points provided they are not separated by more than 3 or 4 inches.   For separation exceeding 4 inches, however, more than one Position or alignment and, possibly, some orienting logically is involved.   At least two Positions would be needed for points as much as a foot apart, and the operator might conceivably use three for even greater distances apart.   The ruler would be located, say, first to the left point, then to the right point, followed by correction of the minor shifting that usually occurs on the left point.   Each of these Positions should be considered separately in allowing time for the sequence.

Positioning the Hand

In some operations, the hand and/or fingers must be located in some definite orientation or dimensional relation to the workpiece prior to actions which follow.   Analysis of this is subject to another MTM rule of thumb stating that positioning of the hand and/or fingers always is a Symmetrical case of orienting.   One example is a typist inserting all the fingers of one hand into the inner hole of an old-style wax dictation cylinder to gain control through finger expansion and thereby permit handling of the cylinder without disturbing the recording cut on its outer surface.   The probable sequence would be an R__D to the cylinder, a P1SE into the hole, and an outward R__A of the fingers to grip against the inner surface of the cylinder.

Most parts can be obtained from a supply by an R__C plus a Select Grasp or, if the part is isolated, by an R__A or R__B followed by a Pickup Grasp.   Minor hand orientation by making Reach-Turns instead, with the Turn time limited out, usually suffices for such events.   However, if an analysis (similar to that above) involving positioning is necessary, it can often be avoided by pre-orienting the parts and/or tools to be obtained.   This time-saving methods improvement could result from parts in slotted trays, fixed tool holders, or parts delivery chutes that always present the part in the same orientation.   Avoiding the hand positioning time in this manner is almost standard design strategy used during workplace layout.

Moves and Positions

The analyst must be careful not to call all insertions Positions.   Many insertions, and approximate locations as well, can adequately be made with only Case B or Case C Moves.   The fit tolerances, mentioned under kinds of Position above and also in the Move chapter, should govern such analysis.

Another frequent question concerns the analysis of Moves following the initial engagement distance of 1 inch covered by Position.   The depth of insertion, minus the 1 inch included in Position, obviously will account for the length of Move in this case.   As to the case of Move, however, this will depend on the conditions in the operation.   The fact that initial insertion has a guiding effect may indicate the use of M–A.   M–B might be justified, particularly if the additional insertion is quite lengthy.   The amount of control to an approximate stopping point may make an M–C correct.   It is even possible that a further Position will occur at the end of the additional insertion, in which event an M–C followed by a Position would be the proper analysis.   Some cases might also require separate and additional applications of pressure.   No definite rules, other than the guiding rules previously given, can be set down for this type of operation.

Placing one part into another does not always mean that a Position has occurred.   An assembly might be disposed to a totepan with a toss, as shown in the Move chapter.   Ordinary stacking of the assembly in the totepan could require M—C.   Precise nesting of the assembly in the totepan, however, would likely necessitate the use of a Position.   Another example where placing a part does not require a Position is that of dropping an object with a conical point into a hole at the end of a Move of the object to the hole.   Correct application of the rules concerning point of initial engagement and fit should make this case obvious.   This illustrates how tapering provided on the engaging part can either reduce the class of positioning or, in some cases, even eliminate the positioning motions.

Multiple Positioning

Note also that when the engaging part is held in a tool, the Position and preceding Move do not differ in analysis from a hand-held case—the tool is essentially an extension of the hand. However, it may increase the classification of the Position since the combination might comprise a more Difficult-to-Handle part.

Many parts require two or more Positions to fully engage them in the hole. For example, a steel shaft with a Woodruff key located more than 1 inch from the end of the shaft might be inserted into the hole of a drive pulley. The end of the shaft would be Symmetrically positioned first, and then a Non-Symmetrical location (positioning) of the key to the keyway would be needed to complete insertion.

The important thing to remember is that multiple Positions must be analyzed in the same order in which they occur. If the parts truly cannot be located with a single Position, the second Position might be influenced greatly by the fact that the first Position limits the amount of additional location needed. A part that must be engaged on both ends might have the Position on the second end reduced in both Class of Fit and Case of Symmetry by the constraining effect of the positioned first end.

In all such cases, attention to the order of Positions and reliance on the rules given for aligning, orienting and engaging parts should suffice to make proper analysis. The operator should also be observed to see the type of motions used in such cases.

INTERNATIONAL DATA COVERAGE

To provide insight into the need for subsequent Position research and development that enhanced its being understood and led to more definitive data, this section describes International Position from a data design viewpoint.

The main characteristics are that: (1) it is *behavioral* as to classification criteria; (2) the times are point values derived directly from a Theory of Position rather than selected values from research data curves; and (3) it covers the full range of fit and symmetry possible for the motion elements included. Each of these points has both advantages and disadvantages in MTM analysis practice.

Figure 6 will assist in discussing these points, for it shows on one plot both the complete International data and the limits of application for each Class of Fit. In summary, the data has nine point values based on the Theory of Position. To these are added nine more values by making the time for a Regrasp—actually an external variable for difficulty in handling—a factor in the data table. No algebraic equation exists to express the time variation, the motions being combined only by a component addition in the theory. These 18 values cover the entire continuum or spectrum of Position from the loosest to the tightest fits at three chosen levels of symmetry.

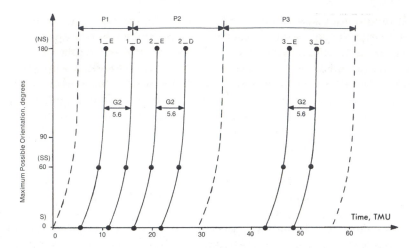

Fig. 6.   Symmetry-time plot of complete International Position data with range boundaries for Classes of Fit.

The time coverage extends to more than two seconds, 61.2 TMU, but is sparsely rather than uniformly distributed over the range. This must be understood in relation to the relative frequency of fits normally met in actual practice. At the loose end of the scale, Class 1 fit has fair coverage for its moderate frequency of occurrence and relatively low time requirement. At the tight end of the scale, the Class 3 fits are rather infrequent in usual handwork; but they still have quite sparse coverage in view of the relatively high time needed for each event. Between these extremes, the vast majority of Positions met in actual practice are Class 2; but the data coverage is far too sparse for optimal distinctions in time values. Remember, all of these Classes of Fit are defined in behavioral terms which are fairly vague as compared to the precision possible with objective dimensional criteria for classification decisions.

Looking at the other axis in the figure, explanation of the scale is in order because its dimensions do not appear in the Theory of Position. In the theory, the case where no orientation is needed is allowed zero Turn, SS is assigned T45, and NS is based on T75. It must be remembered that these Turn degrees are based on observations of the *average* behavioral rotation to achieve geometrical matching when the theory was developed. They do correctly predict the general requirement, particularly when coupled with the helpful criteria about engagement possibilities: only one way for NS, from two to ten ways for SS, and up to an infinite number of ways for S Cases of Symmetry. Field application has revealed that, although all but the first case covers a rather large number of engagement alternatives, three cases are about all the practical differentiations possible between Positions determined behaviorally. On a practicable level it would take objective criteria to permit finer divisions of orientation without being vague.

However, later research disclosed that the symmetry axis really is a strictly geometric *objective* variation expressible as the "Maximum Possible Orientation, degrees" shown as the ordinate. In fact, without such angular degrees of object rotation as a measure, the figure as shown could not even be plotted. In this scale of reference, S is zero, NS is 180, and SS is 0 – 90 with an average of 60 degrees relative object rotation. This concept will be explained completely in the Supplementary Position sub-chapter. For now, the reader should accept it as valid and also accept the assurance that the *entire* physical possibility of orientation is thereby depicted; and, as reflection would reveal, this parameter offers the further possibility of finer symmetry classifications, if needed, with full objective validity.

The three characteristics listed earlier can now be more adequately stated in data design terms. *First,* behavioral criteria can be simpler and more rapidly applied, but are not as clear or consistently interpreted as objective decision rules based on dimensions of the engaging and engaged parts. *Second,* behavioral point values are inherently harder to apply consistently even when well defined, as compared to the accuracy and precision possible if objective algebraic curves were available to predict time variation versus fit and symmetry changes; and, to permit closer time determination, time equations would be superior because the number of time values need not be restricted to a few categories for ease and practicality of application. *Third,* even the sparse coverage of International Position includes the full range of fits although more data is desirable at the tighter fits and decidedly much more data is needed for the medium-fit portion of the range.

All of this illuminates the reasons further Position study was undertaken despite the obvious utility and success of the International data. In fact, the Theory of Position itself pointed the way to progress—separate time data for the actual, verifiable Position components and a reliable way to combine them for working data categories. In the next sub-chapters, it will be seen that the first two areas discussed above were amply advanced. However, because the third area was not fully achieved, the International data must coexist with the newer data until the tighter-fit end of the Position continuum is more fully determined.

DECISION MODELS

The Decision Model for International Position analysis follows as Figs. 7A and 7B.

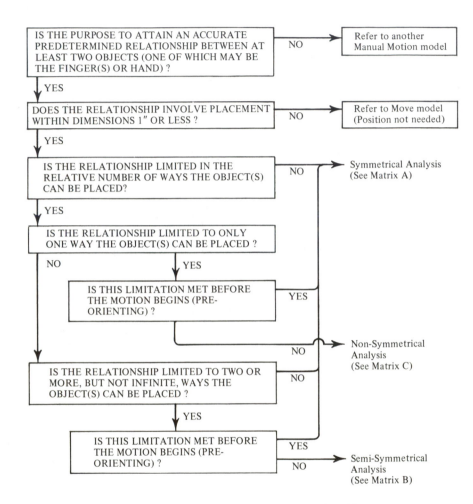

Fig. 7A. Decision Model for International Position.

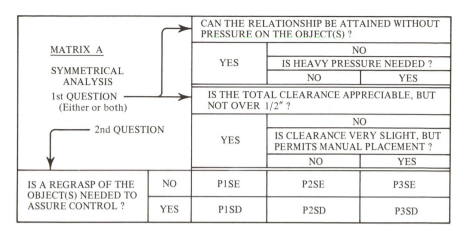

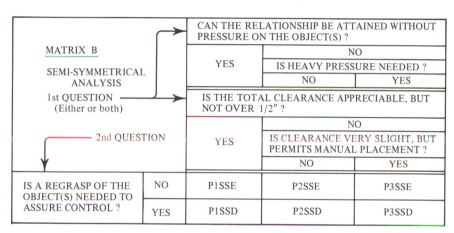

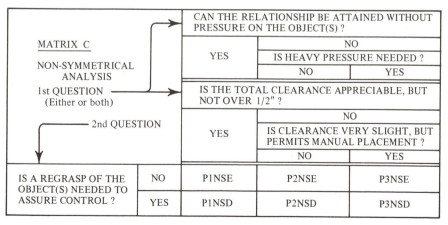

Fig. 7B.   Decision Model for International Position.

12B—SUPPLEMENTARY POSITION

Definition[4]

POSITION MOVEMENTS are those motions necessary to transport an object to a predetermined destination and to seat it in or on this destination in a precise manner.

TOTAL POSITION includes the hand and arm motions which occur from the moment an object is grasped until it is released after being positioned.

POSITION PROPER (POSITION) (symbol P) *includes the motions Align, Orient, Primary Engage, and Secondary Engage which may be required in addition to the basic transporting Move motion. Except for the Orient motion, these components of Position Proper must be performed following the Move.*

> 1. *MOVE Component* (symbol M) *is the transporting motion which brings the positional object to the vicinity of the positioning destination.*
> 2. *PRIMARY ENGAGE Component* (symbol E1) *is the motion in Position Proper which brings the object to the destination surface.*
> 3. *SECONDARY ENGAGE Component* (symbol E2) *is the motion in Position Proper which seats the object in the destination; it must be the last component motion of a given positioning.*
> 4. *ALIGN Component* (symbol A) *is the linear adjustment of the object or tilting motion required to make the axis of the object coincident with the axis of the positioning destination.*
> 5. *ORIENT Component* (symbol O) *is the rotational adjustment of the object or turning motion required to geometrically match the cross-sectional shapes of the object and the positioning destination.*
> 6. *Position COMPONENT COMBINATION is additive in the time sense but is subject to constraints in the method of performance.*

In studying Supplementary Position as just defined, it is assumed that the reader already is familiar with International Position. This is especially true regarding the physical actions as described in Figs. 1 through 3 and the accompanying text, as well as when factual differences are noted in the motion classification and time data. The International Position data is accurate and reproducible as far as it goes, having yielded thousands of satisfactory analyses of motion and time with a relatively direct and simple usage of the data involved. However, analysts frequently must distinguish between positioning movements which require a finer subdivision of motions and times. This is especially true for tasks involving a large number of positioning elements, as often is found in light hand assembly.

[4]From "Application Training Supplement No. 8—Position" by Franklin H. Bayha, James A. Foulke, and Walton M. Hancock, MTM Association for Standards and Research, Fair Lawn, New Jersey, January 1965, revised January 1973.

Accordingly, the MTM Association sponsored basic research[5] into Position which culminated in adding the Supplementary Position data to the U.S.A./Canada MTM Data Card in 1965. This research involved a complete inquiry into the basic nature, true motions and variables, and interaction of the variables in positioning movements. It also developed comprehensive new definitions—as above—and more objective measures of the variables, thereby greatly increasing the understanding of this basic MTM element. Besides the document[4] now considered the basic source, two other Application Training Supplements[6] resulted from an 8-year period of application research and validation which preceded the data card change. The information presented in this sub-chapter is drawn from these sources.[4,5,6]

The Supplementary Position data is based directly on validated laboratory and industrial data with no need for supporting theory other than that included in the research reports. It directly evaluates the methods basis and time requirements of all *Total Position* components demonstrated to exist. This is accomplished by better component definition and more exhaustive analysis of the variables within each component. An important aspect of the research was the verification of the additivity of these components, subject only to methods constraints and the limiting time principle.

POSITIONING COMPONENTS

During the discussion of Supplementary Position data, understanding will be enhanced by referring to Table 2, which summarizes the wealth of detailed statistical information from the research regarding the components and internal variables of Position Proper.

Because of interaction effects, the Supplementary Position data considers integrally all the motions between the completion of the Grasp and the start of the Release of the engaging object. This is the meaning of Total Position as defined. These motions basically include all the Position motions of the International data plus the preceding Move, which requires some additional interaction time above the time it would require when not followed by a Position. Also, the engagement motions have been subdivided into Primary Engage, up to the plane of initial engagement, and Secondary Engage, which describes the penetration of the engaged part by the engaging part. It will help to clarify this to view the

---

[5]R.R. 109 and R.R. 110. (Available from the U.S./Canada MTM Association.)
[6]Published by the MTM Association for Standards and Research, Ann Arbor, Michigan:
a. "Application Training Supplement No. 6—Position," October 1960.
b. "Application Training Supplement No. 7—Position and Apply Pressure," by Franklin H. Bayha and James A. Foulke, June 1964.

Table 2. Variables in Supplementary Position Data

Motion Equation: $P = (M + O) + (A + E1) + E2$
Time Equation: $P = M + O + (A + E1) + E2$

Type of Position	COMPONENT and Symbol	PURPOSE of Component	Qualification Notes	Effect of INTERNAL VARIABLES on Component Time		
				CLASS OF FIT based only upon radial clearance	CASE OF SYMMETRY relative only to geometric cross-sections	DEPTH OF INSERTION from plane of initial engagement
TOTAL POSITION — POSITION	MOVE (M)	Transport object to vicinity of destination	Control is high, normally Case C. Direct effect from both external variables of travel distance and object weight	No effect	No effect if pre-orient completed prior to *Total* Position — Time greater as included orientation is increased. Its total effect is included in the time data for Position *Proper*	No effect
TOTAL POSITION — POSITION	ORIENT (O)	Rotational adjustment of object about the axis of insertion	Not present if object completely pre-oriented prior to *Total* Position	No effect	Direct effect for all degrees of maximum possible orientation	No effect
TOTAL POSITION — POSITION	ALIGN (A)	Linear adjustment of object axis to axis of insertion	Normally limits out Primary Engage (except for P21S and P21SS)	Direct effect is the main cause of time variation. Align time highly sensitive to decreased clearance.	Direct effect is minor except for 90 to 180 degrees of maximum possible orientation	No effect
TOTAL POSITION — PROPER	PRIMARY ENGAGE (E1)	Transport object to surface of destination	Always present, but normally limited out by Align component	No effect	No effect	No effect
TOTAL POSITION — PROPER	SECONDARY ENGAGE (E2)	Transport object into destination	Always the last motion, if present	Direct effect due to highly *controlled* movement	No effect	Direct effect due to highly *constrained* movement.

purpose of the components as either *transporting* motions or *adjusting* motions, as illustrated in Fig. 8.

The *transporting* motions may be considered to sequentially progress the engaging object to the final destination. They include the M, E1, and E2 components, which must occur in that order (see Fig. 8). The Move

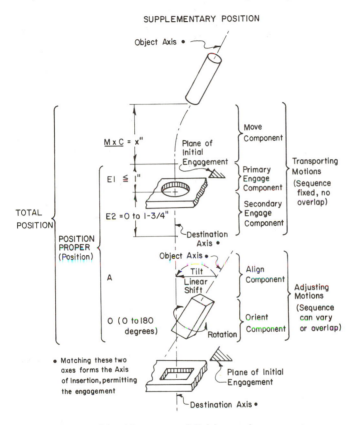

SUPPLEMENTARY POSITION

Fig. 8. Position Movement subdivisions and components.

brings the object to within an inch or so of the destination surface. The more highly controlled Primary Engage completes the travel to the destination surface. Then the Secondary Engage after contact, or potential contact, of the objects involves the final insertion (see Fig. 3) of the engaging part into the target or the destination cavity.

The *adjusting* motions required to accomplish the geometrical matching of the parts are the A and O components (see Fig. 8). Align is a linear shift of the axis of the engaging part, or a tilting motion, or both (see Fig. 1) to cause coincidence of the object axis with the axis of insertion. Orient (see Fig. 2) is a rotation of the engaging part about the axis of insertion to attain the requirements of symmetry between the engaging and engaged objects.

With the significance of the Total Position components in the motion sense thus clarified, the validated components are completely expressed by the following Position Motion Equation:

$$P = (M + O) + (A + E1) + E2$$

The parentheses in this equation indicate the correct usage of the Limiting Principle. In the *motion* sense, the shared terms *may* limit each other, whereas the E2 term must occur independently. Stated another way, the terms sharing parentheses can be performed concurrently. However, there is nothing in this equation which prohibits a Position Proper during which all of the motion terms are performed independently—as though the parentheses were missing—and in any order, provided that the M occurs first and the E2 occurs last. Also, in an actual Position Proper, any given term may be missing, except that E1 must always be present for a *spatial* Position to occur. These distinctions are important in understanding the Supplementary Position data.

However, the Position Time Equation deletes the parentheses around the M and O as follows:

$$P = M + O + (A + E1) + E2$$

This shows how the Total Position time effect of Orient was developed in the research and is used in the Supplementary Position data. Actually, part or all of any necessary orientation time can occur during the Move component, leaving only any remainder to be accounted within the Position Proper time. In addition, the Orient time during Move may not be totally limited out by the Move time; instead, it sometimes has the effect of increasing the Move time above the value normally assigned to a Case C Move of the appropriate distance. To solve this problem of varying M and O mixtures, the Supplementary data reflects the difference between the total Move plus Orient time and the normal Case C Move time as the Orient time assigned within Position Proper. With this simpler approach, the normal Move time data need not be changed to apply the Position data, since any interaction time effects within Total Position are included in the time for Position Proper.

However, if an object is or has been completely pre-oriented before the Move motion begins, no Orient component will be needed during the Position action. This is handled by assigning the Symmetrical case of orienting, so that the orientation time is effectively reduced to zero.

Next observe that the Align and Primary Engage components are subject to being limited by each other. As a matter of fact, the E1 time normally is limited out by the A time except for the loosest fit cases with

Symmetrical and Semi-Symmetrical rotations.   As was true for the Position Motion Equation, any time component except the El time can be missing during a given *spatial* Position.

Finally, all Position Proper motion categories and time categories on the MTM Data Card for Supplementary Position data have been determined in accord with the Position Motion Equation and Position Time Equation.   These have necessarily been confined to the most prevalent combinations of Position Proper components.   For those occasions when an analyst wishes to separately evaluate the various components or use them in different combinations, the component time data are available as presented later.

INTERNAL VARIABLES

With the components of Supplementary Position thus clarified, attention is directed to the internal variables.   One simplification relative to the International data is the removal of the external variables, which have to do with the handling characteristics of the positional objects, from the motion and time included in Position Proper.   Therefore, as shown later, the external variables must be separately analyzed for motion and time requirements in addition to the assignment of Position Proper categories. The basic source[7] for Supplementary Position data again more adequately describes the internal variables. It also includes metric dimensions where appropriate, so that the definitions apply equally to English and metric system data cards.

Definition[7]

    I. *Fit* or *Clearance* is the minimum distance (inches or centimeters) between the positional object and the positioning destination when the object is centrally located at the plane of initial secondary engagement, which is the surface of the destination.

    II. *Depth of Insertion* for Secondary Engagement is the travel distance (inches or centimeters) from the plane of initial engagement to the end of insertion, with a maximum value of $1\frac{3}{4}$ inches; travel beyond this $1\frac{3}{4}$ inch limit should be carefully analyzed and the proper motion so noted.

    III. *Symmetry* cases are a measure of the orientation required by the positional object expressed as the maximum possible orientation necessary to permit final insertion into the positioning destination.   The Cases of Symmetry are judged directly by the geometric properties of the cross-sections of the object and the destination, when projected on a plane perpendicular to the axis of insertion.

    IV. *Object characteristics* prevailing during the research should guide the measurement of actual Positions, especially when unusual conditions are encountered.

Since it is the internal variables which are the actual measurement or
[7]See footnote 4.

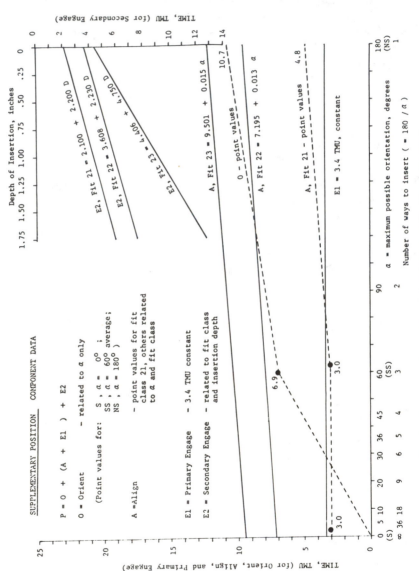

Fig. 9. Behavior and time data for Supplementary Position components.

decision criteria when determining the applicable time for a Position proper, these variables merit careful discussion.   Their interrelationships with each of the components of Position are important to correct usage of the Supplementary Position data.   Again, refer to Table 2 for clarity.   In addition, Fig. 9 shows both the variation of Position Proper components and the equations or point values found to quantify the variation.

Symmetry

Case of Symmetry is the most complicated internal variable of Position in its effects and time interactions.   However, the three cases of S, SS, and NS used on the data card are identical to the International Position data. The contribution of the research was a precise, minute definition of both the motion and time variation caused by the Orient component of Position.   Only Primary Engage and Secondary Engage are unaffected by Symmetry conditions.

The key to understanding Symmetry is the *maximum possible orientation* of the engaging object with respect to the engaged object.   This is a purely geometrical concept amenable to mathematical analysis with no need for experimental proof in the laboratory or shop.   Such an analysis is fully detailed[4] in R.R. 109, where it is demonstrated that the required angular degrees of maximum possible orientation for geometric matching of the parts is a discrete continuous variable.   The discrete steps are easily defined, integral angular rotations which can be computed from the relative shapes of the parts.   The full range of actual variation can be predetermined.   Because engagement can always be completed with an opposite rotation if it is exceeded, 180 degrees is the highest possible value for maximum possible orientation.   Also, the number of ways insertion is possible is related to the maximum possible orientation; if this value is an angle "a," and "n" is the number of ways to insert, then $n = 180/a$ as shown in Fig. 9. The insertion possibilities range from only one way at 180 up to infinite at zero degrees of maximum possible orientation.   In fact, this is the basis of all time plots of Position data as given in this chapter.

However, a simplification results from the fact that the performance time variation is rather limited over the full range of possible angles.   This permits the following measurement criteria for only three standard Cases of Symmetry:

Case of Symmetry	Symbol	Maximum Possible Orientation, angular degrees	Description of Insertion Possibilities	Ways to Insert
Symmetrical	S	0	*Any* orientation	Infinite
Semi-Symmetrical	SS	Between 0 and 90 (average 60 for data card)	*Two* or *several* orientations	2 to 10 practical (3 average)
Non-Symmetrical	NS	180	*Only one* orientation	Only 1

As noted earlier, objects pre-oriented before the start of the Move component require no orientation, and therefore their positioning involves only the S Case of Symmetry which allows zero orientation time. In all other instances, the geometric properties of the parts are the only determinant of the required maximum possible orientation. As the amount of required orientation increases, the Move component time also increases, but the additional time increment is included in Position Proper instead of being accounted for with the Move. The Orient component, being affected only by the maximum possible orientation, requires more time for increased amounts of rotation. However, the time is almost constant between the 90- and 180-degree points because this part of the range is a discrete step in orientation possibility. On the other hand, Align component behavior is practically opposite. Align time increase is almost negligible in the 0 to 90 degree part of the range, but shows a sizeable shift in the 90- to 180-degree portion of the range.

The combined effects of maximum possible orientation are to increase Total Position time as the required angular rotation increases. The same statement is true for Position Proper. However, since Position Proper includes all orientation time *not limited out by the normal Move time component,* the total time increase due to additional angular rotation actually appears entirely within the Position Proper time.

Fit or Clearance

As defined, the fit or clearance is a different measure than was used for the International Position data. Instead of using total lateral clearance to suggest the required force of insertion, or instead of using a drop or gravity test to decide the amount of pressure required, the analyst has objective dimensional criteria in the Supplementary data. The clearance is expressed as *radial clearance,* which is one-half of the total lateral clearance. This is judged at the plane of initial engagement with the axes of the engaging and engaged objects already coincident. Alternately, it is the permissible distance of the engaging object from the axis of insertion which will permit completion of the Secondary Engagement. For example, a 1.0-inch cylindrical shaft has 0.25-inch radial clearance when it is inserted into a 1.5-inch round hole. This distance, measured at the plane of initial engagement, is the sole fit criteria for the Supplementary data.

As before, three Class of Fit categories have been assigned to cover the continuum of radial clearance in discrete steps:

Class 21　　　Inches radial clearance 0.150 to 0.350, inclusive
Class 22　　　Inches radial clearance 0.025 to 0.149, inclusive
Class 23　　　Inches radial clearance 0.005 to 0.024, inclusive

Note the use of two-digit class numbers to avoid conflict with motion patterns based on the earlier data.[8] Also, radial clearance less that 0.005 inch probably involves motions in addition to Position for complete insertion; these must be separately analyzed for the additional motions and time actually required. Of further interest would be a rough comparison of the Supplementary classes with the International classes,[9] although definitive statements regarding this cannot be made due to the difference in basis for each set of data. Approximately, then, P21 corresponds to P1, P22 and P23 represent a splitting of the P2 class, and the P3 which covers almost interference fits is *not covered* by the Supplementary data. Actually, P3 is rare in practice anyway; P22 and P23 are most frequent.

Regarding the effects of radial clearance on the five Total Position components, the following statements apply. Move, Primary Engage, and Orient are not affected. Radial clearance is the principal cause of Align time variation, and the sensitivity is high. The direct effect of radial clearance on Secondary Engage stems from the necessity for higher movement control as the fit decreases.

Depth of Insertion

The explanation of Depth of Insertion is much easier than for the other internal variables. As compared to the International Position data, two main changes are seen. The maximum penetration included as a part of Position Proper was increased from 1.0 to $1\frac{3}{4}$ inches; deeper insertions must be separately analyzed for the motions actually required. Secondly, rather than one depth category, a continuum of penetration distances is recognized by four discrete categories. The category symbols actually key to the nearest $\frac{1}{4}$-inch increment of insertion as follows:

Depth of Insertion, inches	Research Range, inches	Symbol
0	0 (0 to ⅛ inclusive in application)	0
½	over ⅛ to ¾ inclusive	2
1	over ¾ to 1¼ inclusive	4
1½	over 1¼ to 1¾ inclusive	6

Note that the absence of insertion, which is measured from the plane of initial engagement, is uniquely recognized as a zero depth Secondary Engagement to a target. Actually, even finer divisions than those given above, with accompanying time data, can result from utilizing the regression equations in the research reports. The only Position component affected by

---

[8]A preliminary set of field trial data used Classes 11, 12, and 13 which were superseded by these designators.

[9]See this comparison in: Franklin H. Bayha and Walton M. Hancock, "A Comparison Between the Original and New Position Data," *Journal of Methods-Time Measurement,* Vol. XIII, No. 1, January-February, 1968.

Depth of Insertion is the Secondary Engage. The studies show that the direct effect of increasing constraint with deeper penetration results in larger per inch performance times at every distance than would be applicable for unconstrained spatial arm and hand movements of the same distance.

SUPPLEMENTARY POSITION SHORTHAND

In addition to the general symbols for each component of Supplementary Position already given, there are conventions for standardized categories of component combinations. The main set is for Position Proper based on the most frequent combinations in the Position Time Equation. The other set is for Secondary Engage as a separate component following a zero-depth Position to the surface. This is useful for multiple positioning encountered when the parts are chamfered or otherwise contoured at their mating ends to facilitate insertion.

Position Proper

This is a four-field convention, as follows:

Position—A capital P shows occurrence of a Position.

Class of Fit—According to the radial clearance criteria, the Class of Fit is indicated by arabic numerals 21, 22, or 23.

Case of Symmetry—Capital letters after the Class of Fit symbol indicate which Case of Symmetry is present: S for Symmetrical, SS for Semi-Symmetrical, or NS for Non-Symmetrical.

Depth of Insertion—The symbol is an arabic numeral recorded to quarter-inch values shown in the Depth of Insertion criteria, to express Secondary Engage action. Data card time values are available for 0, $\frac{1}{2}$, 1, and $1\frac{1}{2}$ inch midrange values symbolized as 0, 2, 4, and 6. For methods purposes, one could record the nearest quarter-inch symbol next higher than the actual insertion depth measured; but times on the data card are available only for the symbols shown above. In addition, this field is replaced with a capital A when the "Align Only" time on the data card is appropriate. This will differentiate the Align Only situation from a Position Proper with zero Secondary Engage, which requires a different time.

Examples—P21SO, P23NS4, P22S2, and P21SA.

Secondary Engage

This convention is a three-field one, as follows:

Secondary Engage—A capital E shows this component done as a separate motion.

Class of Fit—According to the radial clearance criteria, the Class of Fit is shown by arabic numerals 21, 22, or 23.　This is followed by a hyphen (-) to separate it from the next field.

Depth of Insertion—On the same basis as Position Proper, this is symbolized with one arabic numeral denoting quarter-inch midrange depth. Thus 2, 4, or 6 is shown.　Zero is not needed because then there would be no separate Secondary Engage.　If desired for methods description, one also could write 1, 3, 5, or 7; but, again, there would be no time for these on the data card and the time would have to be for the next higher symbol as given above.

Examples—E21-4, E22-2, and E23-6.

External Variables

Finally, since the Supplementary Position data expresses only the internal variables, all external variables would be separately symbolized and assigned time following the rules given in other chapters.　This holds for AP in binding, G2 for difficult handling, M for Move (or other motions with their symbols) involved in extended insertions, and similar situations.　There are *no* Supplementary Position symbols to cover the external variables.

TIME DATA INFORMATION

As indicated earlier, each Position Proper component has been separately evaluated for its time variation.　The previous discussion, as summarized in Table 2, provides a sound basis for clarity on this time data.　Following the component time data, their combination into the MTM Data Card information is given.

Component Information

On this information, Fig. 9 shows both the graphical variation and the exact algebraic formulas behind the values or tables.

Primary Engage (E1), the minimal motion in a spatial Position, requires a selected constant performance time of 3.4 TMU.　This is unaffected by the internal variables.　Except for the P21S and P21SS cases, it is limited out by the Align time.　Essentially, the Primary Engage is a special motion to the destination surface of very short distance with above-average control.

Secondary Engage (E2), always the last component in any Position if it is performed, is affected by the internal variables of Class of Fit and Depth of Insertion.　Using the symbols presented earlier, Table 3 includes the complete standard time data for Secondary Engage; the equations for any other depths within the allowable range to $1\frac{3}{4}$ inches are in Fig. 9.

Table 3. Supplementary Secondary Engage Time Data, TMU

Class of Fit	Depth of Insertion Symbol			
	0	2	4	6
21	0	3.2	4.3	5.4
22	0	4.7	5.8	7.0
23	0	6.8	9.2	11.5

Orient (O) component time changes only with the Case of Symmetry, and is unaffected by the other internal variables. The three standard performance times found by selecting point values from the time curve for maximum possible orientation are: S = 0.0 TMU, SS = 6.9 TMU, and NS = 10.7 TMU. Due to this strategy of selection, no further variation can be distinguished.

Align (A) normally limits out the Primary Engage time except for P21S and P21SS categories of Postion. While unaffected by Depth of Insertion, Align time does vary with the internal variables of Class of Fit and Case of Symmetry. Based on the previously defined symbols, Table 4 gives the standard performance times in TMU. As shown in Fig. 9, these are a combination of selected and equation values from the regression data. Also, the complete standard Align values appear in the "Align Only" column of the Position Proper table on the data card.

Table 4. Supplementary Align Time Data, TMU

Class of Fit	Case of Symmetry		
	S	SS	NS
21	3.0	3.0	4.8
22	7.2	8.0	9.5
23	9.5	10.4	12.2

Data Card Information

On the 1965 data card, a table for Position Proper was included as Supplementary Data Table 1. This was only slightly revised for the 1972 version, but the latter also includes a separate Table 1A for Secondary Engage.

Position Proper (P) time data results from combining the component times in accordance with the research results and the Position Time Equation for the most frequent situations.   Times beyond these standard values may be derived as shown later in this sub-chapter. Table 5 presents the completed data table for Supplementary Position, using the conventions already given.

Secondary Engage (E2) component data was added on the 1972 data card as Supplementary Data Table 1A for reasons already stated. Note that this table omits the zero depth column, which is all zero TMU and therefore superfluous in the field.   It is given here as Table 6.

Table 5.   Time Data for Supplementary Position Proper [MTM Data Card Table
1—Position—P (Supplementary MTM Data)]

Class of Fit and Clearance	Case of † Symmetry	Align Only	Depth of Insertion (per ¼")			
			0 >0≤1/8"	2 >1/8≤¾	4 >¾≤1¼	6 >1¼≤1¾
21 .150" − .350"	S	3.0	3.4	6.6	7.7	8.8
	SS	3.0	10.3	13.5	14.6	15.7
	NS	4.8	15.5	18.7	19.8	20.9
22 .025" − .149"	S	7.2	7.2	11.9	13.0	14.2
	SS	8.0	14.9	19.6	20.7	21.9
	NS	9.5	20.2	24.9	26.0	27.2
23* .005" − .024"	S	9.5	9.5	16.3	18.7	21.0
	SS	10.4	17.3	24.1	26.5	28.8
	NS	12.2	22.9	29.7	32.1	34.4

*BINDING—Add observed number of Apply Pressures.
DIFFICULT HANDLING—Add observed number of G2's.

†Determine symmetry by geometric properties, except use S case when object is oriented prior to preceding Move.

On infrequent occasions, an analyst may wish to know a positioning component time when only the MTM Data Card is available. Tables 1 and 1A can then be used as follows:

1. *Align* component times appear directly in the Align Only column, Table 1.

2. *Orient* component times can be found by subtracting between the P22SSO and P22NSO values and the corresponding Align Only times in Table 1. This procedure yields 6.9 TMU for SS and 10.7 TMU for NS orienting.

3. *Primary Engage* component time is a constant corresponding to the time for P21SO, or 3.4 TMU, listed in Table 1.

4. *Secondary Engage* component times for insertion are all in Table 1A. To find the remainder of any Position Proper, compare these against Table 1 at the same levels of Class of Fit and Depth of Insertion; and this will check with the zero Depth of Insertion column of Table 1.

Table 6.   Time Data for Supplementary Secondary Engage [MTM Data Card Table 1A—Secondary Engage—E2 (Supplementary MTM Data)]

CLASS OF FIT	DEPTH OF INSERTION (PER 1/4")		
	2	4	6
21	3.2	4.3	5.4
22	4.7	5.8	7.0
23	6.8	9.2	11.5

PRACTICAL WORKING RULES

The application procedure for Supplementary Position data may be described in four steps:

(1) In accordance with the Position Motion and Position Time Equations, analyze the observed or visualized Position for its components and their relative order of occurrence. This includes consideration of restrictions and other practical rules discussed below.

(2) Analyze the internal and external variables affecting the Position under consideration. This includes motion frequency determination.

(3) Symbolize both the Position Proper and any separate Secondary Engage along with the required external motions.

(4) Assign appropriate Supplementary Position times and times for any external motions also caused by the placement.

It is best to proceed in this fashion to assure correct interpretation of the restrictions and exceptions for Supplementary Position data, because they resulted from extensive application research to determine the best format covering the combination of the component data with accuracy.

Restrictions

As required by Definition IV of internal variables, sensitive analysis demands attention to measurement restrictions, which are based on the research samples from the laboratory and industrial studies. These items suggest caution because their effects were not fully explored, thus the data should not be applied without carefully considering actual behavior with objects: (1) flexible or deformable, (2) over 2.5 pounds in weight, (3) with contouring that aids insertion, and (4) which are positioned with both hands together or simultaneously. Conversely, if the data are found by careful analysis to cover particular cases involving these restrictions, a responsible analyst will annotate the facts for potential review.

The *laboratory* sample employed positional objects with six different cross-sectional shapes. The engaging objects were 5-inch-long steel shafts with essentially equal weights of 2.5 pounds. These were positioned into six sets of three hardwood blocks, four inches square by two inches deep and weighing about three ounces each. In the *industrial* sample, the objects included only rigid materials not deformed by either the motions or the act of insertion, although their weights varied. Types of operations studied were drilling, tapping, burring, forming, and assembly.

Some objects may be stretched, bent, compressed, or otherwise deformed while being positioned because of their material, configuration, etc. Since such objects were not covered in the research, analysis should be kept on the conservative—that is, generous time—side of the Position time spectrum. Suggested procedure is to analyze based on the parts remaining rigid and then allow Regrasps and/or Apply Pressures as external variables to fairly account for the difficulty of control and insertion.

Although variable weights were used in the research, none exceeded 2.5 pounds. Thus, even in the Move component, weight factors did not apply to the research times. Naturally, higher weights would justify the usual weight factors applied to the Move component times. It is believed that once the object is in the vicinity of the destination, muscular control requirements are high enough so that weight becomes a relatively minor factor in the total performance time for positioning. However, as an analysis caution, the placement of objects heavier than 2.5 pounds should be examined for any external variables which might be present and/or for unusual methods.

It is a practical fact that when the end of a positional object which must enter the destination has been chamfered or rounded at the edges, it is easier and quicker to insert the object. Likewise, either countersinking or rounding the destination edges will help save some location time. Of course, contouring both the object and the destination should then make insertion the easiest. This time saving due to contouring was not specifically investigated during the additional basic Position research; however, the indications were checked during the application research. The result was a special approach now considered official data which takes advantage of the component data for Secondary Engage. The rule is that *contoured insertion* consists of a zero-depth Position plus a separate, subsequent Secondary Engage. The Position is measured at the plane of initial engagement in accordance with the Class of Fit where lateral motion would cause contact between the objects. The Secondary Engage clearance at the end of the contouring would be the basis for assigning a reduced Class of Fit at the depth of minimal clearance and beyond, to the maximum allowable depth. Since 1972, both the required symbols and times for this total Position analysis appear on the U.S.A./Canada data card.

Although both hands can be involved in placing objects, the available Position times actually are based on single-handed control of the objects. Two-handed Positions can occur in two manners, justifying some comment.

If both hands are involved in positioning the *same object* the analysis could be normal in every respect. For example, one hand could control the engaging object and the other the engaged object, the purpose being to engage the one object. Another possibility is that the first hand does all the basic work of Position while the second hand is used to assist, steady, or guide the engaging object because it is awkward, bulky, heavy, or poorly balanced; normal analysis would present no difficulty in this situation. Such examples fall within the category of assistance, and any time gain due to assistance would probably just balance the time that would be lost if assistance were not used. In general, it is believed that such dual-handed Positions justify the normal analysis. However, if the situation is doubtful, then perhaps a timing device should be used.

The second main kind of two-handed Positions are simultaneous (simo) Positions. As will be explained in Chapter 23, simultaneous motions result in either allowing the time for one or both such motions done together; but the existing simo data for Position is based on the International data. In this case where each hand holds and Positions a *separate object* to a separate destination at the same time, simo is unlikely because of the high control in Position elements. In general, only PlSE can be done simo if the operator has enough practice; all other Positions must be performed independently according to International Position rules. Pending further research, the rule of thumb for Supplementary Position is that the P1 simo data applies for P21, while the P2 simo data applies for P22 and P23; and this means that only P21 can be done simo with practice.

As for the accurate location of the hand and/or fingers for any reason, the analysis is the same as for International Position. The Class of Fit is variable depending on the care and precision needed. However, the rule is that the Case of Symmetry always is Symmetrical for such actions.

Another factor is the effect of exceptional skill or prolonged practice on Position performance. Standard motion patterns and standard elements are for the average operator who has achieved average skill and applies average effort under average conditions. Thus, if the standard pattern is developed by observing actual operators, the analyst must be aware of unusual performance possibilities so that unusual actions will not be specified as standard when they are difficult to attain. Exceptional skill or prolonged practice can save positioning time because higher precision in motions with less need for adjustments definitely and more readily achieves the basic purpose of Position—accurate placement of objects in a predetermined relationship. In terms of Position variables, the component standards are averages or essentially the midpoints of time ranges. Exceptional performance will tend to stay toward the faster end of the individual component ranges, thus the combined components will likewise take less time. For these reasons, analysis based on physical dimensions will be more reliable than when derived from observed behavior.

Internal and External Variables

In most instances, the *internal* variables are analyzed according to the preceding text sections. More information is available, however, for the situation formerly called Align Only; and this replaces the rule-of-thumb analysis for positioning to a line, point, or points used with International Position data.　The key to the revised analysis is the component data.

Note first that Supplementary Position data is provided for analyzing Positions where the Depth of Insertion is essentially zero—actually, by convention, $\frac{1}{8}''$ or less.　A Position of this kind stops when Align, Orient, and Primary Engage have been performed and no Secondary Engage is included.　However, in such a Position, the positional object is spatial or "in the air" and not touching the destination surface as the Position begins; therefore a Primary Engage is needed to cause contact or potential contact, and a "zero-depth" Position includes this.

Opposed to the above situation is the data for "Align Only" in Table 1 of the MTM Data Card; and these data are distinguished by the fact that the positional object is already touching the surface when the Position begins, so that no Primary Engage is needed.　That is, the object is slid along the surface an inch or less to the final destination, at which no penetration is required.　It is also easy to understand how the same time allowance would be appropriate for an Align Only in space, where two objects might need to be precisely located relative to each other and then held for some period of time while a process took place.　For example, an aircraft ramp guider using light wands in directing the airplane to a terminal berth employs fairly precise signals to the pilot formed by spatial alignment of the wands.　In this event also, only Align and Orient are needed.

To summarize, a "zero-depth" Position is when *only* Secondary Engage is absent, but an Align Only ("surface" Position) is when Primary Engage *also* is absent.　In case the geometry of the parts is Symmetrical, Orient also could be absent—resulting in a *pure* Align Only.　For all Align Only, the component time for Align is the only positioning time needed; and this is the reason for direct data card listing.

Since the actual method can be reflected with the Supplementary data, a bit more discussion of analysis is justified.　The clearance for an Align Only will be the permissible $(+$ or $-)$ distance from the target center or target boundary.　This *tolerance* of placement naturally determines the Class of Fit with either the International or Supplementary Position data. For the International data, the Case of Symmetry is always considered to be Symmetrical; but for the Supplementary data, the Case of Symmetry can vary to fit the situation.　The target can be a point, two or several points, one or more lines, an intersection, or the area within any closed figure such as a circle.　It is easy to see that if two points are the target, one Position would not be enough if the points were too far apart.　Thus the general rule states that if the separation is four inches or more, at least one alignment will be needed at each point.　For very accurate placement, or if the points are as much as a foot or more apart, a third Position might be needed to correct any misalignment at the first location caused in achieving the second Position. Also, as seen in a later chapter, some eye motion might be necessary between placements if the distance between points is four inches or more.

The *external* variables are the motion and time occurrences not covered within Total Position; that is, they are analyzed as separate normal motions that happen because a Position was performed but are not part of it. The principal ones are noted on the Supplementary Position data table of the MTM Data Card. For instance, deeper insertion may require a Move (normally Case C because of control, sometimes even Case A), Turn, Apply Pressure, or Regrasps to completely seat the positional object; so the proper symbol(s) and time(s) would be written after the Position analysis. Manually pre-orienting the engaging object before Total Position (sometimes called Pre-Position as with the Therbligs) requires any motions for that purpose to be written before the Position symbol as well as a recognition that the Position Proper will involve only Symmetrical performance. A note on the data card shows that binding upon insertion requires Apply Pressure motions to be separately analyzed, whereas its effects for the International data are included within the Theory of Position. A similar situation holds for Difficult Handling. While the Theory of Position assigned a single G2 to average this variable, the Supplementary data requires separate allowance of the actual number of G2 observed or visualized. Analyst alertness is needed on these items.

Other Analysis Topics

While all Positions must be preceded by a Case C Move (or equivalent for hand and/or finger placement), not all Case C Moves are followed by a Position. The various methods of parts disposal to a totepan affords a good demonstration. Merely tossing parts into the pan, as shown earlier, could be a combined motion of M—Bm with a limited out RL1 "on the fly." If parts are carefully laid into the totepan due to a danger of damaging them, then only a Case C Move with a subsequent RL1 would suffice; and the same applies for orderly, but not precise, placement into the pan. However, if the parts are accurately placed, nested, or stacked in an orderly way into the totepan, then a Position could follow the Case C Move—especially for layers succeeding the first on the bottom of the pan. These hints should help avoid assigning a Position after every Case C Move when it is not warranted. They also should alert its allowance when economy of handling, packaging specifications, operating practices, or other requirements indicate neat disposal to be desirable or necessary— even at the expense of the extra positioning labor.

Many analysts believe that positioning time can be reduced by tilting the engaging object as it enters a hole, based on the supposition that this will reduce the Class of Fit which is gaged at the plane of initial engagement. In general, unless the parts are designed to facilitate tilting, it will save little time; since a Move, Turn, or even Apply Pressure to overcome any binding subsequently must be performed in addition to an otherwise straightforward Position. The analyst should avoid misleading conclusions.

An analyst sometimes must consider multiple Positions with the same parts to achieve the desired final relationship. For example, to insert a long part through a series of flanges separated by less than one inch at entry points, a series of Positions without intervening Moves may be correct; however, the second and succeeding Positions may be reduced to Symmetrical orientation regardless of the Case of Symmetry at the first hole. Also, the Class of Fit possibly could vary at each flange. Again, if the spacing between flanges exceeds the Depth of Insertion limits, the Positions really are not multiple; rather each flange will justify both a Move and Position. Careful analysis of the actual order and sequence will avoid errors in the analysis of multiple Positions.

Depending on its length and shape, a part which must be located at both ends may require two Positions. If it does, the second Position may be lower in both Class of Fit and Case of Symmetry due to the constraining effect of the already-located first end. Remember that multiple Positions must be analyzed in the actual order of occurrence, with the effect of previous Positions of the same part being considered.

Examples

Several examples, if studied in light of the above, will be helpful:

a. Positioning a Difficult-to-Handle object with no Secondary Engage:

	3.4	P21SO
	5.6	G2
	Total 9.0	TMU

b. Positioning an object which on the average binds two times during insertion:

	32.1	P23NS4
	32.4	2 APB
	Total 64.5	TMU

c. Positioning a 10-pound object to a total depth of 3½ inches:

	21.9	P22SS6
	5.8	M2C10
	Total 27.7	TMU

d. Sliding an object less than one inch to Align Only (compare this to the P21SO above):

	3.0	P21SA
	Total 3.0	TMU

e. Bringing an object from space to insert a nominal depth of 1½ inches:

	14.2	P22S6
	Total 14.2	TMU

f. In previous example, end of object or   3.4   P21SO
  hole is contoured to reduce insertion fit   7.0   E22-6
  by one class (compare result):   Total 10.4   TMU

Many of the analysis hints given in this section should seem quite logical if the components and theory are understood, and they will readily become second nature during analysis with sufficient analyst practice.

SUPPLEMENTARY DATA COVERAGE

A previous reference[7] would assist those interested in the data design aspects of the Supplementary Position data. However, additional material beyond that source is offered in this section, together with later technical developments based directly on the approved data content.

Position Proper

The complete Supplementary Position data found on the U.S.A./ Canada MTM-1 Data Card is represented in Fig. 10. This may be directly compared to Fig. 6 since the plotting parameters are identical, and statistical analysis has shown these data to be merely a different "slice" of the same time continuum. In summary, the data has 45 point values based on all the considerations previously mentioned directly from research values. This permits greater definition of Position values than with International Position by a factor of five, excluding the external variables no longer included in Position Proper. The time coverage is somewhat more than one second, 34.4 TMU, and is rather complete and uniform in

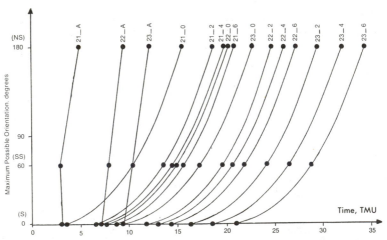

Fig. 10. Symmetry-time plot of complete Supplementary Position data (English).

coverage; however, the spread is about half of the International data because the tighter fits are not covered. In the most frequent range of actual fits, the coverage is considerably better. Most importantly, this is all *objective* data as to decision criteria.

Note several things in Fig. 10. Three of the data points are almost identical in time value, which shows they can be essentially equal even for different positioning methods. Note also the "jog" in the curve for P21___A at the SS value, a tip that the combination of component values at that point should be checked. Finally, notice that the curve direction for the Align Only values behaves differently than those for Position Proper that includes the other internal variables. Otherwise, all the curve slopes are very consistent throughout the range. It also is revealing to notice the spread between curves at the same Depth of Insertion for the three Classes of Fit. Coverage of this data within the selected range of fits in the research is excellent.

Components

The complete component data for Supplementary Position is depicted graphically and algebraically in Fig. 9, together with a legend keying the plots to fixed or selected point values of the variables. Comment on a few items will be helpful on this plot. Cross-comparison to Table 2 is very instructive as to the general trends and interactions between the internal variables. Especially note that Secondary Engage varies completely independently of the geometric parameter, being related only to the Depth of Insertion and the Class of Fit; in fact, it is shown on this plot merely for convenience and to centralize the picture. Primary Engage is a true constant time unaffected by any internal variable.

Align and Orient both depend most directly on the Case of Symmetry, although Orient is independent of the Class of Fit. They share the parameter of maximum possible orientation, which in turn, is an inverse relation to the number of ways insertion can proceed. At zero and 180 degrees, there are definitive end points which are *absolute*. Actually, all other points are in the Semi-Symmetrical range, most of which is covered by the same practical rule of "2 to 10 ways" that applies for both International and Supplementary data. The data table values for SS are based on an average of $a = 60$ degrees (3 ways to insert), but the real range is from zero to 90 degrees in a discrete step; and in the "neglected" part of the range from zero to 18 degrees (infinite to 10 ways to insert) the component variation is almost "flat" except for Orient, for which the time spread is less than 3.0 TMU. In effect, all these technical factors not only support the practices in Supplementary Position, but also confirm those in the International Position.

Extended Values

For the vast majority of MTM applications, the data card values for Supplementary Position will be more than adequate for both method and time discrimination. However, for analysts faced with critical Position times in very short cycles under high-volume, highly repetitive production conditions, occasional need may be found for even finer Position subdivision. Working *strictly* within the established data design boundaries and validated behavior of the variables, Fig. 9 opens the way to developing extended values for Supplementary Position. In fact, the only limitation on the number of values possible—bounded by the established end points of ranges—is the magnitudes assumed for D, *a* (or *n*), and the Position Time Equation as confined to mutual limiting by the A and E1 components.

The Depth of Insertion (D) naturally depends on measurable differences within the zero to $1\frac{3}{4}$ inch range, which is a function of the measurement tools applied and their precision. The degrees of maximum possible orientation (*a*) is restricted to discrete steps, which within the range of 2 to 10 ways to insert would be as follows: 2 - 90°, 3 - 60°, 4 - 45°, 5 - 36°, 6 - 30°, 7 - 25.7°, 8 - 22.5°, 9 - 20°, and 10 - 18°. With hand-held electronic calculators available and these values, there is no hindrance to developing as many Position values as one would find useful. All that is needed are condensed equations as next described, each being a vertical "slice" of Fig. 9.

For S and NS conditions of Symmetry, all possible values exist on the data card except those for Depths of Insertion differing from those found in the columns for that internal variable. For SS conditions, the data card is confined to 3 ways to insert at *a* = 60°, so the symbolism would have to be modified for other values in the SS range. This is easily done by using parentheses in the convention as follows: (SS). Under these provisions, the equations for extended Supplementary Position values will be as follows:

Align Only:

$$P21(SS)A = 3.0 \qquad \text{(a point value)}$$
$$P22(SS)A = 7.195 + 0.013\,a = 7.195 + 2.340/n$$
$$P23(SS)A = 9.501 + 0.015\,a = 9.501 + 2.700/n$$

Zero Depth Insertion:

$$P21(SS)O = 10.3 \qquad \text{(a point value)}$$
$$P22(SS)O = 14.095 + 0.013\,a = 14.095 + 2.340/n$$
$$P23(SS)O = 16.401 + 0.015\,a = 16.401 + 2.700/n$$

All Other Depths:

$$P21S\_\_ = 5.500 + 2.200\ D$$
$$P21(SS)\_\_ = 12.400 + 2.200\ D$$
$$P21NS\_\_ = 17.600 + 2.200\ D$$
$$P22S\_\_ = 10.803 + 2.230\ D$$

$$P22(SS)\_\_ = 17.703 + 0.013\,a + 2.230\,D$$
$$= 17.703 + 2.340/n + 2.230\,D$$
$$P22NS\_\_ = 23.843 + 2.230\,D$$
$$P23S\_\_ \quad = 13.907 + 4.750\,D$$
$$P23(SS)\_\_ = 20.807 + 0.015\,a + 4.750\,D$$
$$= 20.807 + 2.700/n + 4.750\,D$$
$$P23NS\_\_ = 27.307 + 4.750\,D$$

Secondary Engage Only:
$$E21\text{-}\_\_ = 2.100 + 2.200\,D$$
$$E22\text{-}\_\_ = 3.608 + 2.230\,D$$
$$E23\text{-}\_\_ = 4.406 + 4.750\,D$$

Finally, in the above equations, the blank for depth would be filled with the actual depth used; and this could be distinguished from the standard depths by enclosure in parentheses, as (1.125) for 1⅛ inches. Either angle *a* or ways to insert, *n*, can be used in several of the equations. With these equations and a calculator, countless Position analyses are possible.

DECISION MODELS

The Decision Model for Supplementary Position analysis follows as Figs. 11A, 11B, and 11C.   These three sheets cover all data card results.

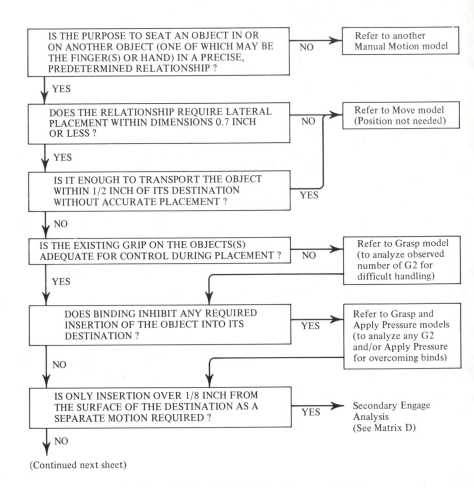

Fig. 11A.   Decision Model for Supplementary Position.

(Continued from last sheet)

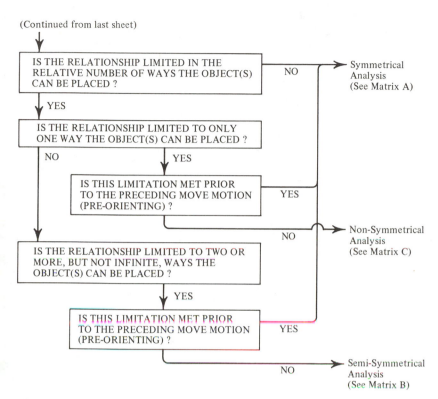

(MATRIX A) SYMMETRICAL ANALYSIS (Find proper intersection)

WHAT IS THE RADIAL CLEAR-ANCE, INCHES, BETWEEN THE OBJECTS AT THE POINT OF ENGAGEMENT ?	IS THE DESTINATION SURFACE PENETRATED AT ENGAGEMENT ?				
	NO	YES			
		WHAT IS THE DEPTH OF INSERTION, INCHES ?			
	Surface Align only	From 0 to 1/8	Over 1/8 to 3/4	Over 3/4 to 1-1/4	Over 1-1/4 to 1-3/4
					Over 1-3/4
					Refer to Move model for the added motion
.350 to .150	P21SA	P21S0	P21S2	P21S4	P21S6
.149 to .025	P22SA	P22S0	P22S2	P22S4	P22S6
.024 to .005	P23SA	P23S0	P23S2	P23S4	P23S6
Under .005	Refer to International Position model				

Fig. 11B.   Decision Model for Supplementary Position.

(MATRIX B) SEMI-SYMMETRICAL ANALYSIS (Find proper intersection)

WHAT IS THE RADIAL CLEARANCE, INCHES, BETWEEN THE OBJECTS AT THE POINT OF ENGAGEMENT ?	IS THE DESTINATION SURFACE PENETRATED AT ENGAGEMENT ?					
	NO	YES				
		WHAT IS THE DEPTH OF INSERTION, INCHES ?				
	Surface Align only	From 0 to 1/8	Over 1/8 to 3/4	Over 3/4 to 1-1/4	Over 1-1/4 to 1-3/4	Over 1-3/4
.350 to .150	P21SSA	P21SS0	P21SS2	P21SS4	P21SS6	Refer to Move model for the added motion
.149 to .025	P22SSA	P22SS0	P22SS2	P22SS4	P22SS6	
.024 to .005	P23SSA	P23SS0	P23SS2	P23SS4	P23SS6	
Under .005	Refer to International Position model					

(MATRIX C) NON-SYMMETRICAL ANALYSIS (Find proper intersection)

WHAT IS THE RADICAL CLEARANCE, INCHES, BETWEEN THE OBJECTS AT THE POINT OF ENGAGEMENT ?	IS THE DESTINATION SURFACE PENETRATED AT ENGAGEMENT ?					
	NO	YES				
		WHAT IS THE DEPTH OF INSERTION, INCHES?				
	Surface Align only	From 0 to 1/8	Over 1/8 to 3/4	Over 3/4 to 1-1/4	Over 1-1/4 to 1-3/4	Over 1-3/4
.350 to .150	P21NSA	P21NS0	P21NS2	P21NS4	P21NS6	Refer to Move model for the added motion
.149 to .025	P22NSA	P22NS0	P22NS2	P22NS4	P22NS6	
.024 to .005	P23NSA	P23NS0	P23NS2	P23NS4	P23NS6	
Under .005	Refer to International Position model					

(MATRIX D) SECONDARY ENGAGE ANALYSIS (Find proper intersection)

WHAT IS THE RADIAL CLEARANCE, INCHES, BETWEEN THE OBJECTS AT THE POINT OF ENGAGEMENT ?	WHAT IS THE DEPTH OF INSERTION, INCHES ?			
	Over 1/8 to 3/4	Over 3/4 to 1-1/4	Over 1-1/4 to 1-3/4	Over 1-3/4
.350 to .150	E21-2	E21-4	E21-6	Refer to Move model for the added motion
.149 to .025	E22-2	E22-4	E22-6	
.024 to .005	E23-2	E23-4	E23-6	
Under .005	Refer to International Position model			

Fig. 11C.   Decision Model for Supplementary Position.

CHAPTER 13

Disengage

Definition[1]

DISENGAGE[2] *is the basic hand or finger element employed to sepa-rate one object from another object where there is a sudden ending of resistance.*

1. *Friction or recoil must be present. Merely* **lifting** *one object from the surface of another would* **not** *be a disengage.*
2. *There must be a noticeable break in the movement of the hand.*

The significant characteristic of Disengage is the hand recoil action. This may or may not be caused by the fit of the parts separated during this action, although tight fit of parts is the most common reason. Recoil is a noticeable break in controlled hand movement and it is the most significant variable to be *observed* in identifying Disengage.

The frequency of Disengage in most industrial work is relatively low for two principal reasons. First, most industry is concerned with building —assembling objects rather than dismantling them—which makes the frequency of Position, for example, higher in expectancy. Secondly, when objects are disassembled or taken apart, the condition of the parts or the motions used may not require the Disengage element defined—even though assembly of the same parts required a Position. Often, more common motions of less complexity will suffice for the disassembly of objects, as is more fully explained later. Typical examples of Disengage are:

[1]From "MTM Basic Specification" (Footnote 1, Chapter 5).
[2]Disengage is another convenient grouping of the more basic motion elements into a more practical, easily utilized category of motions. In Therblig terms, it comprises one type of Disassembly action; but the Therblig Disassembly could encompass more of the basic MTM motions than is the case for Disengage, which consists only of very short motions plus one longer special type of Move that is called recoil.

1. Removing the handle of a socket wrench set from a socket
2. Pulling a plug gage out of a snugly fitting hole in a part being inspected
3. Uncorking a bottle of wine with a corkscrew
4. Pulling a radio tube from its socket
5. Lifting a riveted assembly out of a nesting anvil after the ram of the machine has squeezed a staked component into the contour of the nest.

Position is concerned with the assembly of objects by align, orient, and engage motions. In direct contrast, Disengage concerns taking apart objects previously joined, such that the sudden elimination of resistance between the parts will cause recoil of the finger, hand, and arm muscles used. The requirement for some recoil evidencing resistance to the breaking of contact is basic to fulfilling the definition of Disengage. This recoil is the main distinguishing feature of the type of object separation defined as Disengage. Provided no recoil occurred, one could therefore disassemble objects previously joined by a Position without employing Disengage. For example, only a Grasp would suffice to break contact between a ruler and the surface on which it was previously positioned to two points. Obviously, no resistance to separation of it (other than the minor effect of gravity) from the points or the surface would be present in this example. Recoil action would be conspicuously absent.

Note that only the hand and fingers are capable of gripping and controlling objects in the manner required for the occurrence of Disengage. The action concerned can be performed only by the fingers, hand, and arm working in a concerted fashion at the task of object removal. Disengage can occur, however, when a tool is used to apply the force because tools held in the hand are essentially extensions of the manual members that permit greater leverage, grip, or control than is possible with the bare hands. Disengages with such tools as pliers, tongs, tweezers, etc. used essentially to *pull* apart objects are quite commonly seen. The recoil of a worker's hand in such cases will signal a Disengage just as surely as if the tool were not present. Note also that the same tools used to *pry* objects apart will not involve a Disengage action by the worker, even though a recoil might result; the Grasps, Moves, and Apply Pressures used to pry are not the same in method or combination as those covered in a standard Disengage.

Broadly speaking, Disengage consists of two major actions. One set of motions is first used to break the contact resistance which keeps the objects together, followed by the recoil movement when the object being gripped is suddenly freed from the other object. The first set of motions usually consists of Turns (very small), Regrasps, Apply Pressures, and short Moves (1 inch or less) involving the part being removed. The recoil, as discussed in detail later, is presently understood to be *equivalent* to a Type II Case B Move. Recoil is an involuntary movement, that is, one not under complete control of the operator. The moment complete control has been resumed, Disengage is ended and either motion ceases or a subsequent Move of the disengaged object ensues.

This summary of the basic nature of Disengage serves to introduce the fact that at one time the Disengage data was questioned as to its adequacy to fully describe and allow time for the set of motions referred to above. The very first research report of the MTM Association offered possible alternates to the means of handling Disengage described in this chapter. A *Preliminary Research Report on Disengage*[3] indicated a desire to make further, more complete studies of the present Disengage data or any alternate which might evolve during such studies. Knowing this will help the reader to understand certain parts of the detailed discussion in this chapter.

There is no question about the Disengage description which refers to the "sudden ending of resistance" that causes the recoil. What was questioned is: (1) Of what detailed motions do these actions consist? (2) Is the present system of Disengage categories adequate to cover most occurrences and differentiate between them? (3) Has the set of rules presently governing application of the data yielded sufficiently correct results? The present data was deemed by the originators to suffice for the actions defined in Disengage; and, while further clarification and study may be desirable, the MTM Association deferred any more Disengage research due to its low frequency of occurrence and the generally satisfactory standards possible with the existing data. The time values now on the data card were actual observations validated for the categories presently recognized, but possibly other categories may be needed.

In mentioning the need for additional data on Disengage the authors do not intend to cast doubt on the validity of the present data. It is merely in the interest of being thorough that the following facts are stated: (1) Disengage does not occur too frequently in the average motion pattern; (2) The present data, although based on a minimum of original research, has given adequate results; (3) Exploratory, but inconclusive, research done since the original determination suggests the need for further study to better understand the nature of Disengage; and (4) Because of its relatively infrequent occurrence, it is very low on the priority list for further investigation. The present data can be used with full confidence pending additional developments.

DISENGAGE VARIABLES

A number of variables are present in Disengage. Figure 1 sequentially depicts the breaking of contact, the recoil motion, and the Move of the

[3]R.R. 101. (Available from the U.S./Canada MTM Association.)

object away after the recoil ends.   As will be seen shortly, it is not possible to accurately describe the exact nature of the recoil path, since it is the resultant of two forces—the combined momentum of the object and the involved body members plus the reflex action of the operator's muscles in trying to regain complete control of the freed object.   Thus, the path is not completely under the conscious control of the worker.   Note that Disengage ceases when the recoil ends, the recoil being the only large displacement of the object; very little object relocation results during the earlier task of breaking contact.

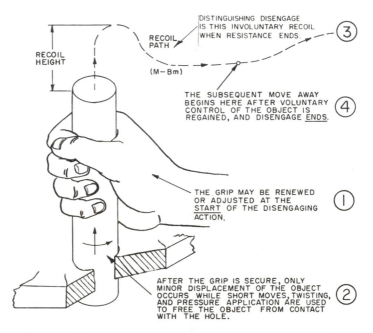

RECOIL HEIGHT

RECOIL PATH

(M − Bm)

DISTINGUISHING DISENGAGE IS THIS INVOLUNTARY RECOIL WHEN RESISTANCE ENDS.   ③

THE SUBSEQUENT MOVE AWAY BEGINS HERE AFTER VOLUNTARY CONTROL OF THE OBJECT IS REGAINED, AND DISENGAGE ENDS.   ④

THE GRIP MAY BE RENEWED OR ADJUSTED AT THE START OF THE DISENGAGING ACTION.   ①

AFTER THE GRIP IS SECURE, ONLY MINOR DISPLACEMENT OF THE OBJECT OCCURS WHILE SHORT MOVES, TWISTING, AND PRESSURE APPLICATION ARE USED TO FREE THE OBJECT FROM CONTACT WITH THE HOLE.   ②

Fig. 1.   Nature of Disengage.   Since recoil is an involuntary motion, the recoil path is rather indefinite and unpredictable but of the general nature here shown.

Unless recoil occurs, Disengage, as presently defined, has not been performed.   It is not always visually perceptible with loose to medium fits, but can always be detected by resistance to separation if actual manipulating of the parts is employed for analysis purposes.

Tightness of Engagement

Since separation of the objects is the first action during Disengage, the intimacy of contact between them will affect the initial motions required and the total Disengage time.   The degree of contact may vary from what may be termed integral, as seen in certain cutting actions, to the ordinary sliding fits found in most Disengages.

Integral contact implies that one portion of the same object is cut or broken away from it with actions that cause recoil of the appropriate manual members to occur. This happens, for instance, when a knife is used to cut twine or string. Several sawing motions, usually analyzed as Case B Moves—possibly with a resistance factor—of short distance, are first employed to sever strands of the cord or string. Following this, a break of contact between the pieces of cord is evidenced by the recoil which signals a Disengage of the cut piece from the remainder of the cord.

When Disengages occur between physically discrete objects, the mechanical fit is a measure of the degree of contact. This may vary from the tightness which would require assembly with a Class 3 Position, an exact fit, to a sliding fit that will cause little or no recoil upon separation. Obviously, from this condition, there is really a continuum of cases during Disengage that would justify a range of performance times in rather finely divided steps. As a practical matter, however, the number of steps must be limited in such a way that the MTM analyst can readily categorize any given occurrence of Disengage in relation to easily recognized differences in the time required. The relative time value for a Disengage to the other motions in a pattern or sequence in which it occurs will also have a bearing on the size of divisions needed for desirable accuracy of the total pattern time. Present MTM data, with three major steps, possibly needs revision into a greater number of categories.

It is now believed that the breaking of contact is actually accomplished by varying combinations of Regrasps, Turns (very small), short Moves, and Apply Pressures. It has been fairly well established that the number of each of these motions is not *directly* related to the amount of initial resistance to Disengage, although the difficulty of separation will dictate their usage in varying degree. At present, the data is qualitative in nature, not quantitative.

Just as application of force inward is needed to Engage objects during Position motions, the employment of Apply Pressure outward would logically be needed to Disengage them. The brief hesitations which identify Apply Pressure are very prevalent during Disengage, particularly where the Disengage does not result in successful separation of the objects. This is not illogical when it is realized that the application of greater force may well have resulted in separation of the same objects. Also, remembering that APB is defined as APA plus G2, the presence of Regrasps during Disengage is acknowledged as probably due to shifting the grip for applying greater force when the initial effort fails to break contact between the objects.

The reader has probably noticed when he has disengaged objects himself that a twisting action is frequently employed. This is often performed as the initial action, with a concurrent Apply Pressure. Apparently, a twist will often loosen objects or at least reduce the friction component in the direction of travel during separation. Essentially, the Turn motion used in this manner overcomes inertia or contact at the boundaries of the parts and is most effective when a slight amount of taper exists between the parts. Also note that the relative location of the hand and arm to the object being separated will determine whether the twist is actually a Turn or only a short Move with heavy force applied. If the Turn is 30° or more or the Move is appreciable, however, separate time should be allowed for them.

Another interesting phenomenon observed during Disengage research was the occurrence of flying starts that seemed to reduce the Disengage time. In flying starts, the hand hits the object being separated with sufficient force to provide momentum of travel that loosens the part enough to permit a quick Grasp and lifting of the object out of its cavity in a continuous motion. The use of this technique appeared in the research films to be associated with highly skilled operators. It was questionable whether average operators could perform the Disengages in the same manner. The conclusion was that the use of flying starts is really a skill factor and should not be considered in establishing Disengage times for normal workers. In addition, Research Report No. 101 pointed out that if the object becomes dislodged purely from the impact of the operator's hand, the absence of recoil would indicate that no Disengage motions are actually involved.

Path of Recoil

MTM Research Report 101 has an excellent analysis and detailed plots of recoil during Disengage. Perhaps its most important contribution to the Disengage data, however, is that it establishes the validity of approximating the time for recoil as an M–Bm. It is worthwhile to discuss the major ideas presented, to clarify recoil action during Disengage.

The idea of recoil, the second major action during Disengage, was described earlier as the muscle action caused by the sudden release or freeing of the object being gripped from the other object. This inherently implies that a measurable displacement of the freed object must occur— indeed recoil is the one essential characteristic to which the definition of Disengage is most firmly connected. If it is absent, even though objects are in some manner separated, Disengage was not the motion used.

The most vexing problem, timewise, in developing valid data for Disengage is due to the fact that the measurable displacement noted above occurs while the operator does not consciously control the object. Recoil is an *involuntary* movement which occurs following the breaking of resistance between engaged objects. It has been stated that the recoil time can be *approximated* by a Type II Case B Move of appropriate distance. The time measurement problem consists of (1) determining the exact shape of the recoil path, and (2) obtaining a distance measurement practical for use by the MTM analyst who does not have films to view in detail.

Employing plots of vertical distance versus horizontal distance, Research Report 101 shows seven different types of recoil paths that can occur. Only the essential differences will be discussed here to highlight the problem of determining the shape of the recoil path. Refer to Fig. 2 as this is discussed. Remember that *as soon as conscious control is regained, the recoil—hence the Disengage—has ended* and the next motion, if any, begins. The small circle shows this on the diagrams, and small arrows show the direction of pull at the start of the recoil.

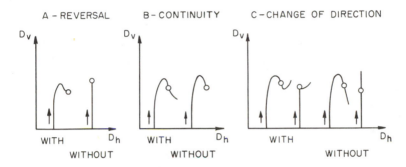

Fig. 2. Path of recoil. D_v is the height of recoil, D_h is the lateral recoil travel.

A. *Reversal* of path during recoil. A recoil may be straight or curved, depending somewhat on the magnitude of the forces of the Disengage. Reversal causes compensating changes in the velocity of recoil which tend to equalize the time for equal heights of the two cases shown. With reversal, a slight hesitance during which the regaining of control is realized following recoil balances the higher velocity associated with the forces that cause reversal; whereas there is extra uncontrolled acceleration of the object when reversal is absent. Analysis in Research Report 101 of the velocity-time plots, where the area under the curve is proportional to the distance traveled, shows both cases to be adequately approximated by a Case B, Type II Move.

B. *Continuity* of motion following recoil. After control is consciously regained, the operator may further Move the object or stop its travel. If a Move follows, it is probably a Type II in motion at the start. If the object is stopped, a deceleration time occurs. These actions also cause differences in the velocity-time plots, but the results seem to average out for the two cases so that an M–Bm again suffices.

C. *Change of Direction* in connection with a subsequent Move. The path of this Move may continue tangent to the end of the recoil or veer off to a new direction. The velocity-time compensations associated with the regular analysis of Type of Motion in Reaches and Moves would apply in this case so that any time differences associated with a Move following Disengage would not be a part of the problem of Disengage time.

D. *Dampening* of the recoil. It is possible for a recoil to be less uncontrolled if special effort is exerted to dampen it. This will alter the behavior seen with regular Disengages. Special rules of thumb have been developed for Disengages where restrictions are present as a result of analysis of dampened recoils. The restrictions and rules are discussed later.

Having seen the kind of difficulties connected with the shape of the recoil path, it is easier to understand the problem of distance measurement for recoil. While the length of recoil path is proportional to the initial resistance overcome during Disengage, it is not directly determined by the force of separation. The nature of grip on the object and the relative

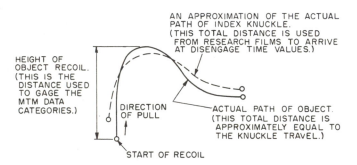

Fig. 3. Measurement of recoil length.

body position of the operator also affect the path length. For example, a longer recoil will occur when the object is at arm's length than would be true if the arm was close to the body of the worker.

Measurement of the distance of recoil may be clarified with the aid of Fig. 3. Remember that, in the analysis of films to assign MTM times, the reference point for distance measurements is the index knuckle. The manual time will depend on the *total* index knuckle travel during displacement motions. This travel path does not coincide with the recoil path of the object, although the distance difference is small. To attempt accurate measurement of either of these paths would be difficult in practical MTM usage, if not impossible, especially since the recoil path may vary for successive Disengages of the same kind of objects.

Since the actual path length is difficult to measure, a practical compromise was used to set up the MTM categories for Disengage. The observed recoil *height*—obviously an approximation under even the best conditions— is taken to be a gage of the total path, and thus becomes one of the factors in establishing the time for the basic action. This makes for ease of measurement with relatively small resultant error, in most cases, although in some cases the error could be large. Depending on the engineer to observe the operator and estimate the extent of recoil is perhaps a minor inconsistency in determining the time allowance and does increase the difficulty of writing an accurate MTM standard through visualization.

Nature of Grip

The kind and tightness of grip can affect Disengage time. For example, a tighter grip will increase the amount of force it is possible to exert and could thus increase the length of recoil. Also, when the hand does not grip an object in a manner permitting the application of force readily, Disengage will be delayed until the fingers and/or hand are adjusted properly. Such corrective action may range from a minor finger movement to a full G2 Regrasp; the type of Regrasp labeled "occasional" in the Theory of Grasp occurs frequently during Disengage.

In addition to the type of grip improvement just noted, Regrasps due solely to the difficulty of handling the part can occur at the start of Disengage; this type of Regrasp was discussed at some length in the Position chapter. When present during observed Disengages, this type of Regrasp is revealed by finger shifts or hesitations at the start of Disengage. The same Easy and Difficult classes of handling as were assigned for Positions are used in the MTM data for Disengage to account for the nature of grip, and they are judged in a similar manner.

Restrictions

Disengage time will also vary when restrictions are placed on the conditions under which the worker performs a Disengage. Basically, restrictions have the effect of limiting the amount of pull and/or the recoil path (shape and distance) the operator can allow to occur. The nature of the grip could also be altered.

Proximity of nearby objects can restrict the Disengage by either preventing travel in certain directions or beyond certain distances. Either the objects in proximity or the object being removed may endanger the worker's hand due to fragility, burrs, or sharp cutting edges. The objects could likewise damage each other upon contact resulting from an uncontrolled recoil. Disengaging a metal object from a tight clamp used to hold it partially submerged in a caustic chemical plating solution, for instance, would be an operation requiring a worker to prevent a recoil path that would splash, drip, or throw the caustic on himself or his fellow workers. Also, removing a gear keyed to a shaft mounted inside a rough cast housing would present a Disengage problem, especially if the projections on the inner surface of the housing were sharp.

Another type of Disengage restriction exists when care to prevent damage of the object or injury to the worker is needed during the action prior to recoil. Pulling a broken lamp from a bayonet type socket would require caution to protect the fingers and hand from cuts. Less force would be applied to Disengage a highly polished shaft than a rough one merely to avoid marring the finish. In both of these examples, the break contact action during Disengage would be altered, even though the recoil might be unrestricted.

Binding, or resistance to separation other than that associated with the fit or friction, may place restriction on Disengage from the standpoint that extra motions not normally needed are required to break the bind. These might be the renewal of grip, reapplication of force, or both, in such manner as to permit the Disengage to continue. Binding may be caused by cocking of the part in the hole, compression of the part against the hole sides due to the pulling, or snagging of projections of the part. A familiar example is a lock key snagging in the tumblers of the lock. Corrective motions in such cases permit the Disengage to proceed.

Restrictive variables during Disengage are generally handled by rules of thumb. Another approach is to determine statistically the frequency of occurrence, as explained in a later chapter, and then use a percentage modification of the normal basic time for the motion.

KINDS OF DISENGAGE

As was true for other MTM motions, categories for Disengage were assigned to aid classification of the variables present in a way that permits application of predetermined times which reflect the average occurrence for that situation or set of motions. This aim was perhaps not as well met for Disengage as for other motions, in that the large number of variables just discussed are categorized into relatively few classes that might fail to permit adequate time differentiation. In addition, the observer is expected to evaluate an object-muscle reaction identified as recoil that at best is only a reasonable approximation.

The present data lists six classes or kinds of Disengage. Further research might well reveal new data justifying additional Classes of Fit, accounting separately for the break contact and recoil portions of the action, and more explicitly expressed restrictions. It is helpful to note that Class of Fit and Ease of Handling, which comprise the two lumped variable categories of Disengage, are expressed in terms similar to some of the classes found in Position.   These bear no direct relation, however, but are similar only on the basis of being defined limits.   The classes are arbitrary breakpoints even though Disengage times actually vary on a continuum that may be more finely approximated in future data editions.

Class of Fit

The three classes of fit for Disengage each account for the break contact and recoil actions present in terms of either the effort required or the height of recoil.   The force and recoil height are more easily judged than are the detailed separating motions and the total path length of the recoil.

Class 1—Loose Fit

Definition: A Class 1 Disengage requires only *slight effort* to break contact and results in separation with *minimal recoil* that blends smoothly into the subsequent Move.

Although some resistance or constraining effect must be present—or else only a normal Move would be needed to disassemble—it is not easy to see the recoil that distinguishes a Class 1 Fit.   *Minimal infers that it will be no more than one inch.*   Likewise, little or no pause would be discernible to an observer when control of the disengaged object was regained. To verify his doubts as to whether a Disengage actually occurred, the analyst should perform the motion himself; in this event he can feel the resistance and sense the recoil directly.   For example, it is common to find that parts which were Positioned with a Class 1 Fit will seldom require more than a Class 1 Fit to Disengage.   If no resistance or recoil is experienced, however, the motion observed should correctly be assigned a regular Move category and time.

Class 2—Close Fit

Definition: A Class 2 Disengage requires *noticeable effort* to break contact and is followed by *moderate recoil* denoting separation.

The height of *recoil suggested for a Class 2 Disengage is over one inch and up to 5 inches.*   The contact can be broken with very little pressure application.   This kind of Disengage occurs with the highest frequency of all classes.   Reversal of recoil is often associated with this class, as well as some pause or deceleration as control is regained.

Class 3—Tight Fit

Definition: A Class 3 Disengage requires *considerable effort* to break contact, and is readily evidenced by the hand *recoiling markedly.*

The use of Apply Pressure is obviously necessary, for considerable effort must be exerted to effect the separation.   The *recoil is unmistakable, being over 5 inches and up to 12 inches high* and sometimes suggesting a violent muscular reaction to the pull exerted to break contact. The analyst will have no difficulty classifying this kind of Disengage without trying the motions himself.

That the Classes of Fit, as just described, do not adequately cover all the Disengage variables they are intended to cover may be a valid criticism. That more research study is needed on the element is frankly admitted; possibly future research will result in an approach more similar to that used for Position. The advantages of explanation and differentiation found in Position would then prevail for Disengage, perhaps permitting a Theory of Disengage that would directly yield useful time values.

Ease of Handling

In essence, Disengages are classified as to the nature of the part being removed in terms of whether additional grasping motions beyond the initial Grasp are needed to permit continuance of the separating motions. Odd shapes, flexibility, unusual surface texture, and the like might be properties of the part necessitating adjustment of the Grasp. It was also indicated that additional grasping is often needed due to proximity restrictions. It is relatively easy for the analyst to decide, however, which of the two handling categories is appropriate.

Easy to Handle

Definition: Parts which can be grasped securely and disengaged without changing the original Grasp are Easy to Handle.

Difficult to Handle

Definition: Objects which cannot be readily grasped securely enough to permit their disengagement without correction of the Grasp are Difficult to Handle.

THE SHORTHAND OF DISENGAGE

Only six combinations of coded symbols are used to cover the kinds of Disengage.

Disengage

The capital letter D is written to symbolize the occurrence of a Disengage.

Class of Fit

The Classes of Fit encountered in Disengage are denoted by Arabic numerals 1, 2, or 3 following the letter D to correspond to the defined categories of fit.

> *Example:* D2 shows a Disengage, Class 2 Fit, where noticeable effort and moderate recoil are involved.

Ease of Handling

The use of either the capital letter E or D following the numerical designation of the Class of Fit will identify the ease or difficulty of handling during a Disengage.

Examples: D1D is the convention for the disengagement of a Difficult to Handle part with a Class 1 Fit, which shows little effort or recoil.

D3E indicates that an Easy to Handle object was forcibly disengaged with considerable effort and marked recoil, a Class 3 Fit.

DATA CARD INFORMATION

The MTM-1 data card information shown in Table 1 is self-explanatory. The supplementary information added on the 1972 revision, showing Care in Handling and Binding analysis, will be explained later.

Table 1.   Disengage Time (MTM-1 Data Card Table VII—Disengage—D)

CLASS OF FIT	HEIGHT OF RECOIL	EASY TO HANDLE	DIFFICULT TO HANDLE
1—LOOSE—Very slight effort, blends with subsequent move.	Up to 1″	4.0	5.7
2—CLOSE—Normal effort, slight recoil.	Over 1″ to 5″	7.5	11.8
3—TIGHT—Considerable effort, hand recoils markedly.	Over 5″ to 12″	22.9	34.7

SUPPLEMENTARY		
CLASS OF FIT	CARE IN HANDLING	BINDING
1— LOOSE	Allow Class 2	———
2—CLOSE	Allow Class 3	One G2 per Bind
3— TIGHT	Change Method	One APB per Bind

THEORY OF DISENGAGE

The time values for Disengage appearing on the MTM-1 data card were all taken from research films.[4] A theory later was synthesized to try to explain the motions involved during Disengage in terms of the times observed. This theory in its present form is summarized in Table 2.

[4] For those interested in time equations for this data, only three data points for each equation is a questionable basis. However, they do show a smooth curve when graphed. Analysis of the two curves was used to develop, correct within 0.1 TMU, these two equations:

$$\text{TMU} = 0.21\, x^{1.82} + 3.68 \text{ for Easy to Handle, and}$$
$$\text{TMU} = 0.45\, x^{1.68} + 5.22 \text{ for Difficult to Handle,}$$

where x is the height of recoil in inches.

In contrast to the Theory of Position which is considered so adequate that the original data card times for that element were developed directly from the theory, the present Theory of Disengage is primarily a rationalization of the observed data. This rationalization is evident from the motion significance notes at the bottom of Table 2. While it is an aid to understanding what happens motionwise during Disengage, the use of this theory for synthesis of non-standard time values should be limited. One of the few features about it rather well validated at present is the approximation of the recoil movement by a Type II, Case B Move.

Table 2. Theory of Disengage

Case of Motion	Motion-Time Synthesis f = fractional inches or less than 1″	Time Value, TMU		
		Theory	Experiment	Data Card
D1E	M3.5Bm = 3.95	3.95	4.0	4.0
D2E	M8.5Bm = 7.55	7.55	7.5	7.5
D3E	APA + M15.5Bm = 10.6 + 12.45	23.05	22.9	22.9
D1D	MfB + M3.5Bm = 2.0 + 3.95	5.95	5.7	5.7
D2D	MfB + RL2 + RfA + G5 + M8.5Bm =2.0 + 0 + 2.0 + 0 + 7.55	11.55	11.8	11.8
	RfA + G5 + MfB + RL2 + RfA + G5 + M8.5Bm = 2.0 + 0 + 2.0 + 0 + 2.0 + 0 + 7.55	13.55		
D3D	APA + APA + M15.5Bm = 10.6 + 10.6 + 12.45	33.65	34.7	34.7
	G2 + APA + G2 + M15.5Bm = 5.6 + 10.6 + 5.6 + 12.45	34.25		

Motion Significance:

RfA	finger reach with automaticity
MfB	minor twisting movement of object while still engaged
G2	regrasp of the object (or its equivalent) prior to the exertion of force or pressure
APA	application of force or pressure to assure grip and/or apply pull to object to break engagement contact
M–Bm	synthetic equivalent of recoil motion with length of recoil path shown. In actual practice, the motion cases are judged as to recoil distance breakpoints by height of recoil as follows:

Motion Case	Height of Recoil	Recoil Path Length
D 1	Up to 1″	3.5″
D 2	Over 1″ to 5″	8.5″
D 3	5″ to 12″	15.5″

PRACTICAL WORKING RULES

It is important to stress that resistance to separation must be present for a Disengage to occur. The recoil is the main clue in considering whether the removal of one object from another should be classified as a Disengage. A Disengage does not *necessarily* occur when disassembling objects merely because they were engaged with a Position, although Disengage *usually* takes place when friction and/or close spatial tolerance are involved in their separation.

The Class of Fit for a Disengage does not necessarily indicate the same tolerance or clearance as the Class of Fit for Position having the same numerical case. Also a part that was Difficult to Handle during Position might be easily handled while subsequently being disengaged. The threading of a needle would likely entail a Class 3 Position, and the thread is Difficult to Handle; however, the thread could be removed from the needle with a Class 1 Disengage and no difficulty in holding the thread firmly—the actions of the hands and fingers are quite dissimilar in the two cases. It is even more likely that the thread would be removed with a simple Case B Move, rather than requiring a Disengage, since recoil is not usual for this task.

There are several rules of thumb that facilitate practical application of the MTM data for Disengage. These are based primarily on the behavior of the motions observed in film studies of the element and relate to Care in Handling and the occurrence of Binding. These will be recognized as the other two Disengage restrictions previously mentioned but not covered directly in the standard Disengage categories. Their usefulness was acknowledged by listing as supplementary information on the 1972 revision of the MTM-1 data card.

Care in Handling

If damage to the part or injury to the worker is likely during Disengage unless caution and care are exercised, the slowing of the motion is usually recognized by making the following adjustment in the motion analysis and time allowance:

1. If the Class of Fit normally would be Class 1, the need for care is accounted for by assigning a Class 2 Fit to the Disengage.
2. If the Class of Fit normally would be Class 2, a Class 3 Fit will be allowed to cover the need for care in disengaging.
3. Where care is required on what normally would be a Class 3 Disengage, the Disengage should be analyzed for possible elimination from the work. The data card indicates for this case a "Change Method" as the correct analysis. If this cannot be done, the extra

MTM motions actually observed to be necessary to avoid injury or damage should be allowed in addition to the normal Disengage time.

In effect, this rule of thumb raises the Class of Fit to the next higher when Care in Handling is an additional factor in an otherwise normal Disengage.

Binding

When Binding action results in an unsuccessful attempt to Disengage, the Disengage motion ceases and new, additional motions are made to overcome the Binding.   These extra motions should be allowed in addition to the Disengage motion that would normally be appropriate if the Binding had not happened.   Binding is not likely to occur with a Class 1 Disengage, but for the other two classes the following rules of thumb apply:

1. For Binding in conjunction with a Class 2 Disengage, a Regrasp (G2 of 5.6 TMU) should be allowed each time Binding occurs.   This presumes that the Binding can be broken by obtaining a new grip on the part.
2. For Binding during a Class 3 Disengage, an Apply Pressure (APB of 16.2 TMU) should be allowed each time Binding occurs.   The assumption here is that, besides obtaining a new grip, the operator will need to re-apply the force normally used to break contact in a Class 3 Fit.

It is important to note when applying the Disengage data that the object being removed *has been assumed to be gripped once before the Disengage begins*. Obviously, any other MTM motions such as Reaches and Grasps performed prior to the beginning of Disengage must be analyzed in the usual manner in addition to analysis of the Disengage action.   To remove a long bolt from a closely fitting hole, for example, would require that the operator apply a wrench to the bolt to loosen it, lay aside the wrench, Reach to the head of the bolt, Grasp the head of the bolt, unscrew the threads with a series of Turns and Reaches or Moves and Reaches, and only then—for removal from the hole—will the Disengage action begin. Even this latter motion will not be a Disengage unless it results in recoil of the operator's hand.   Binding could occur if the axes of the bolt and hole became misaligned during removal of the bolt and would require revision of the class of Disengage assigned.   If another object were in close proximity to the bolt, the Disengage could be considered as Difficult to Handle even though the bolt itself is not hard to grip firmly.   Finally, any reason for the exercise of care would justify additional G2 or APB motions in the motion pattern if the Disengage were other than a Class 1.

DECISION MODELS

Disengage requires the Decision Model for analysis shown in Figs. 4A and 4B.

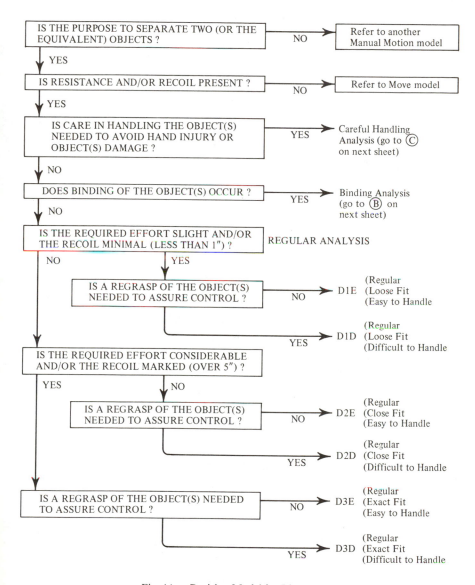

Fig. 4A. Decision Model for Disengage.

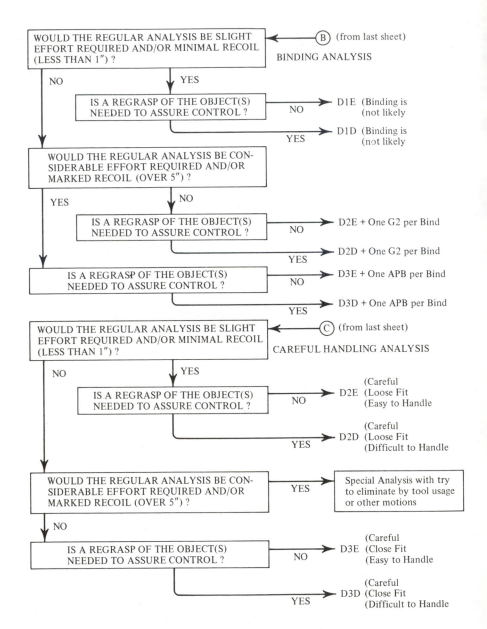

Fig. 4B.   Decision Model for Disengage.

CHAPTER 14

Eye Usage

In connection with the discussion of the Methods-Time Measurement definition in Chapters 3 and 4 the major body subdivisions were considered in relation to the motions used in the MTM system. These included: (1) the manual members; (2) the sense organs· of the head, notably the eyes; and (3) the heavier limbs and body trunk associated with body motions. The motion chapters to this point have elaborated greatly regarding the motion capabilities of the manual members and the MTM means of analyzing and measuring them. Motions of the body will be treated in the next chapter. This chapter contains information concerning the motions of which the eyes are capable, how these intermingle with other motions, and the MTM treatment of several other topics that include major usage of the eyes. Thus, this transitional chapter is another link in a complete study of the normal human capabilities. After the body motion chapter, the text will show how all of these data can be used to depict a worker's smoothly coordinated actions during actual productive sequences.

The eyes are used almost constantly as the body performs its work; eye usage is obviously a vital part of human performance—work study must necessarily include analysis of this function. Ample proof of this is evident from observing blind workers; they use different motions than employees with sight and must rely on non-visual senses for the information needed to achieve their tasks. On the other hand, it is equally evident that the eyes do not usually control the total time required for a task—particularly when factory work, as opposed to office clerical work, is being analyzed for motion and time content.

To achieve intelligent understanding of eye motions, it is necessary to consider the circumstances under which eye time is the limiting factor during the performance of work.

MTM data already given includes reference to the fact that the use of the eyes while performing other motions has been recognized as one of the control factors that tend to prolong the performance times of those motions.

For example, the main reason a Case C Reach takes longer than any other Reach is because during the Reach the operator is searching and selecting one object from a jumbled pile and will next employ a G4 to achieve control of the chosen object. These motions involve the use of the eyes in a very direct manner. Obviously, the motion picture films from which the total time for these motions was obtained included the time for visual control, so that the time needed for eye action in this example is included in the Reach and the Grasp times reported on the MTM data card. It is thus unnecessary to allow additional eye time for such cases as these two motions, or others like them, when the required eye time has already been included with the over-all motion during which it occurred. Attention to this fact is essential for the practitioner who mistakenly tends to allow excessive eye time in setting MTM standards.

Contrarily, the eyes are used in certain work sequences in a manner that precludes performance of other motions until the eyes have finished their tasks. In such instances separate eye time must be allowed. Typical operations that either cause or tend to cause the operator to cease other motions while the eyes accomplish their distinct and separately required work appear often during inspection routines, the reading of instructions and other printed matter, and the visual perception of data from instruments and measuring devices. It is in such cases that the MTM analyst must recognize and allow time for eye motions.

Although it does not determine the total time required, eye time comprises an important element in the task of printing and writing. It can easily be proved that printing and writing involve the use of the eyes as a prime control factor if the reader will attempt to write or print with ease and legibility with closed eyes. While the coordinated use of the manual members controls the allowed time for printing and writing, and not eye time, treatment of these topics so closely allied to eye time is included in this chapter as a matter of convenience since the MTM data on printing and writing is a combination of manual motions that does not readily lend itself to discussion in either earlier chapters or as a separate topic.

The following rule summarizes the preceding discussion:

Eye time is allowed only when it occurs during a complete lapse of other operator motions or limits out other simultaneous operator motions, with the specific provision that the eye motions in question are necessary for the worker to complete his task or before the next manual motion can be performed.

There are many places in this chapter where this rule could appropriately be restated but, with the emphasis just given it, this should not be necessary.

To provide a unified concept of eye time, this chapter begins with basic data on eye motions and rules governing their time requirements; sections on inspection and reading are next in order; and the subject is concluded with a discussion of printing and writing.

EYE MOTIONS

The original MTM text by Maynard, Stegemerten, and Schwab did not include any specific discussion or data on eye motions but did note a time value for visual inspection that appeared in the original research and later proved to be the MTM value for Eye Focus. Isolation of Eye Travel and Eye Focus as distinct MTM elements, and determination of the time required for Eye Travel, was accomplished later by research sponsored by the originators. It is instructive to note that the originators had diligently accounted for all of the most frequent MTM motions before they released their data; the relative infrequency of instances in which eye time controls the working time in the more usual manual tasks partially explains why eye time data did not appear in their text.

Eye Travel and Eye Focus, the only two actions of which the eyes are capable in work performance, were investigated soon after the formation of the MTM Association. The resultant data on eye time added to the MTM-1 data are reported here.

However, as this Third Edition is being prepared, the MTM Association currently sponsors a research project on eye behavior at Georgia Institute of Technology. While additional knowledge of eye usage is being sought, it is too early to report this project here; nor have any indications been released as yet. The interested reader is advised to watch current MTM Association literature for developments on this topic. They may or may not change the material given here, and possibly might only clarify it.

Eye Focus

DEFINITION:[1] *EYE FOCUS is the basic visual and mental element of looking at an object long enough to determine a readily distinguishable characteristic.*

1. *Eye focus is a hesitation while the eyes are examining some detail and transferring a mental picture to the brain.*
2. *The line of vision does not shift during the eye focus.*
3. *Eye focus is a limiting motion only when the eyes must identify the readily distinguishable characteristic before the next manual motion can be started.*
4. *Eye focus is not the normal control over the reaches, moves, positions, grasps, and other motions; eye control affects the time for these motions and this time is included as an integral part of the motion.*

[1]From "MTM Basic Specifications" (Footnote 1, Chapter 5).

The eye muscles accomplish a focus by altering the contour of the lens of the eye and thereby changing its focal length. Several important conditions are essential for the eye to achieve clear perception of an image. Since they are all included in the above definition, adequate analysis of Eye Focus requires taking account of them.

Note that the definition covers a specialized manner of "looking" at an object, not the ordinary usage of the eyes in viewing objects without sharp distinction of details.

First, the eyes cannot focus *as defined* while they are in motion; the eyeballs must be stationary long enough to permit the necessary adjustment of the lens. After aiming of the eyes at the particular image or object characteristic being examined, they must be stopped long enough for completion of the focus before the target can actually be seen. The eyes can *maintain* their focus on an object (moving object), however, so long as the focal length required does not change significantly.

To see details of objects in motion, the eye adjusts to its initial focus and then holds it ("locks on") while the head is moved in synchronization with ("tracks") the motion of the object. This can succeed only if the object's velocity is low enough so that the eyes will not lose their point of aim while the initial focus is being made; at higher speeds, the eyes only isolate a blur because the initial focus cannot be made successfully before the target area passes from view. Also, focus cannot be achieved initially or maintained if the object distance from the eyes varies greatly while focusing action is occurring. Therefore, the eyes cannot achieve initial focus on an object in motion until the *relative* motion and distance between the eyes and the object becomes essentially fixed.

Secondly, the eyes cannot "see everything at once." They must search and aim at specific target areas. To get the total image of a large object, the eye must scan the object with a series of individual focuses. Additional factors are the amount of illumination present and the distance of the object surface in respect to the focal capability of the eyes.

The object area included by one focus will vary somewhat with the individual and especially with the degree or ease with which the target characteristic(s) being sought can be differentiated from other surrounding characteristics on the object being examined. Obviously, the Area of Normal Vision (ANV) must be defined with respect to the persons involved. The research data was developed with persons having the accepted norm of 20/20 visual acuity. The time values, therefore, properly apply only to persons with this degree of vision or to those with correction to this norm by eyeglasses.

During MTM research, using persons with 20/20 vision, the average area covered per Eye Focus was found to be that within a cone from the eye to a 4-inch diameter circle at a distance of 16 inches from the eyes.

This was then defined as the Area of Normal Vision (ANV) and is diagrammed in Fig. 1. As previously mentioned, this area can be affected by the size of the characteristics or features of the object being examined. The area defined as normal is average for situations ordinarily encountered. The Area of Normal Vision, incidentally, is an important factor in the MTM system for judging whether certain motion combinations are possible for the normal operator; this is discussed in Chapter 23.

The last part of the Eye Focus definition refers to the necessity that the focus be maintained on the target "long enough to distinguish it." As soon as the image is clearly perceived, the Eye Focus is completed. That is, the purpose of the focus has been accomplished as soon as the mind quickly notes the nature or characteristic of the image seen. Note that Eye Focus, therefore, inherently includes a minimal amount of mental processing time. This is important in analysis of inspection operations. The mental time is adequate to complete recognition (perception) of the image and its nature. Eye Focus time does not, however, include time for complex judgments or extensive reasoning. A single Eye Focus is complete as soon as the minimal mental action required has occurred. Beyond that, either additional focuses are required to clarify the impression received or else thinking time not covered by any MTM motion begins, this being a prime example of "process time" requiring stopwatch study.

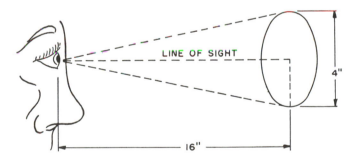

LINE OF SIGHT

4"

16"

Fig. 1. Area of Normal Vision (ANV).

The preceding information should help the reader to realize that the definition for Eye Focus is a precise one that fully accounts for the eye action involved. A simple but typical instance involving analysis of Eye Focus occurs when an operator is required to check the legibility and uniformity of the impression he has just made on the identification tag of a product with a rubber stamp. He knows what to expect, being familiar with the stamp and quality requirements; the mental decision is quick and easy. If a defect appears in the stamping, however, the image will not be normal and the fact that an error is present will be recognized immediately. He may next focus again to classify any obvious defect prior to rejecting the tag and/or segregating it into a pile to be restamped. He may even require thinking time to isolate the cause of the defect so he can take corrective action in stamping further tags.

Eye Focus is symbolized in MTM by the capital letters EF. The time for a normal operator is a constant value of 7.3 TMU. This is about ¼ second or 0.00438 minutes. This value was confirmed by independent sources cited in MTM Research Report No. 102[2]; an average value of 7.35 TMU for EF in reading was obtained by a number of psychologists. Because the eye (or more particularly its lens) takes that long to accommodate a clear image, it is noteworthy (as mentioned in a previous chapter) that *no motion whose complete performance takes less time than an EF can actually be seen and identified by the naked eye.* The effects of the motion on objects can, of course, be seen and also the observer of a situation including the motion knows logically that it must have occurred if certain following conditions are observed. Remember that "see" as used here refers to the clear perception, identification, and mental recognition required by the MTM definition of EF; what has been said may not hold for casual glimpsing, ordinary visual sensing of movement, or other conditions of eye usage which do not involve manual motion identification of the type required by MTM.

Eye Travel

DEFINITION:[3] *EYE TRAVEL is the basic eye motion employed to shift the axis of vision from one location to another.*

1. Eye travel is a limiting motion only when the eyes must shift their axis of vision before the next manual motions can be started.

Just as the eyes must pause to allow Eye Focus to occur, they must be in motion during Eye Travel; these are the only useful actions of the eyes in work performance. Adding the necessary variable Eye Travel time to the correct number of Eye Focuses multiplied by their constant time value will yield any work cycle time *under sole influence of the eyes.* As is true for other MTM motions, correct use of the times depends on clear understanding of the precise nature of the motions in question.

The aiming of the eyes to actually "see" an object, or a given characteristic of the object, is analogous in many respects to the aiming of a gun at a target. Indeed, the point at which the eyes are directed is designated as the target in many accepted treatises on the subject of vision. The nature of the viewing area and what happens after the eyes are aimed at the target has already been fully explained under Eye Focus above.

---

[2]R.R. 102.
[3]From "MTM Basic Specifications" (Footnote 1, Chapter 5).

Needing clarification here is the meaning of "axis of vision" in terms of motions and their measurement.

In discussing vision, reference is commonly made to the concept of the line of sight, this being an imaginary line from the eye directly to the target. This useful concept is adequate to explain the action of each eye individually, since the eye can readily focus and perceive the image when the line of sight has been established. It does not serve as well when considering the combined action of both eyes[4] because of the additional element of depth perception. For the operator to ascertain depth (or distance) of the target from himself, his eyes must converge on the target. When this occurs, there is a line of sight from each eye to the target, and these lines intersect at the target as seen in Fig. 2. The "axis of vision,"

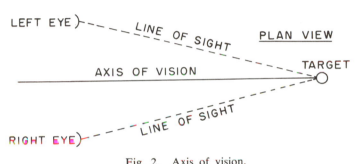

Fig. 2.   Axis of vision.

the displacement of which is the measure for Eye Travel, then consists of an imaginary line equidistant between the eyes and passing through the intersection of the lines of sight.

Eye Travel occurs only when the axis of vision moves in space. When the eyes are not moving, they can only focus; but to perceive more than a single point image, the eyes must execute a series, succession, or pattern of travels and focuses dependent on the nature of the total area which a succession of targets will encompass. The axis of vision may be displaced by (1) the eye muscles positioning the eyes, (2) the eyes being stationary in their sockets while the head is rotated, or (3) a combination of head rotation and eye movement. The latter case is most frequent when long distances between the areas of vision as defined under Eye Focus are involved. The first case occurs quite often, however, when close distances between target points and careful discernment of the images is a requirement of sight.

Because head assistance invariably occurs with Eye Travel beyond certain distance limits, the MTM procedure has placed a maximum time

---

[4] This is also known as "binocular" vision.

value of 20 TMU on Eye Travel. Regardless of the theoretical answer produced by either of the two formulas to be given shortly, *no single Eye Travel will be assigned more than 20 TMU*. The MTM eye time procedure must thus permit a way to find the time for Eye Travel between values limited out by other motions and the maximum value of 20 TMU.

MTM research has shown the exact Eye Travel time to be 0.285 TMU per degree of angular sweep between the initial and final points on the target. Note that the maximum travel time of 20 TMU thus imposes a mathematical limit of about 70 degrees (exactly 70 degrees, 10 minutes, 31 seconds) of angular sweep without head assistance as shown in Fig. 3.

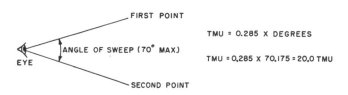

Fig. 3. Exact Eye Travel time.

For actual applications, estimating the degrees of sweep could be an awkward way to evaluate Eye Travel. The inventors of MTM therefore developed an approximation formula and a table for use by analysts. The formula, which appears in Table VIII on the MTM-1 data card, is given in Table 1. This data also includes Supplementary information on the Area of Normal Vision and the Reading Formula, which were added in the 1972 revision of the U.S.A./Canada card.

Table 1.   Eye Travel and Eye Focus Times
(MTM Data Card—Table VIII—Eye Travel and Eye Focus—ET and EF)

Eye Travel Time = $15.2 \times \frac{T}{D}$ TMU, with a maximum value of 20 TMU.

where T = the distance between points from and to which the eye travels.
      D = the perpendicular distance from the eye to the line of travel T.

Eye Focus Time = 7.3 TMU.

SUPPLEMENTARY INFORMATION

— Area of Normal Vision = Circle 4″ in Diameter 16″ from Eyes

— Reading Formula = 5.05 N Where N = The Number of Words.

The tabular data is given in Table 2. Although based on the *exact* Eye Travel formula, the table is entered by the T and D values used in Table 1 and the approximation formula given in Fig. 4. It should be noted that the eye times for industrial operations are different from the reading times, which are discussed later.

Table 2.   Eye Travel Time—TMU

Formula: ET = 0.285 × degrees

This table gives the *exact* ET time in terms of T = inches travel between target points and D = perpendicular inches between eyes and line of travel between points.

The maximum allowed ET time of 20.0 TMU is used to the right of the heavy line.

D	1	10	11	12	13	14	15	16	17	18	19	20	22	24	26	28	30
10	1.63	15.1	16.4	17.6	18.8	19.9											
11	1.48	13.9	15.1	16.3	17.4	18.5	19.5										
12	1.36	12.9	14.0	15.1	16.2	17.2	18.2	19.2									
13	1.26	12.0	13.1	14.1	15.1	16.1	17.1	18.0	18.9	19.8							
14	1.17	11.2	12.2	13.2	14.2	15.1	16.1	17.0	17.8	18.7	19.5						
15	1.09	10.5	11.5	12.4	13.4	14.3	15.1	16.0	16.8	17.6	18.4	19.2					
16	1.02	9.9	10.8	11.7	12.6	13.5	14.3	15.1	16.0	16.7	17.5	18.2	19.7				
17	.96	9.3	10.2	11.1	11.9	12.8	13.6	14.4	15.1	15.9	16.6	17.4	18.8				
18	.91	8.8	9.7	10.5	11.3	12.1	12.9	13.7	14.4	15.1	15.9	16.6	17.9	19.2			
19	.86	8.4	9.2	10.0	10.8	11.5	12.3	13.0	13.7	14.4	15.1	15.8	17.1	18.4	19.6		
20	.82	8.0	8.8	9.5	10.3	11.0	11.7	12.4	13.1	13.8	14.5	15.1	16.4	17.6	18.8	19.9	
21	.78	7.6	8.4	9.1	9.8	10.5	11.2	11.9	12.6	13.2	13.9	14.5	15.8	17.0	18.1	19.2	
22	.74	7.3	8.0	8.7	9.4	10.1	10.7	11.4	12.0	12.7	13.3	13.9	15.1	16.3	17.4	18.5	19.5
23	.71	7.0	7.7	8.3	9.0	9.6	10.3	10.9	11.6	12.2	12.8	13.4	14.6	15.7	16.8	17.9	18.9
24	.68	6.7	7.4	8.0	8.6	9.3	9.9	10.5	11.1	11.7	12.3	12.9	14.0	15.1	16.2	17.2	18.2
25	.65	6.4	7.1	7.7	8.3	8.9	9.5	10.1	10.7	11.3	11.9	12.4	13.5	14.6	15.7	16.7	17.6
26	.63	6.2	6.8	7.4	8.0	8.6	9.2	9.7	10.3	10.9	11.4	12.0	13.1	14.1	15.1	16.1	17.1
27	.60	6.0	6.6	7.1	7.7	8.3	8.8	9.4	10.0	10.5	11.0	11.6	12.6	13.7	14.7	15.6	16.6
28	.58	5.8	6.3	6.9	7.4	8.0	8.5	9.1	9.6	10.2	10.7	11.2	12.2	13.2	14.2	15.1	16.1
29	.56	5.6	6.1	6.7	7.2	7.7	8.3	8.8	9.3	9.8	10.3	10.8	11.8	12.8	13.8	14.7	15.6
30	.54	5.4	5.9	6.4	7.0	7.5	8.0	8.5	9.0	9.5	10.0	10.5	11.5	12.4	13.4	14.3	15.1

Note: The second column gives times for $T = 1$. To compute values when T is between 1 and 10 inches, multiply by T.

Geometry and trigonometry were employed to obtain the approximate formula used for calculating Eye Travel time.   How this was done is shown in Fig. 4.   The arc distance S was assumed approximately equal to the chordal distance T and the resulting formula, based on the exact formula, was then corrected for the average per cent error to obtain the constant of 15.2.   The time values yielded by the formula $ET = 15.2 \times T/D$ will show

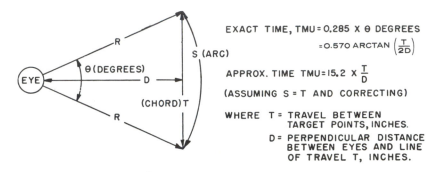

Fig. 4.   Approximate Eye Travel time.

less than 1 TMU error in the usual ranges of the T/D ratio. For example, if θ = 50 degrees and D = 10 inches, then $T = 2D \tan \theta/2 = 2 \times 10 \times 0.4663 = 9.326$ inches.   The approximate Eye Travel time is thus

$$15.2\left(\frac{9.326}{10}\right) = 14.176 \text{ TMU}$$

which is in error by only 0.074 TMU (since the exact Eye Travel time is $0.285 \times 50 = 14.250$ TMU).

The easiest measurements to make or estimate in practice are the distance of travel between focus points (denoted by the symbol T) and the perpendicular distance from the eyes to the line of travel (denoted by the symbol D).   In fact, the Eye Travel symbol convention consists of ET for Eye Travel, followed by a fraction that shows the ratio between T and D which is needed in the use of the approximation formula.   For example:

ET $^{10}\!/_{16}$ shows Eye Travel between points 10 inches apart at a distance of 16 inches from the eyes.   The time allowed would be $15.2 \times {}^{10}\!/_{16} = 9.5$ TMU.

Application of Eye Motions

Eye motion symbols usually are written in the right-hand column of the motion analysis form.   Occasionally, one might wish to call attention to the leftward direction of an Eye Travel by writing the symbol in the left-hand column; but normally this permissible refinement is not done. The only other reason for departing from the usual practice would be when the right-hand column is overcrowded and entry in the opposite column would emphasize the eye usage.

If the analyst watches a worker's eyes, the focuses and travels can be seen as distinct pauses and sweeps during performance of eye motions. When visualizing the motions, however, he must rely on knowledge of the eye motion data and then examine the method he would use himself. In any case, it is important to remember that *eye time is not allowed unless it is limiting.*

ET is seldom limiting over other MTM motions when T is less than 10 inches, nor is ET or EF allowed unless there is a recognizable pause between other MTM motions, provided the eye motions are necessary to complete the task or before the next motion can be performed.

The tendency of the MTM neophyte to allow excessive eye times can be overcome to some extent by reiterating that normal MTM motions do contain, to some degree, eye time and mental choice; the eye time discussed in this chapter is only that additionally demanded by observed methods.

If an operation is being studied for standards purposes *before* the performer has achieved full learning of the work, more than the usual number of limiting eye motions will be observed by the analyst. This is due to the learning effect as found in sponsored MTM Association research,[5] to be fully explained in the projected second book by the present authors. Such excess limiting eye usage will be reduced as the worker gains practice on that particular operation. In fact, reduction of limiting eye usage to the required minimum shown in a proper MTM analysis indicates that the operation is fully learned. This means that when an operator lacking ample practice on an operation is studied for writing an MTM standard, the observed excess eye usage should not be included in the prescribed motion pattern. The standard always should be written for average, fully learned operators.

It is not necessary that ET occurs or is allowed for each and every EF that is assigned. Oftentimes, the ET is accomplished during other motions, and is therefore limited out. In this case, the EF following might be the only eye time permissible; this is more likely to be true if the simple mental decision time included in EF must be expended before later manual motions can properly ensue. One or more EF are often assigned when visual checking is needed while other motions cease; these often do not involve ET. In addition, the eyes may have been already directed at the target because of the need for visual control in accompanying

[5]R.R. 113A.

manual motions (e.g. a Case C Move) preceding, in which event they can easily focus as required without any ET preceding the focus.

Such occurrences are fairly frequent before and after critical Positions, when motions involving care to prevent damage or injury are made, and in the reading of measuring instruments. For instance, from two to four EF would reasonably be allowed when reading a 1–inch micrometer to the nearest thousandth; one for the tenths of an inch, another for the nearest 0.025-inch mark, and one or two to decide which vernier mark is nearest to the index line. The same reasoning could apply to reading a vernier scale to thousandths. The eyes would not shift sufficiently in these cases to necessitate allowing ET.

On the other hand, aligning a ruler very accurately to several points (in addition to any Positions included) might involve an EF at the first end, an ET to the other end, a short Move and a Position, a second EF, an ET back to the first end, and a final EF to assure that the align has been completed correctly. In this example, the alignment of the second end of the ruler cannot be properly made until the eyes are brought to that end, so that the need for ET is clearly indicated.

Remember that MTM sets time values for average, fully learned operators. This naturally assumes normal vision or vision corrected to normal by proper eyeglasses. This point should not bother the analyst in visualizing, but he might have difficulty in reconciling some stopwatch times to the indicated MTM eye motion times if the observations are not on an operator having normal vision.

While other MTM motions are usually best observed from alongside or behind the worker, it is usually easier to see eye motions by standing in front of the worker. At times, particularly when the operator's head is bent over the work, it might even be necessary to stoop to observe the eye actions for accurate analysis. This type of observation requires tact and finesse because employees who do not mind having their hands observed are often more self-conscious with a work analyst peering intently into their faces and eyes.

INSPECTION

One of the more obvious and especially important applications for the foregoing eye motion data is in inspection operations. While no official MTM-1 rules or detailed data exist for this type of work, a number of application rules and suggested procedures have been sanctioned by the MTM Association pending future research. To aid the reader wishing to apply eye time data to inspection, as done successfully by many practitioners in a wide variety of industries, the presently available MTM-1 material on the subject is included here.[6]

The MTM eye time data are valid for inspection work that involves only relatively simple judgments or reasoning similar to that included in making rapid "yes" or "no" decisions. This was explained under Eye Focus. To any focus motions deemed necessary under the rules previously given and any additional rules given below must, of course, be added the necessary Eye Travel time between the pauses which signal focuses. Observation of an inspector will readily reveal the complex series of "shifts" and "pauses" in eye action during an inspection operation. Such direct viewing of the working inspector is much more reliable for determining the inspection time by MTM than is the visualization method, since frequencies and kinds of both eye and body member motions during examination of parts are of paramount importance in achieving correct time standards for inspection.

It is commonly thought by many persons dealing with industrial inspection that eye time and mental decision time comprise the major portion of inspection routines. Many MTM applications engineers have found, however, that this is not usually so. Pure inspection work as practiced in industry often averages from 85 to 90 per cent manual and body elements as limiting motions, with the eye and mental time accounting for the balance. The latter do occur constantly as the parts are examined; but, as this is also true of all other MTM motions, the real problem in inspection standards is only when eye and mental elements must be completed during a cessation of other necessary motions or when they limit out other motions in respect to the elapsed time for the work cycle. The eye time can be handled with MTM; the mental time—if not covered—can only be adequately evaluated by stopwatch study or dependence on previously accumulated time formulas. Thus, to write inspection patterns, all of the necessary manual and body motions must be considered as well as eye and mental time. Failure to do this has been the principal reason why some efforts to time inspection with predetermined standards did not succeed—a major error was committed.

Since Eye Travel time between focuses during visual inspection is relatively easy to measure and account for after the correct number of Eye Focuses has been determined, the latter becomes the central problem in the use of eye time data for such work. The variation in the types of inspection, kinds of objects to be examined, nature and size of defects, and operator differences obviously creates a wide difference in the total Eye Focus requirement of such operations. No rules or procedures

[6]One of the higher-level data systems, MTM-GPD (a seldom used and no longer promoted), does contain specific information and data on inspection operations. Since this will be included in the projected companion text by the present authors, it is not given here.

could, therefore, possibly substitute for the skill of a good analyst in arriving at valid times.

A knowledge of the true nature of the inspection process is a great aid in analysis. Basically, inspectors must (1) recognize a defect and (2) dwell on it long enough to "classify the defect." The word "defect" implies that an object is being compared to an established standard or mental concept of another object which possesses no flaws, or at least does not have the flaws which are of concern at the inspection station. It is absolutely necessary to have a standard of comparison that has been defined by instructions to the inspector and that he in turn has translated this into a mental image or concept; that many industrial inspection operations fail in this respect has been attested frequently by responsible parties. However, this requirement is necessary for satisfactory and successful inspection and many aids to this communication process exist. These include data tables, photographs of standards of comparison, standard samples, procedures, and oral instructions (least reliable) by the persons setting the original quality requirements. No system of timing could possibly be valid unless the inspector's concept of the acceptable part was correct and only the more simple decisions are required. A system like MTM requires that such conditions exist. The use of such inspection instruments as gages, magnifying lenses, profilometers, etc., are of additional help in setting up an inspection station amenable to successful establishment of time standards.

The inspector must exercise sufficient "vision" to allow sighting of the defect and then apply further EF as required to classify, if necessary, what variation from the norm exists. Each of these looks requires at least one Eye Focus.

The practical significance of the foregoing is as follows. Inspectors do not normally search out every possible defect in each part, but instead they scan the part or inspection instruments by shifting the eyes from point to point. This sequence will occur without additional pauses until a defect by which the part deviates from the mental image of the acceptable part appears or the indicators show the part out of acceptable tolerance. The number of Eye Focuses for the scanning action will depend mainly on the area which can be covered per Eye Focus, as discussed next. The presence of defects which prompt closer scrutiny will then require additional Eye Focuses. The time standards problem can be solved, therefore, by determining the number of Eye Focuses required for both scanning and defect classification and adding to this the Eye Travel time between focuses.

The area covered per Eye Focus apparently is a function of the ease with which the potential defects can be seen. For such flaws as deep, open cracks in a large casting, one EF could easily cover 2 square feet. Conversely, the checking of minute scratches on a dull surface might dictate only 2 or 3 square inches per EF. An MTM investigation reported at the first MTM conference *suggested* that an operator can normally check an area 16 times the size of the expected defect per EF needed. The limits of the Area of Normal Vision provide another means by which to gage the area covered by each EF. The work analyst must determine, with these two guides, how many Eye Focuses are needed to cover the toal area of the part and the average distance between the points of focus.

The number of additional EF's required for defect classification can be best judged by the frequency with which the various attributes or charac-

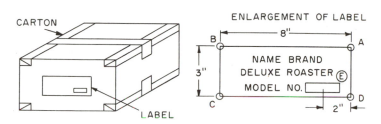

Inspection Requirements

Cartons pass the inspector seated at a roller conveyor so that his eyes are about 20 inches from the side bearing the label. He must check adhesion to the carton at all four corners of the label. He must also examine a model number stamp for inclusion and legibility. Rejects are defined as the amount of error allowed. All cartons with rejects are shunted to a side branch conveyor.

MTM Analysis for Fig. 5.

	EF Req'd.
Eye Focus:	
Scanning—Allow one EF for each pattern point as shown by the letters on the sketch above	5.000
Defects—One extra EF for each loose corner found to see if it is a reject. Maximum 7% per carton from inspection records	.070
One EF for absence of model stamp. Average missing has been 1 for each 200 cartons sealed since revision of the packing method	.005
One EF for poor legibility of stamp. Occurs on 10% of packs at the most	.100
Total EF allowed per carton	5.175 = 37.8 TMU

Eye Travel:

Eyes to Carton, ET $^{30}/_{20}$	20.0 TMU (Max.)
Label Corners, 2 ET $^{8}/_{20}$ + ET $^{3}/_{20}$	
= 2 (6.6) + 2.5	= 15.7 TMU
Model Stamp, ET $^{2}/_{20}$	= 1.6 TMU
Total ET allowed per carton =	37.3 TMU
Total TMU inspection time allowed per carton	75.1 TMU

This amounts to 0.04506 minutes average per carton.

Fig. 5.   Example of inspection allowance for eye motions.

teristics that define the defect appear. The normal frequency of occurrence for a given type of defect must, therefore, be known or estimated for proper analysis leading to a time standard. Size, shape, and color might be three inspection criteria, for instance; the MTM engineer must decide or know how often defects in each of these categories will be found in the part being inspected. Inspection and quality control records will greatly aid the analyst in this; but it is more common that he must determine the frequency himself from observation.

Figure 5 provides a hypothetical illustration of one manner in which an inspection operation can be analyzed by MTM for the EF and ET involved. Note that the ET times are taken from Table 2; they are exact. In this case, specific points are checked for specific kinds of defects that may be expected to occur with frequencies as discussed in the example. For many other practical situations, the allowable EF and ET may be found to depend more directly on area, with the type and frequency of defect unknown. Such inspection proceeds on the basis that any defect whatever is caught and classified by the inspector as it arises, from full coverage of the concerned area. While analysis of the latter situation is more involved, the same principles involved in the example and the rules given earlier will apply.

READING

Another specialized use of the eyes during work sequences sometimes occurs with sufficient frequency to justify specific MTM information on the matter. This is the nature of eye usage in reading and the time required for reading. Operators must read instructions, check material tickets, read labels on the product and on packages containing materials or the product. Inspectors must check quality specifications, read information from blueprints, and examine data written by operators or test personnel. Clerks and timekeepers obviously spend much time reading various forms and other printed matter. Other examples where a work analyst might wish to include reading time in a pattern containing other manual, body, and eye motions could be cited abundantly.

The second research report of the MTM Association was an analysis of *Reading Operations*.[7] While the results there reported are especially applicable to the reading of printed non-technical prose at normal speeds, the data is a guide to the expected reading time for other types of writing and printing (including handwriting and hand printing) at other speeds. More important, the research accumulated in a single source a number of interesting facts and variables concerning reading and its nature.

[7]R.R. 102.

The time for reading actually varies greatly, mainly because the methods employed, in terms of eye motion, can be greatly divergent. Three main variables affecting the process of reading are:

(1) Complexity of the material read. Reading speed commonly ranges from 500 words per minute for easy prose to 150 words per minute, or less, for highly technical or scientific expositions. For the general run of reading, however, 330 words per minute is the *norm* accepted by many authorities. Of course, people trained in "speed reading" can greatly exceed these values.

(2) Length of line printed. There exists an optimum line length which facilitates the eye motions to minimize the reading time; longer and shorter lines require more reading time. The accepted optimum line length seems to be 3.2 inches, or slightly less than the width of two standard newsprint columns.

(3) The style and size of type in the printing. Examination of a printer's font will reveal the large range of styles available in any given size of type; and the combinations are further magnified when the range of sizes is also included. As to size, 10 to 12 point type is apparently easiest to read, but no agreement exists on the easiest style. Style seems to be a matter of taste and habit, suggesting that the reader will read easiest the style of type he is accustomed to seeing most.

How does the eye behave during reading? Basically, of course, it can only focus on a word or section while not moving, and then perform travel to the new focus point. For inspection operations, the eyes were said to "shift" and "pause" in a series of jumps that cover the area being scrutinized. The MTM system has adopted special terms used by psychologists for the Eye Travel and Eye Focus; they are *saccades* for the "shifts" and *fixations* for the "pauses." The reason is that a number of special eye motion variables are found in reading that differ somewhat from normal eye usage, as discussed below. The problem in establishing reading time is to determine the number of saccades and fixations required to read a given number of words. With this known, the number of words can be counted or estimated and the corresponding time computed.

Very simply, the number of words per fixation averages 1.56 over a wide range of researches and types of reading. This means that division of the number of words by 1.56 will yield the number of Eye Focuses required for the passage by an average operator reading with average speed under normal conditions. The more complicated analysis of saccades is next discussed.

The purpose of all saccades (or saccadic movements) is the same as for Eye Travel—to permit the next focus. Three varieties have been isolated that describe the task performed by the travel in relation to the total task of reading the material at hand. The three types of saccades are:

(1) Forward Interfixation. This kind of saccade includes the Eye Travel between fixations in the same line. Since it is possible to have one or more fixations per word, for example three in a long three-syllable word, the saccadic distance can vary greatly. It is always very short, however, in relation to the distances associated with other usages of Eye Travel; which is one reason the term saccade is used for reading rather than simply Eye Travel. A further interesting fact is that any given individual cannot consciously vary his saccadic speed, although normal Eye Travel is subject to a good degree of conscious control; variation of saccadic speed between people is very noticeable, however. Only by remedial reading exercises and practice at increasing reading speed can a person significantly change his normal rate of saccadic movement.

(2) Regressive Interfixation. Oftentimes, due to lack of concentration or unfamiliarity with the material, a reader wishes to reread some portion of the material he has already covered. As his eyes travel back, they perform a regression or regressive saccade. Some regression is necessary in reading highly technical or complex matter. Also, for instance, when a word is hyphenated to carry another syllable to the next line, the meaning might be obscured without a regression. The distance chargeable to this saccade would be highly unpredictable in advance, except for the more patently necessary regressions.

(3) Return Sweep. As the eye finishes reading the end of one line, it must rapidly sweep back to the beginning of the next line to continue the reading pattern—this is the performance of return sweep saccades. The length of these travels will approximate the length of line being read.

Another variable which affects all saccades is the distance from the eye to the printed material. For persons with normal 20/20 vision, most physicians and oculists recommend that the reading be held from 12 to 16 inches from the eyes, with the proviso that the amount of lighting might affect this distance. If industrial operators must read as a normal part of their cycle, it is thus important that illumination be adequate.

MTM Research Report 102 showed one way to find the total time for the saccades required in reading a certain passage. It involves laborious computation to account exactly for the degrees of Eye Travel for the saccades and then applies a variable Eye Travel time found from a curve. A less cumbersome method was evolved based on the fact that, for a wide range of reading material, the saccade time will average about 8 per cent of the fixation time. Since the fixation time depends directly on the time for an EF and the number of words read, a practical formula for average reading time is:

$$\text{TMU reading time} = \text{Fixation time} + \text{Saccade time}$$

$$= 1.08 \times \text{Fixation time}$$

$$= \frac{1.08\,(7.3)\,N}{1.56} = 5.05\,N$$

where N = the number of words read, and 1.56 = average words per fixation. This formula for reading average material is included as Supplementary Data on the 1972 revision of the MTM-1 data card.

Other ways of stating the same result are: 5.05 TMU per word, 0.003 minute per word, or 0.182 second per word. A check of these constants will show that the formula corresponds to the accepted average reading rate of 330 words per minute. The reading time data can be applied in several ways. Figure 6 shows the method of determining the time required. The exact time[8] is found by evaluating the fixations and the three kinds of saccades, with the regressive saccades being obvious from the numbers that appear out of normal forward reading order above the words.

In actual normal usage, however, the analyst merely counts the total *words* in the passage, with allowances for more than one "fix" for extremely long words, and then applies the simplified reading formula given above. Alternately, he may only estimate the total words instead of counting them. The most widely used method, however, unless the passage is short enough to count the words easily, is to count the words in several lines and find the average number of words per line; count the number of lines in several pages; and then multiply with the number of pages to obtain the approximate total number of words. This figure is then used in the simplified reading formula to compute the allowed time.

WRITING AND PRINTING

Because the eyes are in constant use during printing and writing, even though the motions controlling the time for the operation are all hand and finger motions, this subject has been included in this chapter. Another reason is that the mental activity connected with such operations is not too dissimilar to the type of "yes" and "no" decision noted for Eye Focus.

---

[8]The details of this are not shown here, but the calculation is exemplified in R.R. 102.

In every line of the passage below, each word is given a sequential number to identify a suggested order in which the eyes might see the material—some regressive saccades are thereby illustrated. It is suggested that the reader first check himself against the average time of 16.9 seconds to see that the computed TMU rate will be close for most readers.

Reading Passage	Forward Fixations	Regressive Fixations	Return Sweeps	No. of Words
2 1 3 4 5 6 7 8 9 10 11 "While it is true that a good work analyst must be	10	1	0	11
2 1 3 4 5 6 7 8 9 10 a good observer, keen to note and succinctly clas-	9	1	1	9
2 1 3 4 5 6 7 8 9 sify minute motion details, it does not follow that	8	1	1	8
1 2 3 4 5 6 7 8 all work analysts possess this necessary trait. By	8	0	1	8
1 2 3 4 5 6 7 8 9 the same token, some work analysts fail to qualify	9	0	1	9
1 2 3 4 5 6 7 8 9 for commendation of their work purely because they	9	0	1	8
1 2 3 4 5 6 7 employ slipshod analysis, circumventing good prac-	7	0	1	6
2 1 3 4 5 6 7 8 9 10 tices of work study, in the mistaken belief that a	9	1	1	9
1 2 3 4 5 6 7 8 rapid and "practical" answer will endear them in	8	0	1	8
1 2 3 4 5 6 7 8 9 10 11 the eyes of those who depend on their work for data	11	0	1	11
1 2 3 4 5 6 permitting management decisions which prove correct."	6	0	1	6
Totals	94	4	10	93

Exact: TMU reading time = 435.2 TMU fixation + 24.6 TMU interfixation
                               + 10.7 TMU return sweep = 470.5 TMU or 16.9 sec.
Normal: TMU reading time = 5.05 × 93 =   469.7 TMU or 16.9 sec.

Fig. 6.   Examples of exact and normal reading analyses.

Usually, a person setting down his thoughts or data is well aware of what he wishes to do; the control required is therefore principally limited by muscular, rather than mental requirements.

Typical examples in which a worker might need to write or print are: noting production data on time tickets, signing for tools, requisitioning items needed for his work, noting facts about his job as required by process instructions, recording data and informative details on paper work accompanying the task, initialing production tags, making minor arithmetical computations for his own needs, etc. Only when such activity is essential and necessary to the task at hand should the work analyst allow time for it in the motion pattern. Oftentimes, companies prefer to cover such activity with a miscellaneous allowance against a base standard from which writing operations are omitted; this data would not then be needed.

Consider printing and writing as a total operation. The writing implement—pen, pencil, electric etch-writer, chalk, or crayon—is a tool adequately held and controlled by the worker, and is subjected to a series of motions by the hand which can be described and timed by MTM elemental motions. The time required for obtaining the implement prior to use and laying it aside when finished can be determined by the usual Reach, Grasp, Move, etc. data as found in ordinary MTM analysis. It is only the act of writing or printing itself that is described in this section. Several factors concerning the use of writing implements should be clarified before considering the MTM method of handling these topics. Of prime importance is the fact that methods deviations between operators can be extreme. Few people write or print exactly like anyone else. It is, therefore, impossible to predetermine accurately for all persons exactly what time they will use to write certain words or the precise motions they will employ. While this is true mainly in regard to frequency with which they apply the two basic motions of which printing and writing are composed—Moves and Positions—it is still true that the motions themselves have been isolated and can be accounted for timewise where they occur in the analysis.

Skill and effort, especially as they affect legibility of the script, are important variables in writing and printing. A person's writing skill cannot be consciously varied in a short span of time, but he can vary his effort at will. Poor writing methods that usually accompany poor skill result in waste of time and lack of legibility; the reverse is also true. Because of these factors, any MTM standard on writing and printing must be properly understood as being strictly an average estimate for the situation, not a precise determination for any given individual who may write.

The MTM motions presently assigned (this is application practice, not research data) for the use of writing implements to produce either letters or numerals (specific examples of which will be given later) are as follows:

(1) POSITION. An Easy-to-Handle, Symmetrical Position with the Class of Fit appropriate to the type of script produced (P–SE) is allowed:

    (a) Each time the implement touches the paper or makes a dot, as when a period is placed at the end of a sentence, the writer dots an "i", or a "t" is crossed. For touching the paper, the implement must first be given a short M–C in accord with the regular MTM rule for Moves preceding Positions; this Move will be allowed, for example, each time the implement begins a new word after having been lifted from the paper at the completion of the preceding word.

(b) Each time an *abrupt* change in the direction of writing or printing occurs. By abrupt, the analysis implies that a change of direction appreciable enough to cause a cessation and resumption of motion has occurred. It reflects the presence of deceleration and acceleration in the writing motion. In general, the direction change must exceed 90 degrees for a Position to be valid, unless a sharp corner can be clearly seen in the writing.

(c) Each time the ending must be clearly defined or limited. This accounts for the action in accurate lettering where the implement is brought to a definite stopping point before lifting from the paper. Blueprint drafting, for example, requires sharp lettering. Ordinary writing and printing do not require this kind of Position, since the implement is merely lifted from the surface in motion at an approximate location, without a definite deceleration.

(2) MOVE. All connecting Moves in writing will logically be M–C because basically the implement progresses from Position to Position on the surface. The distance will be the average of many lengths of motion that appear in a given size and style of writing. It is from fractional—symbolized MfC—to no more than 1-inch long for ordinary-sized writing and printing. Writing large words, as for example with chalk on a blackboard, might indicate an average Move length of 2 inches or more. The time standard will be incorrect if the shorter distances are used. Moves are allowed for:

(a) Each time the implement approaches the surface. Beginning to write (after the initial M–B to bring the implement to the general area on the paper), starting of the next word, and just prior to Positions as in (1) (a) all justify counting a new Move.

(b) Each time the direction of motion changes enough to require control exceeding that for which no velocity change occurs. This will, in general, be true when the path departs more than 45 to 90 degrees from its original course. Whether the direction change is sharp or blended does not alter the need for control; this merely affects the finished appearance of the penmanship.

(c) EXCEPTION. No Move is allowed for the lifting of the implement away from the end of a writing action. This is due to the fact that the Move always allowed for starting to write as in (2) (a) above will tend to average out as the implement is brought to the next word or letter. If the Move in question is the one after the writing is finished, it will actually become a part of the Move normally used (much longer) to lay aside the writing implement, and no additional time is needed.

In terms of application, this data will be described in relation to four usual kinds of writing. Examples of each can be seen in Fig. 7, A and B. Each example is displayed in normal size and the larger specimen shows the MTM method of analyzing the Moves and Positions involved. Solid arrows show the Moves on the paper and dotted arrows show the Moves off the paper. Small circles show the positions, plain for P1SE and blackened

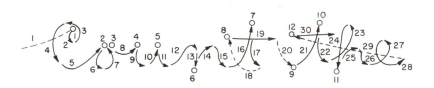

A - ORDINARY WRITING

Count the motions.

```
30 MfC  X  2.0 = 60.0 TMU
12 P1SE X  5.6 = 67.2 TMU
TOTAL TMU REQ'D. 127.2
```

B - FREEHAND PRINTING

Count the motions.

```
30 MfC  X  2.0 = 60.0 TMU
14 P1SE X  5.6 = 78.4 TMU
TOTAL TMU REQ'D. 138.4
```

Fig. 7A. Examples of writing and printing.

C-CAREFUL LETTERING

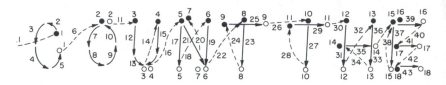

43 MfC   X   2.0 = 86.0 TMU
18 P1SE X   5.6 = 100.8 TMU
18 P2SE X 16.2 = 291.6 TMU
TOTAL TMU REQ'D. 478.4

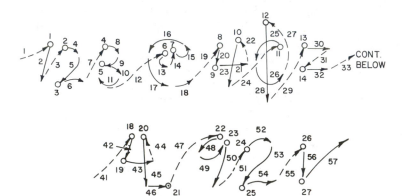

D -TYPICAL SPECIAL MARKS AND SYMBOLS

$123 @ 4¢ = \$4.92 ✓$

57 MfC   X 2.0 = 114.0 TMU
27 PISE X 5.6 = 151.2 TMU
TOTAL TMU REQ'D   265.2

Fig. 7B.   Examples of writing and printing (*cont.*)

for P2SE.   The Moves and Positions are numbered to show the sequence of motion and also the motion frequency needed to compute the MTM time value as at the right.

(1) ORDINARY WRITING. Use MfC = 2.0 TMU and P1SE = 5.6 TMU. Count the actually observed or apparent Moves and Positions in accord with the rules given above.   It will be found that this kind of writing consists of relatively few Positions and a large number of Moves.

(2) FREEHAND PRINTING.   Use MfC = 2.0 TMU and P1SE = 5.6 TMU.   This kind of lettering is not precise.   It differs from ordinary writing only because each letter is separated from all others, which increases the number of positions necessary and takes longer. Words are not continuous pencil lines as found in writing.

(3) CAREFUL (DRAFTING) LETTERING. Use MfC = 2.0 TMU unless the letters are large enough to justify longer Moves, such as found in most poster size printing. In many cases, the Position time can be covered by assigning a P2SE = 16.2 TMU at the start of each letter or separate pencil line and a P1SE = 5.6 TMU at the end of each letter or pencil line. These are "rules of thumb" sometimes used. It is obvious that Positions will tend to equal or even exceed the Moves for this kind of printing, with the attendant time allowance matching the greater time requirement. It is also obvious that the use of lettering guides or devices will require other MTM analysis than that given here.

(4) SPECIAL MARKS AND SYMBOLS.. When check marks, division signs, and other special marks and symbols appear in the script, the analyst must decide which kind of Positions and what length of moves are appropriate. No special rule can be given, except that the Moves and Positions tend to match those found in the context where the symbols and marks occur.

In MTM patterns, the usual way to record data on writing and printing is to show the symbol for Moves and Positions in the left- or right-hand columns and the number counted in the frequency column of the analysis form. Sometimes, an analyst may wish to sequentially list each motion separately for certain reasons, but this is wasteful of analysis forms and requires more analysis time.

A little practical judgment is also required. For example, the average name on a timecard might be written, say, with twenty moves and eight positions. Some names could be much less and others vastly more. With judgment, the analyst might decide to write the frequency for the longest name of the persons who will usually perform that operation, since all other people obviously will have enough time permitted for their signatures. The same would be true with regard to the number of words normally needed to perform such operations as manually marking a shipping carton with a destination. An MTM analyst who must record writing analyses will thus find frequency of occurrence, both of the content and of the individual Moves and Positions, the major problem in his efforts.

DECISION MODELS

Only one Decision Model, Fig. 8, is needed for Eye Motions. This covers only the basic EF and ET analysis, not the allied topics in this chapter—which should be reviewed separately.

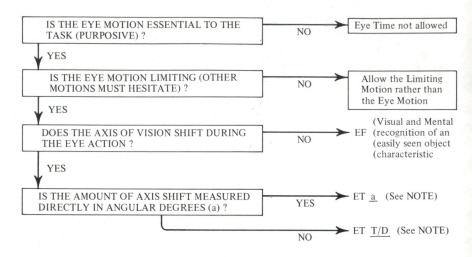

NOTE: Substitute angle a or linear ratio T/D in the Eye Travel
         symbol to complete it as measured, with these variables:

         a = angle of shift, degrees, for the axis of vision.

         T = distance between points from and to which the eye travels.

         D = perpendicular distance from the eye to the line of travel T.

Fig. 8.   Decision Model for Eye Motions.

CHAPTER 15

Body, Leg, and Foot Motions

Body, Leg, and Foot Motions are by nature the most noticeable and easily recognized human motions because their relatively high time requirement assures ample opportunity for their recognition. Similarities in these motions, plus the fact that the reader is by now conversant with MTM concepts, permit them to be discussed together in this chapter.

Body, Leg, and Foot Motions may, in general, be defined as all usual motions which the operator might perform that are not included in the definitions for the basic manual (finger, hand, arm) and eye motions in earlier chapters. Since these latter motions have been presented previously in detail, the data in this chapter will complete consideration of the motions involved in human performance.

GENERAL CONCEPTS

Body, Leg, and Foot Motions in work cycles are used to: (1) locate the hands and arms to permit their proper functioning in performing work elements; (2) orient the hands and arms with respect to the materials and/or the workplace layout involved in a work process; and (3) perform directly required elements of the work.

Since body motions ordinarily require a different kind of control and accuracy of movement as compared to manual motions, they are often termed *gross* motions and are, normally, affected by fewer variable factors. For example, most Foot and Leg Motions are made in conjunction with mechanical guides so that any control needed to operate the equipment is small.

Body, Leg, and Foot Motions will usually account for only a small proportion of a time cycle because they are often limited out by other simultaneous motions. Recall again that the time to perform any MTM motion is allowed only when it is a necessary part of the work cycle that controls

327

the performance of the work being accomplished. However, since the times for Body, Leg, and Foot Motions tend to be large in magnitude, the observer must not overlook or neglect to properly analyze them in a motion sequence—especially when they are not limited out and/or if the frequency of the body motions is unusually high or if many different types of body motions appear in the same sequence. In other words, errors in time allowance can readily occur if the Body, Leg, and Foot Motions have been missed in analysis.

The data card time values for these motions were obtained by a combination of film analysis and conventional timestudy techniques. Although future research and refinements may require a change in values, they have provided satisfactory and reliable results in thousands of applications.

In order to justify classification of human actions by MTM motion designations and, accordingly, to allow assignment of the appropriate time values, it is essential that the actions truly correspond to the defined motion classifications. The MTM engineer must be especially careful not to confuse *body motions* with the many forms of *body assistance* previously discussed. When distinct body motions occur, there are specific MTM rules to aid classification. In the case of body assists to manual motions, the time for the body movement is limited out by the time for the concurrent manual motions; and this is not *of necessity* true for body motions as defined below, although it *might* be true. Also, the principal effect of body assistance on the classification of the accompanying manual motion is to decrease the chargeable path length of the manual motion, while a body motion usually is performed for other reasons than to produce a similar effect on any simultaneous manual motion.

Proper placement of the correct symbols for Body, Leg, and Foot Motions on the study form is a distinct aid in helping avoid errors in the calculation of pattern times. Just as left- and right-handed motions are noted in their respective columns when writing symbols for manual actions, the same scheme is used for the lower limbs. Actions with the left leg and/or left foot are noted in the left-symbol column, while the right Leg and Foot Motions are listed in the right column on the form. Because the trunk lacks the attribute of handedness, however, it is the practice (as was true for eye motions) to always *show general body motions in the right-hand column.* With this convention, the analyst will be less likely to make errors in regard to combined and/or simultaneous motions that could result in incorrect times for individual motions.

BODY MOTIONS

The predominant purpose of body motions is to move the trunk or torso in certain ways or to locate it in a specified manner so as to facilitate subsequent action of the limbs; any attendant actions of the limbs are incidental to this predominant purpose. Often body motions are executed to preserve body balance during use of the limbs for other specific purposes.

Because the bulkiest, heaviest parts of the body are involved, these motions take place with less precision, speed, and control than those of the upper limbs. They are gross motions with large time values which are subject to very little variation.

In the discussion which follows, body motions are grouped into categories which depend on the kind of trunk displacement or relocation achieved. This should help clarify the reader's thinking in regard to classification and/or analysis of the body motions as contrasted to situations discussed in earlier chapters where only body assistance was involved. This grouping will further aid an analyst in deciding, during visualization, which body motions will be needed to attain certain desired results in a motion sense, thus avoiding confusion and saving analysis time in assigning the correct body motion to a pattern. The logical possibilities of trunk displacement are as follows:

1. Vertical displacement to and from the sitting position. This body action is described under Sit and Stand.
2. Lateral displacement of the trunk to either side with accompanying leg and foot action. The MTM motion accounting for such body movement is Sidestep.
3. Pivotal or rotational displacement with the aid of the legs and feet. Action of this sort is categorized as Turn Body.
4. Vertical displacement without movement of the feet, but with accompanying forward or backward action. Such motion occurs when Bend, Stoop, Arise from Bend, and Arise from Stoop are performed.
5. Vertical displacement including foot movement and concurrent forward and backward location of the trunk. This is what happens when the kneeling motions are made: Kneel on One Knee, Kneel on Both Knees, Arise from Kneel on One Knee, and Arise from Kneel on Both Knees.
6. Lateral displacement forward and backward with accompanying leg and foot action. The MTM motion accounting for this movement is Walking and is discussed as a sub-topic under Leg and Foot Motions for reasons later described.

The various actions just indicated are discussed below in detail, together with the definitions, isolation, and discussion of the variables, and the delineation of measurements connected with each kind of body motion. They comprise a rather complete coverage of the motions by which the location of the trunk or torso of the operator can be affected.

Sit and Stand

DEFINITION:[1] *SIT is the motion of lowering the body from an erect standing position directly in front of the seat and transferring the weight of the body to the seat.*

  1. *At the completion of sit, the weight of the body is supported by the seat.*
  2. *Sit does not include such motions as stepping in front of the chair or shifting the position of the chair.*

DEFINITION:[2] *STAND is the motion of transferring the weight of the body from the seat and raising the body to an erect standing position directly in front of the seat.*

  1. *Stand does not include such motions as shifting the position of the chair or stepping to the side of the chair.*

MTM Symbols: Sit   SIT   Time Values: 34.7 TMU
                 Stand   STD   43.4 TMU

Neither SIT nor STD includes moving or locating the chair or seat; any motions required for this purpose must be analyzed separately. Also, any motions needed for the operator to place himself in front of the seat prior to SIT or remove himself from near the seat after STD must be considered independently of SIT and STD. Note further that during performance of either SIT or STD the feet will remain fixed on the floor. With all these qualifications, it should be obvious that SIT and STD motions include only vertical displacement of the trunk or torso to and from a resting position on the seat.

SIT begins when the body descent starts and ends when the lean back has been completed. While it is difficult to perform any precise motions during the descent time of SIT, it is possible to Reach to and Grasp large objects such as a chair arm. More precise motions can be performed during the lean back portion of SIT. However, the SIT time will usually limit out the major portion of this latter type of action, so that the analyst needs to allow time for other motions only from the point at which the lean back is completed.

MTM research has resulted in several tentative conclusions regarding the true motion variables in SIT as outlined in an MTM Bulletin dated February, 1953. The Lean Back portion apparently requires 10.3 TMU. The Descend time depends on the arc distance traversed by the hip joint. An arc distance of 15 inches or less consumes the minimum time of 20.2 TMU, while each extra inch of travel will increase the time by 1.5 TMU; the average hip traversal during the study was 18 inches for the complete

---

[1]From "MTM Basic Specifications" (Footnote 1, Chapter 5).
[2]*Ibid.*

Descend action.   Using this data, the average time for SIT should include 20.2 + (18–15) (1.5) = 24.7 TMU for Descend, which added to the Lean Back time of 10.3 TMU gives an average time for SIT of 35.0 TMU. This value compares well with the data card time of 34.7 TMU.

STD begins with the start of the lean forward to place the weight on the feet, continues as the weight is shifted from the seat to the feet, and is completed as soon as the rising action has brought the body to an erect stance in front of the seat.   As was true for SIT, it seems to be difficult to perform any overlapping manual motions during the portion of STD in which the main relocation of the trunk takes place; this occurs during the arising part of the STD motion.   Simultaneous manual motions can be made during the initial stages of STD, however, while the Lean Forward and Shift Weight are proceeding.   In other words, overlapping motions can be made during the start of STD or during the finish of SIT motions.

Referring again to the tentative research results, STD variables were evaluated as follows.   The Lean Forward time averages 10.3 TMU. Shift Weight, which approximates an APA in nature, was found to require about 11.0 TMU.   The arc distance of hip joint travel again appears during the Arise portion of STD, with an average of 20 inches being the measuring point for time evaluation.   However, the Arise time varies depending on whether the operator completes the STD as a Type I motion or a Type II In-motion-at-the-end.   This latter case can occur when walking follows the STD and the worker does not pause before the first step.   For the Type I Arise, distances of 15 inches or less take the minimum time of 25.9 TMU, with 1.0 TMU per inch being required for extra distance.   For the Type II Arise, the time for 15 inches or less travel is 20.8 TMU; for 20 inches it is 23.1 TMU; and for 25 inches it is 25.6 TMU.

Using the data just given, the data card time for STD can be checked as follows.   The average Type I STD requires 10.3 TMU for Lean Forward, 11.0 TMU for Shift Weight, and 25.9 + (20–15)(1.0) = 30.9 TMU for Arise; this gives a total of 52.2 TMU for a Type I STD.   The average 20 inch Type II STD takes 10.3 TMU for Lean Forward, 11.0 TMU for Shift Weight, and 23.1 TMU for Arise; the total in this case is 44.4 TMU. Therefore, the data card time of 43.4 TMU that was found to be average in the original MTM research checks out as being closer to the "in motion" time found from the additional research.

In applying the MTM information to analysis of SIT or STD, the engineer can either apply the data card times or make a more detailed analysis based on the times developed in the later research.   Which course is justified depends on the relative frequency of these motions in the pattern being developed, and also on the purpose of the time standard resulting.   If SIT and STD occur rarely and the job is not too repetitive under daywork conditions, the data card time will suffice.   However, if they are very frequent in a highly repetitive condition under incentive payment limitations, it would be wise to make the more detailed analysis. Remember that even the average person is very proficient at either SIT or STD, so overlapping motions should normally be anticipated.

From the methods standpoint, it is desirable to minimize the need for SIT and STD time. This might involve requiring the operator to maintain one or the other condition during his work. Other production factors, such as personnel policies, reduction of boredom and fatigue, and psychological satisfaction of the employee have traditionally dictated that where possible the choice be left to the worker himself—and provision is generally made for him to either SIT or STD at his work and to change as it suits him. Usually a change in position under such circumstances is ignored, however each situation needs to be analyzed before deciding what is proper.

Figure 1 shows the combined Decision Model for Sit/Stand.

Sidestep

DEFINITION:[3] *SIDESTEP is a lateral motion of the body, without rotation, performed by one or two steps.*

1. The body moves directly to the side without any noticeable raising or lowering or rotation.

While there are two basic kinds of Sidestep, the fact that two directions—right or left from the operator's original body location—are possible makes a total of four combinations of Sidestep possible. The discussion of the variables below will be based on a *right* Sidestep, but it is obvious that changing of certain words would make the analysis cover left Sidesteps; this will shorten the presentation.

Analysis of Sidestep during MTM research has shown that it is seldom limiting unless the trunk has been displaced more than 12 inches to the right or left. Shorter Sidesteps occur rather frequently while the manual members perform Reaches or Moves, in which case the latter motions will *usually* be limiting. Sidesteps under 12 inches thus provide another form of body assistance, in effect, which reduces the allowable Reach or Move distance by the distance of the Sidestep. For example, an R—B to an object 20 inches to the right of the operator which is assisted by a 7-inch Sidestep will be noted as an R13B and allowed 13.7 TMU. However, if a Sidestep of less than 12 inches is necessary for *body displacement,* and no Reach or Move occurs simultaneously, the correct Sidestep distance should be noted; but the assigned time will be that for a 12-inch Sidestep, which is the minimum Sidestep time value for all distances up to and including 12 inches.

Sidestep, Case 1 (to the right) proceeds as follows:

1. The entire body weight is shifted onto the left leg and foot.

[3]From "MTM Basic Specifications" (Footnote 1, Chapter 5).

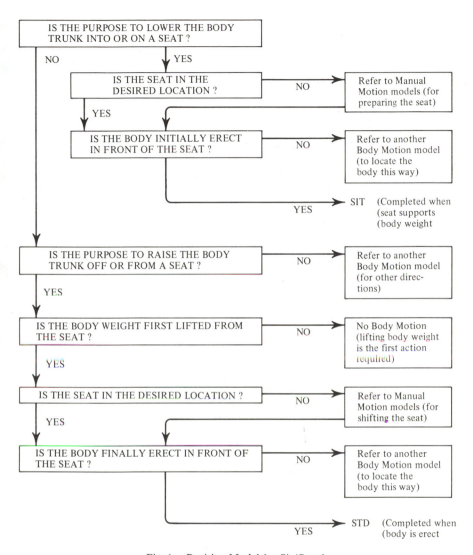

Fig. 1. Decision Model for Sit/Stand.

2. The right leg and foot are raised from the floor and moved to the right.
3. The right foot is placed on the floor.
4. About one-half of the body weight is shifted onto the right leg and foot for body balance.

Note that the left foot and leg remain in place, and that they carry all the weight of the body, while a right Case 1 Sidestep occurs. Since the feet after a Case 1 Sidestep will be more than the normal distance apart, the body of the operator may not be in the most comfortable working position. For this reason, Sidestep Case 1 may be considered as a temporary expedient to facilitate the operator's work.

Having completed the right Sidestep, Case 1 (note that the body weight will then be approximately equally distributed between the feet), the operator must use another Case 1 Sidestep to relocate either leg. However, actual operations ofteń involve the performance of a Case 1 Sidestep in which the weight shift is continued to the point where the leading foot essentially assumes all of the weight. For this occurrence it is believed that the time of performance is practically identical, since the readjustment of weight between the feet does not occur. In this situation, it is possible to bring the lagging foot alongside the leading foot at a later time by using another Case 1 sidestep which may or may not be limited out.

Sidestep, Case 2 (to the right) proceeds as follows:

1. The entire body weight is shifted onto the left leg and foot.
2. The right leg and foot are raised from the floor and moved to the right.
3. The right foot is placed on the floor.
4. The entire body weight is shifted onto the right leg and foot.
5. The left leg and foot are raised from the floor and moved to the right.
6. The left foot is placed on the floor near the right foot.
7. About one-half of the body weight is shifted onto the left leg and foot for body balance.

During a Sidestep, Case 2 the right and left lower limbs each obviously support the full body weight in turn. Note further that after completion of à Case 2 Sidestep, the final standing position achieved can be maintained for any length of time; return motions need not of necessity appear later in the motion pattern. If and when a return to the original location is needed after a Sidestep, Case 2 was employed, a full Sidestep, Case 2 in the opposite direction can be used. It would also be possible to make a Sidestep, Case 1 in the return direction temporarily and later on use another to return the right leg and foot to the starting point.

From the discussion just presented, it should be obvious that the two cases of Sidestep differ in the body action involved, the time required, and the ways the operator can return to his original starting position. Also note that when complete control of body balance is desirable after a lateral displacement, the analyst should, in fairness to the worker, allow the Case 2 Sidestep in his motion pattern.

One additional relationship between the cases exists: Case 1 Sidestep must be followed by Sidestep, Case 1 before another Sidestep in the same direction can occur.   The MTM data card uses the term leading leg for the first leg moved and the term lagging leg for the movement of the other leg.   To move in the return direction from a Case 1 Sidestep, the first step can be only another Case 1 Sidestep.   After a Case 2 Sidestep is completed, another Sidestep of either case can be made in either direction.   Therefore, only one step at a time can be taken with a Case 1 Sidestep, whereas any number of sidewise steps can be made successively as long as only Case 2 Sidestep is employed.

Nothing has been said yet as to how the distance of Sidestep is measured. The method of measuring should, of course, correspond to that used in the MTM research on which the data card times are based if the time allowance is to be valid.   The data card times are based on considering the lateral distance moved as the *displacement of the trunk,* not the distance moved by the legs and/or feet.   Hence, present MTM application rules state that the sidewise movement should be measured at the centerline of the trunk—the belt buckle or the base of the spine would be convenient measure points, as they correctly reflect centerline displacement.   It would also be possible, of course, to measure from any other point on the trunk.

It has been recognized that manual motions can be combined with Sidesteps.   Any such concurrent motions will be symbolized, limited out (where proper), and assigned times in the same manner as was discussed for other combined motions in the Turn chapter.   Since the legs and feet are involved in Sidesteps, the symbol should be written in the left- or right-hand column on the study form as appropriate.   Sidesteps to the right are shown in the right-hand column, those to the left in the left-hand column.   If concurrent manual motions with either of the hands occur, they will likewise be shown in the proper column along with the Sidestep.   Therefore, these concurrent motions can be either combined or simultaneous motions as discussed in the Motion Combinations chapter and the conventions on writing of symbols would be followed accordingly.

Sidesteps are symbolized as follows:
    1st —Numeral(s) for the paces taken (if more than one)
    2nd—Capital letters SS for Sidestep
    3rd—Numerals for the inches of travel distance (centerline)
    4th —C1 or C2 to denote the case of Sidestep performed

Examples of the complete symbols are:
    SS16C1 which shows a Case 1 Sidestep involving 16 inches trunk travel
    SS20C2 illustrating a 20 inch Sidestep, including both legs acting
    2SS18C2 symbolizing two Sidesteps, Case 2, each covering 18 inches of sidewise motion

As explained previously, it is seldom necessary to allow time for a Sidestep less than 12 inches long, but assign the 12-inch time when shorter Sidesteps must be allowed. At 12 inches the performance times for the two cases are:

SS12C1 — 17.0 TMU   and   SS12C2 — 34.1 TMU

For each extra inch of travel over 12 inches, the additional time allowance will be:

SS–C1 — 0.6 TMU per inch   and   SS–C2 — 1.1 TMU per inch

Computation of the times for the three illustrative examples given will demonstrate the use of this data:

$$\text{SS16C1} = 17.0 + (16 - 12)\,(.6) = 17.0 + 4\,(.6)$$
$$= 17.0 + 2.4 = 19.4 \text{ TMU}$$

$$\text{SS20C2} = 34.1 + (20 - 12)\,(1.1) = 34.1 + 8\,(1.1)$$
$$= 34.1 + 8.8 = 42.9 \text{ TMU}$$

$$\text{2SS18C2} = (2)\,(34.1) + (2)\,(6)\,(1.1) = 68.2 + 13.2$$
$$= 81.4 \text{ TMU}$$

Also, where x is the inches of centerline trunk displacement with a minimum value of 12 inches, net formulas for Sidestep would be:

$$\text{SS}\,x\,\text{C1, TMU} = 0.6x + 9.8, \text{ and}$$
$$\text{SS}\,x\,\text{C2, TMU} = 1.1x + 20.9.$$

Figure 2 shows the Decision Model for Sidestep.

Turn Body

DEFINITION:[4] *TURN BODY is a rotational movement of the body performed by one or two steps.*

1. In performing the turn body, the steps are made with the feet turning in the same direction as the body.

There are two basic cases of Turn Body, as was true for Sidestep, with the additional possibility of either clockwise or counter-clockwise rotation. Only *clockwise* Turn Body need be used in this discussion and only a change in directional wording is necessary for describing counter-clockwise Turn Body.

Simplification of the Turn Body discussion results from knowing at the start that MTM research showed it to be seldom limiting when the body was turned less than 45 degrees. Lower degree Turn Body action often seen during concurrent limiting manual action can also be regarded as equivalent in effect to the type of body assistance earlier detailed as radial assistance; the rules for that kind of assistance should suffice to enable the

[4]From "MTM Basic Specifications" (Footnote 1, Chapter 5).

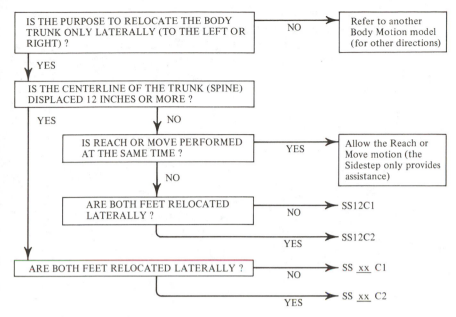

NOTE: Substitute the distance of lateral spine movement in inches,
         xx , to complete the Sidestep symbol.

Fig. 2.   Decision Model for Sidestep.

MTM analyst to assign the correct net distance for Reach or Move aided by any Turn Body less than 45 degrees.

The reader should not confuse Turn Body, however, with rotation of the shoulders to provide radial assistance, since the latter action occurs with the feet kept stationary   To say that Turn Body under 45 degrees often has the net effect of radial assistance does not imply that the body actions are identical.

The true variables in Turn Body are the degree of rotation and the state of body balance resulting.   The balance will depend on whether one or both feet are moved as the Turn Body is performed, which is accounted for by the Case of the motion.

Turn Body, Case 1 (clockwise in direction) proceeds as follows:

1. The entire body weight is shifted onto the left leg and foot.
2. The right leg and foot are raised from the floor and moved angularly clockwise.
3. The right foot is placed on the floor.
4. About one-half of the body weight is shifted onto the right leg and foot for body balance.

The left leg and foot remain in place at all times during a clockwise Case 1

Turn Body. Since after completion of Turn Body, Case 1 the working position will not be comfortable for long periods of time, the motion tends to be a temporary means of facilitating the worker's task. To bring the feet together again will normally require another Case 1 Turn Body; the rules will determine whether it is limited or limiting. Another possibility following Case 1 Turn Body is to start walking with the lagging foot taking the first pace.

Turn Body, Case 2 (clockwise in direction) proceeds as follows:

1. The entire body weight is shifted onto the left leg and foot.
2. The right leg and foot are raised from the floor and moved angularly clockwise.
3. The right foot is placed on the floor.
4. The entire body weight is shifted onto the right leg and foot.
5. The left leg and foot are raised from the floor and moved angularly clockwise toward the right leg and foot.
6. The left foot is placed on the floor near the right foot.
7. About one-half of the body weight is shifted onto the left leg and foot for body balance.

The right and left lower limbs thus alternately fully support the body weight, and the degree of body balance at completion of the Case 2 Turn Body is normal—permitting even the most difficult manual motions. A full Case 2 Turn Body in the opposite direction will restore the original location of the worker. Otherwise a Turn Body, Case 1 in the return direction, followed by a subsequent motion in the same direction, will achieve the original body position. The differences in Case 1 and 2 should now be apparent. Fairness to the operator will also dictate assignment of a Case 2 Turn Body when complete control of body balance must exist after the rotation is finished.

The performance of either case of body turn implies that the degree of rotation will be between 45 and 90 degrees. There will be few occasions for a worker to Turn Body more than 180 degrees, since this amount of rotation could be achieved by a Turn Body of smaller angle in the opposite direction. To cover the range from 90 degrees to 180 degrees, several MTM rules—which are merely logical extensions of the actions for the two cases above—delineate how to assign the motions and decide limiting action. When the turn is over 90 degrees, the first Turn Body will be a complete Case 2. If the remaining rotation desired does not exceed 45 degrees, the idea earlier discussed of manual motions limiting further action will hold. When the remaining angle exceeds 45 degrees, however, an additional Turn Body of either case will be performed; the case will depend on the desired state of body balance. This covers the range of motion possibility for Turn Body.

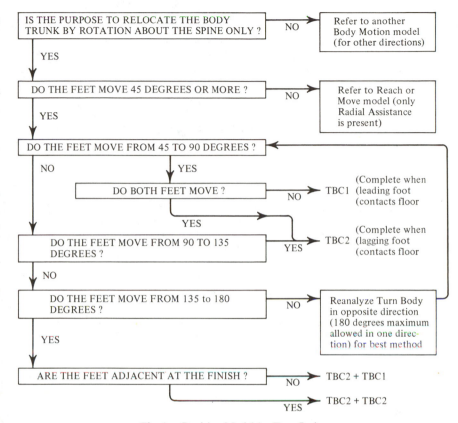

Fig. 3.   Decision Model for Turn Body.

Distance has no bearing on the measurement—the body pivots about the spine and is not essentially displaced by the rotation. By reference to this point it is relatively easy to analyze whether a Turn Body or some other motion occurred. The analyst needs merely to estimate whether the 45 degree breakpoints discussed have been swept and for case determination to observe whether one foot or both feet were used. The spine is the correct pivot point about which to gage the number of degrees, although the difference between the initial and final locations of the leading foot also provides an indication.

The comments on concurrent manual motions—either combined or simultaneous—made under Sidestep apply equally well to Turn Body. This is also true for the conventions regarding the columns in which the symbols belong and whether the manual action can be considered as combined or simultaneous.

Turn Body symbols consist simply of:

1st—TB in capital for Turn Body

2nd—C1 or C2 for the case, depending on the rotation and leg action.

The only possible examples, with their time constants are:

TBC1   18.6 TMU   and   TBC2   37.2 TMU

Turn Body analysis by Decision Model will use Fig. 3.

Bend, Stoop, and Arise

DEFINITION:[5] *BEND is the motion of lowering the body in a forward arc, from a standing position, so that the hands can reach to or below the level of the knees.*

1. *Bend is performed with little or no rotation of the body or flexing of the knees.*
2. *Bend is controlled by the back muscles and leg muscles.*

DEFINITION:[6] *STOOP is the motion of lowering the body in a forward arc, from a standing position, so that the hands can reach to the floor.*

1. *Stoop is performed by bowing forward at the hips and at the same time lowering the entire body by bending at the knees.*
2. *Stoop lowers the hands further than bend through a simultaneous "bend" and knee bend.*

DEFINITION:[7] *ARISE FROM BEND is the motion of returning the body from a bend to an erect standing position.*

DEFINITION:[8] *ARISE FROM STOOP is the motion of returning the body from stoop to an erect standing position.*

MTM Symbols:	Bend	B	Time Values:	29.0 TMU
	Stoop	S		29.0 TMU
	Arise from Bend	AB		31.9 TMU
	Arise from Stoop	AS		31.9 TMU

These four motions are characterized by vertical displacement of the trunk with the feet remaining fixed on the floor at all times during the motions. The usual purpose of these body motions is to locate the manual members or the head in a more useful position, using conscious control of the back and leg muscles for balance of the body during and after lowering or before and during rising.

The principal difference between these motions and body assistance to Reach or Move is that, for the assistance referred to, only the upper trunk to about the waist level is involved. Assistance motion is not consciously done, although the performer can notice it if he is paying particular attention. The body motions discussed in this section must, however, be

[5]From "MTM Basic Specifications" (Footnote 1, Chapter 5).
[6]*Ibid.*
[7]*Ibid.*
[8]*Ibid.*

consciously controlled because some of the larger body muscles are used and just maintaining balance is enough to require attention.

Bend begins as the shoulders start their descent forward, normally proceeds while the arms swing naturally forward and down, and is technically completed as soon as the hands could or do pass the knee level. While it is difficult for preceding motions to blend into a Bend, the hands can smoothly continue a Reach or Move at the end of Bend—in fact, they can be considered as reaching or moving all during the Bend. All the hand motion prior to the time the trunk stops moving, however, would be limited out by the bending action which controls the time. For these reasons, it is common to see Type II Reaches and Moves follow Bend, with the hands being in motion at the beginning of the separately shown Reach or Move. Note that the slight amount of bending action present during body assist to Reach or Move differs mainly in the amount of trunk displacement and the degree of conscious control as compared to Bend.

Stoop can be conceived as a Bend with a simultaneous leg action that brings the trunk and hands lower. The leg action is completely limited out, which explains why the time values are equal for both motions. During Stoop, the thigh muscles share the work load with the back muscles to a greater degree than is true for Bend. The worker would naturally, therefore, find Stoop more fatiguing even though it would require no greater time. This is a good example of why fatigue allowances are used.

Arise from Bend or Stoop starts as soon as the ascent is initiated and is finished as soon as the body has been brought to a fully erect standing position. In opposition to what was true for the downward motion, preceding motions can blend into Arise; while following motions tend to begin only after the Arise is completed. The time to Arise is about 10 per cent greater than descent because the body balance is achieved in a different manner and the pull of gravity must be overcome. Since the straightening of the legs required by Arise from Stoop is concurrent with the body action, the time is equal to that for Arise from Bend.

Handedness has no connection to these body motions, so the MTM symbols are always indicated in the right-hand column of the study form. Consistency in this convention will help avoid analysis errors and the assigning of incorrect times.

The Decision Model for these motions is combined with that for Kneel in Fig. 4.

Kneel and Arise

DEFINITION:[9] *KNEEL ON ONE KNEE is the motion of lowering the body from erect standing position by shifting one foot forward or backward and lowering one knee to the floor.*

[9]From "MTM Basic Specifications" (Footnote 1, Chapter 5).

1. At the completion of kneel on one knee, the weight of the body is supported on one knee and one foot with the other foot helping to maintain balance.

DEFINITION:[10] *KNEEL ON BOTH KNEES is the motion of lowering the body from erect standing position by shifting one foot forward or backward, lowering one knee to the floor, and placing the other knee adjacent to it.*

1. At the completion of kneel on both knees, the body is supported by both knees with the feet helping to maintain balance.

DEFINITION:[11] *ARISE FROM KNEEL ON ONE KNEE is the motion of returning the body from kneel on one knee to an erect standing position.*

DEFINITION:[12] *ARISE FROM KNEEL ON BOTH KNEES is the motion of returning the body from kneel on both knees to an erect standing position.*

MTM Symbols:

Kneel on One Knee	KOK	Time Values:	29.0 TMU
Kneel on Both Knees	KBK		69.4 TMU
Arise from Kneel on One Knee	AKOK		31.9 TMU
Arise from Kneel on Both Knees	AKBK		76.7 TMU

The general purpose of these motions is the same as that stated for Bend and Stoop. The principal difference, motionwise, is the concurrent leg and foot action which provides a final body position that is firmly balanced and may be maintained for a greater length of time. Note that the final body position for Bend or Stoop, while balanced, is much less steady and cannot be held for long periods. Maintaining Bend and Stoop is also more fatiguing than is true for kneeling. On the other hand, no reason would exist for kneeling motions unless the work requirements justified allowing the operator to expend more time than he would for Bend or Stoop. As for all other MTM motions, intelligence, judgment, and understanding by the analyst will be needed; it would be prohibitive of clarity and practicality for any predetermined time system if every minute choice were rigidly bound into a plethora of rules and exceptions.

KOK is performed with the following progression: the weight is shifted off the leg and foot to be lowered; the trunk and knee descend simultaneously while the supporting leg bends at the knee and its calf and thigh muscles carry the load; the lowered knee is placed near the stationary foot on the floor; most of the weight is shifted to the knee; and body balance is regained with the other leg and foot exerting a stabilizing force on the body through a horizontal thigh. Most people also tend to rest their elbow

[10]From "MTM Basic Specifications" (Footnote 1, Chapter 5).
[11]*Ibid.*
[12]*Ibid.*

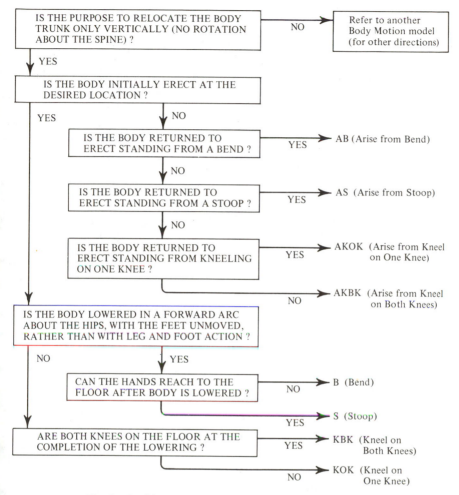

Fig. 4.   Decision Model for Bend/Stoop/Kneel (and Arise).

near the knee on the level thigh as a further assurance of adequate balance. All of this muscular action is obviously not simple, as most body members are involved in some manner; this makes difficult the performance of other purposeful motions during KOK without great skill or much practice.   The KOK, when achieved, is almost always the limiting motion.

KBK may actually be performed with two distinct sequences.   Completion of a KOK followed by a separate leg action to place the other knee on the floor; shifting of an equal load to this knee; and regaining of body balance comprises one method.   The alternate method includes a downward trunk movement without shifting the position of the feet; lowering the trunk until almost sitting on the heels of both feet; swinging the trunk forward to bring both knees to the floor; and regaining of body balance. Because the motions are less blended and balance must be shifted and reassumed an additional time, it is evident why KBK requires more time than KOK. The muscular control requirements are also greater.

With an understanding of the KOK and KBK motions just outlined, the reader can readily extend the remarks concerning Arise under Bend and Stoop to cover the motions known as AKOK and AKBK. The body motion which consumes the greatest single motion time is AKBK.

To record the MTM symbols and other motion conventions, the appropriate right- or left-hand column (based on the knee lowered) should be used for KOK and AKOK; to show KBK and AKBK, however, only the right-hand column can properly be used.

One Decision Model, see Fig. 4, covers Bend, Stoop, Kneel, and the four Arise motions connected with them. This emphasizes their basic similarity in downward or upward—vertical—body displacement.

LEG AND FOOT MOTIONS

In the performance of various body motions previously discussed, actions of the legs and feet were included with the total motion. Such usage of the lower limbs as was cited in those cases was incidental to the predominant purpose of effecting displacement of the trunk in a given manner. The legs and feet are, however, often used in an independent manner that does not essentially involve the body trunk; the predominant purpose of employing the lower limbs in this manner is to perform effective labor or to locate the legs or feet as an essential part of the work cycle. Such separate action of the lower limbs is analyzed as Leg and Foot Motion, with attendant motion analysis and time values as discussed in this section.

As a matter of convenience, Walking is also discussed in this section. It is the remaining type of major body displacement—forward and backward—not discussed under body motions earlier. The Walking data fills in this gap concerning actions directly involving the trunk or torso of a worker during productive labor. The basic reason for deferring it until now is the nature of the motions by which Walking is accomplished. The *purpose* of Walking is trunk displacement; but the *mechanics* by which it is accomplished depend almost entirely on the action of the lower limbs.

Remember that Leg and Foot Motions, as defined here, do not include any purposeful motion of the trunk in any direction—the lower limbs are used for other purposes under this motion category. Walking, however, includes displacement of the body forward and backward as an accompanying effect to displacement of the lower limbs in those directions. In other words, when the legs and feet displace the body in other than the forward and backward directions, some other category of body motion rather than Walking is being performed.

Leg Motion

DEFINITION:[13] *LEG MOTION is the movement of the leg in any direction with the knee or the hip as the pivot, where the predominant purpose is to move the foot rather than the body.*

1. *Leg motion may be made while either sitting or standing.*
2. *Leg motion made while standing usually has the hip as the major pivoting point.*
3. *Leg motion made while sitting usually has the knee as the major pivoting point.*

Leg Motion as defined above—to locate the leg and/or foot rather than facilitate body action—is frequently performed in order to allow the leg or the foot to accomplish a task directly. The foot can be used, following Leg Motion to get it into position, to: operate pedals, shove objects along the floor, brush objects aside (if not too heavy), and do similar tasks. The leg can also perform useful labor after a Leg Motion has placed it in contact with an object to be activated by having it: make a further Leg Motion to operate the knee lever of an electric sewing machine or rotary ironer; operate the water valve of a surgical washbasin by a pressure application (AP) and another short Leg Motion; shove open swinging doors; or to perform holding or restraining action on objects that would move into the worker's way without the leg action. In general, it is good methods practice to employ the legs and feet as much as possible when they free the manual members for more productive tasks or even make some job sequences with the hands possible at all.

A standing operator can pivot the leg in any desired direction—forward, backward, or sideways—and usually does so about the hip joint. The leg can also be swiveled about the knee while standing. When sitting, however, most leg action results from pivoting about the knee. For example, Leg Motions can be used to roll a chair with casters while sitting; this probably will require some AP's. Large movements of the leg while sitting, as when crossing one's legs, will ordinarily display pivoting about the hip as well.

In any of the cases cited, the measuring point for Leg Motion is the ankle or instep of the leg moved. This distance is either scaled or estimated to the nearest inch. It has been found that a constant minimum time of 7.1 TMU will be required for all Leg Motions up to and including 6 inches distance. Each extra 1 inch of travel will require 1.2 TMU additional time. Thus, a net formula for Leg Motion is: TMU = $1.2x - 0.1$, where

[13]From "MTM Basic Specifications" (Footnote 1, Chapter 5).

x is the ankle movement, in inches, with a minimum value of 6 inches.

The MTM symbol for Leg Motion consists of the capital letters LM (the initials of the motion name) followed by a numeral that shows the travel distance. Some analysts always record a 6 for all distances equal to or below 6 inches, although this is not an official MTM rule—they use it to remind themselves of the minimum time.

An example of the symbol and time calculation is: LM14 designating a Leg Motion of 14 inches that requires 7.1 + (14–6) (1.2) or 7.1 + 9.6 = 16.7 TMU to perform. The symbol should be shown in the left- or right-hand column of the methods analysis form as appropriate for the leg acting.

LM is easy to recognize and measure, but care must be taken to check whether it is limited out by concurrent manual and/or eye motions. Many of the LM performed will be part of a combined motion with the work of the hand; the usual rules for combined motions will govern the time allowance for such situations.

Minor LM are sometimes made that have no bearing on the allowed time; such cases are often not even written into the motion pattern. They are essentially balance or reflex actions that are not consciously made as required by the LM definition. In using the LM data, therefore, an analyst may often include or omit it in the pattern on the basis of whether it is an essential action to task achievement.

The net analysis for Leg Motion is shown in Fig. 5, in a Decision Model including Foot Motion.

Foot Motion

DEFINITION:[14] *FOOT MOTION is the movement of the ball of the foot up or down with the heel or the instep serving as a fulcrum.*

1. Motion of the toes of the foot generally is 2″ to 4″.

2. Foot motion, with pressure, includes a hesitation for the application of force directly by the foot or a transfer of body weight in conjunction with the foot motion.

MTM Symbols:·

Foot Motion	FM	Time Values:	8.5 TMU
Foot Motion with Heavy Pressure by the leg muscles	FMP		19.1 TMU

FM or FMP usually occurs in activating pedals, levers, and switches. They are more easily executed when the operator is seated. Whether seated or standing however, the operator will be fatigued by excessive repetition of FMP. Its use to relieve the hands of possible tasks should, therefore, be judicious. The symbols for Foot Motions obviously should be shown in the left- or right-hand column of the analysis form to show which foot is acting.

---

[14]From "MTM Basic Specifications" (Footnote 1, Chapter 5).

The direction of foot motion may be either vertically up or down or it can occur sideways.   Note that the added motion and time when pressure is used is an APA taking 10.6 TMU.

The distance of movement, in any case not involving LM as well, will be small—from 2 to 4 inches long; it is therefore only a minor variable and does not require measurement and notation on the MTM analysis. The two time constants given are based on the maximum condition of motion and therefore no analysis of the motion except for recognition of its occurrence is required.

The Decision Model in Fig. 5 covers both Leg Motion and Foot Motion.

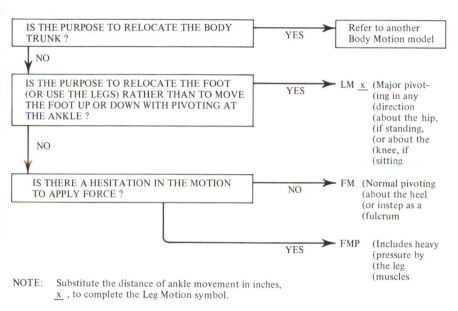

NOTE: Substitute the distance of ankle movement in inches,
 x , to complete the Leg Motion symbol.

Fig. 5.   Decision Model for Leg/Foot Motion.

Walking

DEFINITION:[15] *WALKING is a forward or backward movement of the body performed by alternate steps.*

1. Walking does not include stepping to the side or turning around.

According to this definition, the act of Walking is essentially an alternate series of LM by each leg and foot for the purpose of conveying the trunk to a new location.   The direction of travel is restricted, however, to forward and backward only; any lateral or rotational movement of the trunk will be done with either SS or TB as previously discussed.   While

[15]From "MTM Basic Specifications" (Footnote 1, Chapter 5).

Walking seems to be a simple act (probably because almost everyone is highly experienced and skilled at it) the relatively large number of variables associated with it in industrial situations will be included in this section. Remember while considering this data, however, that the MTM motion and time values basically apply to *purposeful* Walking during work sequences—it would be improper to use it to compute the time for a leisurely stroll. The MTM Walking data may not compare too well to walking rates generally accepted for other purposes.

Besides the data given in the MTM procedure, variously published rates that have been well established for purposeful walking (or marching) are given and these tend to corroborate the results reported from rather extensive MTM research based mainly on time study.

1. AMERICAN ARMY. The parade standard of 128 paces per minute with a 30 inch stride can be converted to *3.64* miles per hour, effort being noticeable.
2. BRITISH ARMY. An hour's march is defined to include the coverage of 3.00 miles in 50 minutes followed by 10 minutes rest to overcome fatigue prior to continuation of the march. The steady marching rate would be based on the actual time to cover the specified distance, or $3.00/50 = 0.06$ miles per minute. The hourly rate would then be $(60) (.06) = 3.60$ miles per hour. The effort, again, would be at an average or higher level.
3. A rather exhaustive investigation by United States Steel Corporation resulted in a standard of *3.57* miles per hour.
4. Walking standards from other sources range from *3.00* to *4.05* miles per hour. The conditions specified were not all equal, however.
5. MTM RESEARCH. As described in the MTM text by the originators, time study under controlled conditions with no loads being carried showed an average velocity of *3.57* miles per hour with a ·34-inch pace. This will correspond to rates of *5.3 TMU per foot* or *15.0 TMU per pace* for no load walking over an unobstructed path; this is the MTM walking data.

In addition to the data just given, other tentative time values for carrying loads and accounting for the hindrance offered by obstructions are part of the MTM data.

Maynard, Stegemerten, and Schwab discuss in their text the effect of Walking variables and provide graphs and tables to show this variation. Since the MTM analyst ordinarily should choose an average operator for observation, however, this data is informative rather than essential in nature.

It is instructive, on the other hand, to record here some of the variables and how they influence the time for Walking. Men tend to walk faster than women—particularly when carrying loads. Younger and older walkers apparently walk slower than those between 17 and 24 years. The ideal weight for Walking is 170 pounds. Physical condition and health will affect Walking ability, since they will be reflected in both the pace and effort. Heavy shoes, or high heels, will also slow down the walker. A smooth, hard, level surface will result in better velocity than will be true with a less desirable foothold; this is further discussed under obstructed Walking below. Objects that interfere with a clear path result in extra steps or more caution to avoid stumbling and thus slow the Walking rate. Effort will affect not only the time per pace, but also the length of pace employed. Low effort reduces the length of stride and increases Walking time per foot; high effort lowers the per foot Walking time because of an increased stride.

A *tentative* time value of 17.0 TMU per pace has been used with success for obstructed Walking. Obstructions may be defined as physical objects in or near the usual pathway or as surface conditions differing from those in the original research just noted above. Walking on sand, loose dirt, oily or slippery floors, ice and snow, railroad ties, and similar impeding objects are all instances of obstructed Walking that will slow the rate. Besides reducing the pace or causing increased caution, obstructions result in an increased length of path to get around objects in the way. For such cases, the 17.0 TMU value may be used cautiously. Even though still a tentative value, it was added to the U.S.A./Canada MTM-1 data card in the 1972 revision.

The *tentative* 17.0 TMU per pace value also is taken to cover Walking with loads over 50 pounds (with the pace also shortened) and the pushing of loaded hand trucks or heavy objects on wheeled dollies.

The MTM Walking standard is stated in connection with a definite length of pace. Short striding walkers tend to exert greater effort, so that their per pace Walking rate tends to equal that of long striders in spite of a shorter pace. Operators with other than average build normally use a pace length appreciably different than the average and therefore require a different number of paces to cover equal distances as compared to average operators. Also, operators using other than average effort tend to alter their pace length and therefore the number of paces required to travel a given distance. Note in all of these cases that the MTM time per pace is essentially constant, for a given Walking condition. The number of paces used, however, depends on the pace length. Backward Walking seems to require the same time per pace but the pace length is reduced from what it would be for forward steps. Similarly, heavier loads seem to reduce the pace length with little change in the per pace rate.

All of the preceding data and discussion can be combined into a practical table (see Table 1) which applies to Walking by average operators. While this table includes several tentative time values, it also includes the official MTM-1 Data Card information as indicated.

Walking up and down stairs presents another analysis problem. Essentially, the 15.0 TMU per pace value is applied and the number of paces must be found. In addition, such Walking should be examined for the use of extra leg motions and even, at times, the use of AP by the feet. A stairway consists of vertical pieces called risers and horizontal members called treads or steps; when these are of average dimensional proportions, one pace per tread is the usual requirement. If a long flight is climbed frequently during a work sequence, however, an extra fatigue allowance will be in order. Risers over 12 inches high, or extra deep treads, or whether the treads are enclosed rather than open present further need for

Table I. Walking Data for Average Operators

WALKING CONDITION		UNOBSTRUCTED		OBSTRUCTED	
LOAD, pounds	PACE, inches	TMU per Pace	TMU per Foot	TMU per Pace	TMU per Foot
0– 5 incl.	34	15.0*	5.3*	17.0*	6.0
Over 5–35 incl.	30	15.0	6.0	17.0	6.8
Over 35–50 incl.	24	15.0	7.5	17.0	8.5
Over 50	24	17.0*	8.5	17.0	8.5

* Official MTM-1 Data Card information.

practical analysis that probably will be best resolved by using time study.

The conventional MTM symbols should always be shown in the right-hand column of the analysis form. They are as follows:

1st. A capital W for Walking
2nd. A numeral for either the paces taken or the distance in feet traveled
3rd. The capital letters for paces—P—or for feet—FT, as appropriate
4th. A capital O when obstructed conditions are present

Examples, including the time values are:

1. W5P (Walking 5 paces)   75.0 TMU
2. W20FT (Walking 20 feet)   106.0 TMU
3. W5PO (Walking 5 obstructed paces)   85.0 TMU

The MTM engineer must choose between analyzing Walking by the per pace or per foot methods, since times are available for both symbols. Factors influencing the choice are indicated in the following:

1. PER PACE. The time value per pace is essentially constant for given Walking conditions. It applies even for the somewhat shortened paces at the start and finish of Walking. A pace should be counted

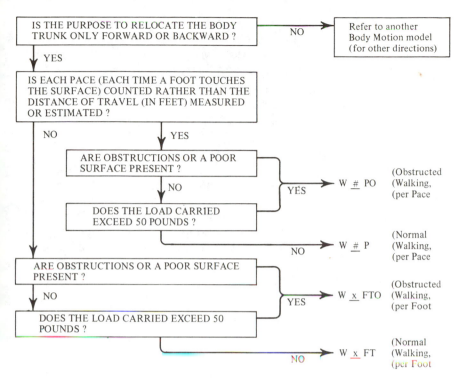

NOTE: Substitute either the number of paces taken, #, or the
linear travel in feet, x, to complete the Walking symbol.

Fig. 6.   Decision Model for Walking.

each time a foot strikes the floor, and its length will be from the
initial to the final location as found on a convenient measuring point
such as the ankle or heel.   It is easier to record paces, rather than
measure the path, when single paces or short distances are walked.
With the originators' data, it is possible to adjust the allowed number
of paces from these actually done by an observed off-average operator
or one who does not walk with average effort.   For most Walking
conditions, workers tend to use the average pace.   When they do not,
a pace correction can be made with the following formula:

$$\frac{\text{Actual Paces} \times \text{Actual Stride, inches}}{34 \text{ inches Standard Stride}} = \text{Standard Paces}$$

The analyst can pace the distance himself if necessary.   When writing a visualized analysis, however, the per pace method is harder to
use than the per foot method.

2. PER FOOT.   When analyzing motions based on layouts, it is easier to

determine the number of feet covered than the number of paces. It is also easier to specify distance to walk than the paces to take. The per foot Walking rate is, however, not really constant because it does not include prorated start and stop time. Visualizing is easier done when it is based on the feet to be traveled rather than on a per pace basis. The distance specified or allowed should be the actually necessary path, not a straight line measurement. This makes the per foot method harder when obstructed conditions are encountered, as the true path length is harder to determine.

While the MTM data for Walking is far from complete, the discussion given here will permit establishment of usable times for the average Walking situation. More research may be desirable. A final point is that the Walking data should not be applied where unusual conditions prevail or when the operator does not fully control his Walking rate. Such *process time* indicators suggest that stopwatch time study rather than MTM data should be used.

The Decision Model for Walking appears as Fig. 6.

DATA CARD INFORMATION

MTM-1 time data has been given with the discussions of Body, Leg, and Foot Motions. In practice the analyst will ordinarily use the data as compiled on the official MTM-1 card as shown in Table 2. The extended

Table 2. Body, Leg, and Foot Motions
(MTM-1 Data Card—Table IX—Body, Leg, and Foot Motions)

TYPE		SYMBOL	TMU	DISTANCE	DESCRIPTION
LEG–FOOT MOTION		FM	8.5	To 4″	Hinged at ankle.
		FMP	19.1	To 4″	With heavy pressure.
		LM__	7.1	To 6″	Hinged at knee or hip in any direction.
			1.2	Ea. add'l inch	
HORIZONTAL MOTION	SIDE STEP	SS__C1	*	<12″	Use Reach or Move time when less
			17.0	12″	than 12″. Complete when leading
			0.6	Ea. add'l inch	leg contacts floor.
		SS__C2	34.1	12″	Lagging leg must contact floor before
			1.1	Ea. add'l inch	next motion can be made.
	TURN BODY	TBC1	18.6	——	Complete when leading leg contacts floor.
		TBC2	37.2	——	Lagging leg must contact floor before next motion can be made
	WALK	W__FT	5.3	Per Foot	Unobstructed.
		W__P	15.0	Per Pace	Unobstructed.
		W__PO	17.0	Per Pace	When obstructed or with weight.
VERTICAL MOTION		SIT	34.7	——	From standing position.
		STD	43.4	——	From sitting position.
		B,S,KOK	29.0	——	Bend, Stoop, Kneel on One Knee.
		AB,AS,AKOK	31.9	——	Arise from Bend, Stoop, Kneel on One Knee
		KBK	69.4	——	Kneel on Both Knees.
		AKBK	76.7	——	Arise from Kneel on Both Knees.

data and special values previously discussed in the text normally will be reserved for special demands and are, of course, less convenient than the data on the card.

It will be noted that the card has been arranged in condensed fashion, primarily by grouping motions that require the same time values and share the same predominant direction of motion. Note also that tentative time values, such as obstructed and loaded Walking times, have been added to the 1972 revision of the card. This does not preclude data changes which might result from clarifying research.

GENERAL APPLICATION RULES

Rules of application pertinent to each of the Body, Leg, and Foot Motions have been given in the discussion of each. Most of these rules have been formulated by considering the facts that: (1) The body motions are gross movements involving the heavier muscles of the worker; (2) the times for body motions are usually longer than for any manual motions that can be made concurrent with body motions; and (3) a distinct manner of using the body occurs when heavier loads are handled by the worker.

Regarding the first point, use of the heaviest muscles obviously will require the greatest expenditure of energy per unit of effective output. This will result in greater fatigue per unit of working time. No reasonable management expects their employees normally to incur more than a nominal amount of fatigue during the working hours; continued over-fatigue can only result in a net loss of efficiency for the company itself with attendant higher costs. The day when it was believed profitable to unfairly overburden workers has passed; in addition the actions of those who might want to be unfair are now restricted in this matter by years of labor negotiation and social legislation. The logical outcome of the foregoing facts is that either excessive use of body motions is avoided through providing adequate material handling and other worker aids, or else some means for recovery from fatigue is permitted for jobs in which a large amount of body motion is unavoidable. The latter takes the form of rest periods, special cycle allowances at stated intervals, and permission for the worker to refresh himself during the working day.

In view of the foregoing, the methods analyst has a clear responsibility to specify methods, where possible, that minimize use of the body motions that are especially fatiguing. In this way he will implement more effective work patterns as well as meet the moral and legal requirements assigned to management. Armed with knowledge concerning the body members acting and the required time for each body motion, he is ideally equipped to achieve these aims. He can, sometimes, devise methods which employ less fatiguing body motions to attain equal work results. An example of this is provision of a roller conveyor to relieve the worker of the Bends, Stoops, Kneeling, and Walking usually found when moving heavy tote boxes.

The use of pedals and leg-actuated buttons in place of hand operated mechanisms is another example of a fatigue reducing measure; however, as pointed out previously, they must be used judiciously. Heavy pressures or light short movements can be applied easily by the foot—often easier than by hand.

The second point mentioned alludes to limiting body motions and over-lapping body motions that could or might be made concurrently with manual motions. The MTM analyst either should not require the use of such combinations of motions or else be sure that he has fairly allowed time for the true occurrences he finds. Short Reaches and Moves are almost always limited out by body motions; although it is possible for them to blend into body motions or begin while a body motion is ending. In the latter event, only that portion of the Reach or Move which has not been overlapped should be written separately, usually as a Type II motion, and time allow-ance permitted. For instance, the Reach (or part of it) to a part made during the last step of Walking forward toward it can reasonably be limited out.

Body motions may themselves overlap and limit one another. Bend and Turn Body can be performed together if weights are not being carried. A step may be taken so as to overlap with either bending or arising from a Bend. A Leg Motion is often used to bring the feet together as a worker performs a Sit; Sidestep can also be used in this manner. However, care must be taken in such analysis. A Leg Motion would not be limited out, for instance, when used to cross the legs after sitting, to return the body to a place prior to sitting, or to shove a chair with casters to a required location. Another factor is that, although any given body motion might be limited out, it will nevertheless provide a degree of assistance to the long Reach or Move that is limiting.

When the analyst is in doubt as to whether manual or body motions limit, or which of several concurrent body motions predominates, he should generally perform the motions himself and note the ease or difficulty —particularly the state of body balance—with which they can be combined. Remember, always, that the *normal* operator at *average performance* levels is the standard against which the difficulty of motions or their combinations is gaged, and body balance must be preserved during per-formance of all *normal* body motions.

Finally, the special manner of handling heavy loads should guide the MTM engineer in assigning body motions. Any manual Moves connected with this topic will, of course, involve the use of weight factors as discussed in the Move Chapter. In addition to this, the heavier loads require making the Moves separately from the body motions, in other words, such Moves are seldom overlapped or combined with body motions. They occur both before and after body motions required during handling of heavy weights. The worker tends to Move the heavy object to within a few inches of his body before he is balanced enough to make the body motion. He will then Move the object away from his body before he performs further motions with it following completion of body motions. Both of these Moves with weight must be separately allowed. Also, it is hard to avoid using some variety of body motion if the weight is moved more than very short distances; such weights are seldom moved as far as light objects without either body assistance or a definite body movement to aid the action.

No adjustments are made in the description or time for individual body motions used in handling weights; nothing similar to weight factors for manual motions pertains to body motions. The handling of heavier loads will dictate the use of different kinds of body motions rather than noticeable changes in the elemental motions themselves. The reason is that the body or trunk muscles are the heaviest, strongest parts of the body, but are less adapted to different kinds of actions than are the manual members. If any doubt exists in the MTM analysis of heavy transports, the best approach is to use stopwatch time study instead of MTM body motions. For long range solution of such work study problems, probably the best approach would be to take a motion picture and analyze the film for accurate general data which will augment the available MTM data.

Finally, Fig. 7 concisely indicates the directional aspects of body displacement as a good summary and review. Note that the Leg and Foot

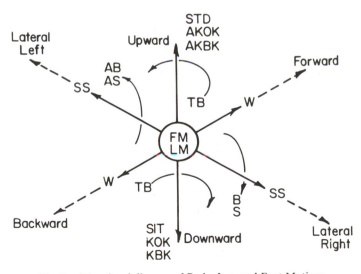

Fig. 7.   Directional diagram of Body, Leg, and Foot Motions.

Motions are shown at the center of the diagram because, in addition to their productive capability when acting alone, the legs and feet are involved in most of the other body motions. The vertical, lateral, and fore-aft axes, of course, are mutually perpendicular. Horizontal movements are: Walking (W) forward or backward, Sidestep (SS) laterally to the left or right, and Turn Body (TB) in either direction about the vertical axis. Vertical movements are: Sit (SIT), Kneel on One Knee (KOK), and Kneel on Both Knees (KBK) downward; Stand (STD), Arise from Kneel on One Knee (AKOK), and Arise from Kneel on Both Knees (AKBK) upward; Bend (B) or Stoop (S) down and forward about the lateral axis; and Arise from Bend (AB) or Arise from Stoop (AS) up and backward about the lateral axis.

CHAPTER 16

Motion Combinations

Chapter 5 through 15 presented detailed descriptions of the basic motions defined in the MTM-1 system and traced their development from origin to present status. These are the "building blocks" of the MTM system; the next task for the student is to learn how to connect them into a unified whole. This must be accomplished before any useful results can accrue from this study of the entire Methods-Time Measurement system. It is the purpose of this chapter to set forth the connections existing among the basic human motions.

The reader's problem is to learn how to obtain time values for complete motion patterns by coordinating the mass of somewhat discrete data concerning each motion of the MTM system. Intimate knowledge of the subject of Motion Combinations covered in this chapter will enable a skilled analyst to write valid MTM motion patterns, assign appropriate time values, and evaluate the effectiveness of the sequences and/or tasks involved in the majority of work cycles. This procedure is valuable, especially for manual work, even if the standard time aspect of MTM is not used. Such analysis is a powerful tool for management purposes, applicable to a whole gamut of industrial engineering problems.

To gain full appreciation of the highly logical and practical MTM data on motion combinations, the reader must be able to quickly and accurately: (1) recall and apply the individual motions and all factors connected with them; (2) structurally unite such motions with due account for the variables affecting their union; and (3) interpret the resulting methods-time data for the many uses to which a finished MTM analysis can be put. When lacking full confidence with respect to Step (1), the reader should *review and re-examine* the individual motion chapters. How to take Step (2) is the direct subject of this chapter. Application chapters to follow will delve more deeply into Step (3) so that the reader's understanding will become well-rounded and reliable for work measurement purposes.

THE LIMITING PRINCIPLE

The reader is familiar with the concept of a performance time required and allowed for each of the individual MTM motions. Also, the concept of combined motions, which was introduced in the Turn chapter, explained the meaning of "limiting" motions and times as well as the converse of "limited" motions and times. The method and logic by which time data is assigned to the many possible motion combinations is yet to be presented. The first step is to extend the reader's knowledge of the Limiting Principle.

Maynard, Stegemerten, and Schwab stated this very important and basic principle, closely related to the Laws of Motion Economy (see Chap. 26), as follows:

In performing industrial operations, it is usually undesirable to have only one body member in motion at a time. Two or more body members should usually be in motion simultaneously if the most effective method for doing the job is to be used . . . . If two or more motions are combined or overlapped, all can be performed in the time required to perform the one demanding the greatest amount of time, or the limiting motion.[1]

Although the general character of motion combinations can thus be described rather simply, the study and evaluation of their occurrences are the most complex aspect of any work measurement system, including time study and motion study in the sense known before the advent of predetermined times. A myriad of variables and subtle differences in both the motions and their attendant time requirements makes difficult both the isolation and systemizing of data into a reliable form.

It has been generally recognized by most authorities in the work measurement field that the MTM system contains the most complete, practical, and research-based data on motion combinations that exists. Indeed, most other systems are woefully weak in this aspect of motion-time data; and useful data on motion combinations is, after all, the focus about which successful methods work and accurate time determination revolves in almost a primary manner. Some attempts to approach this most basic problem have been included in some of the other predetermined time systems. Whether their data rests on reliable research or on pure judgment and estimation is difficult to ascertain without more *public* information on their motion combination rules and how they were developed.

The main source of MTM motion combination data is *A Study of Simultaneous Motions*,[2] which reports the results of over two years of intensive laboratory and industrial study of the problem. The reference includes a discussion and analysis of other predetermined time systems'

---

[1] By permission from *Methods-Time Measurement* by H. B. Maynard, G. J. Stegemerten and J. L. Schwab, New York: McGraw-Hill Book Co., 1948.

[2] R.R. 105. (Available from the U.S./Canada MTM Association.)

handling of motion combinations which the reader will find worth investigating. However, this research and others[3] indicated a wide area yet to be investigated in this subject. Although the information in this chapter is relatively comprehensive, it is by no means offered as final or complete data.

Not only is it evident that additional studies should be made to completely investigate the effect of one motion on another in a combination, but it is also true that any inaccuracies in defining basic motions and/or their characteristics could affect motion combinations time-wise. The latter is true because the time for any motion combination will be that allowed for the individual motion in the combination which is limiting. For an example of possible inaccuracies in basic motion data, the following statements concerning motions and motion times have been given previously:

(1) The performance time of a Reach or Move exhibits proportionality between the time and the distance traveled.
(2) The acceleration time for a motion equals the deceleration time.
(3) Between the periods of acceleration and deceleration, the velocity of a motion achieves a relatively stable, constant value.

These are known not to be completely precise and exact statements of what occurs, and further research might change them slightly. Certainly, these ideas contribute an additional effect which complicates the study of motion combinations with statistical and analysis problems of a high order. However, the data presented in this chapter is practical and has proved reliable in thousands of applications.

The practical significance of the motion combination data is easy to state—it would be almost impossible to write the method in terms of predetermined motions without this information. Even with the mass of data to be presented here, the analyst will at times find unanswered questions which he can solve for the present only by using his best judgment based on full knowledge of the variables here discussed and clarified. Perhaps one of the major contributions of MTM to the field of predetermined times consisted in pinpointing, for the first time, exactly what the variables are and providing practical rules or procedures by which a fairly adequate accounting of them can be made in motion patterns. No substantial information of this nature existed until the MTM Association work was first published.

Let the problem of motion combinations be stated as follows—it is possible, during the same time span, for the left members of the body (or the entire body itself) to perform one, two, or even more basic motions while the right members complete only one single motion; the reverse is

[3]R.R. 108 and R.R. 112. (Available from the U.S./Canada MTM Association.)

also true for cases when the left members perform the limiting motion. In such cases, the procedure is to apply the relatively simple rule to allow or assign on the pattern only the longest time combination of basic motions. However, this practically always is complicated by the question of whether a particular combination can be performed by the operator under the imposed conditions with the presence of major degrees of motion overlap.

The MTM system has practical rules based on defined breakpoints which were somewhat arbitrarily selected from the mass of data evaluated in the research—this being a practical necessity because of the limits of present knowledge. The MTM analyst's thinking, therefore, should not be confined or restricted by the limits imposed by these categories, since a multiplicity of analytical approaches can exist to data which actually varies on a continuum. With due caution, the analyst should rely instead on full understanding of the variables discussed here and exercise independent judgment only *when required*.

While categorical definitions for the prevalent types of motion combinations are given in the next section, together with rules for determining the "limiting" and "limited" motions and time, the analyst should be guided by the following general definition paraphrased by the authors from R.R. 105:

A motion combination may be defined as a complete single motion or a sequence of motions by one or more body members accompanied by a complete or partial motion or sequence of motions by other body members or by the complete absence of motion by all other body members.

Consecutive motion consists of a succession of unitary motion combinations; while combined, simultaneous, and compound motions as later defined are comprised of two or more concurrent motions involving either one or more body members. Because of this complication, no practical time results will be possible unless certain categorized rules given in this chapter are carefully applied and followed whenever they are applicable.

TYPES OF MOTION COMBINATIONS

The possibilities include three distinct categories, together with a fourth that is actually a composite of the second and third.

Consecutive Motions

*CONSECUTIVE motions are individual, complete motions **performed in sequence** by the same or different body members without overlap or pauses between motions.*

1. *The effects of the motions on objects must be logical and physically reproducible.*
2. *Each motion in a consecutive series may be connected to the preceding and following motion by constraints in basic motion definitions.*

Perhaps the greatest error a work analyst can make in dealing with consecutive motions is to regard them as discrete entities that lack any essential connection, and can therefore be strung together like a necklace of beads at random. Such is not the case. A consecutive sequence of motions, when properly understood, will be considered as a special combination of motions subject to limitations no less important than the limits governing other motion combinations.

The basic supposition of all predetermined times systems is: *The time required by a series of consecutive motions is the sum of the times for the individual motions in the series.* The validity of a total time based on this axiom depends on meeting the following necessary connections between individual motions:

(1) that the motions assigned really match both the individual motion definitions and the motions actually being performed or desired,
(2) that the individual motions conform with the limitations and implications of the basic definitions which account for the preceding and following motions, and
(3) that the resultant effect of the motions on the objects involved is physically logical and reproducible.

The first requirement is obvious, but necessary. As an example of the second requirement, recall that the definitions for Reaches and Grasps included restrictions on the kind of Reach that must precede a given type of Grasp. For instance, full accounting for the muscular, mental, and visual control involved in selection time from a jumbled pile can be made only by the use of an R–C prior to the G4 Grasp employed. Another example of where a definition connects motions is the necessity for an M–C to precede the positioning of an object; the object must be brought very close to the final destination before the short motions comprising Position really begin. Many other such connections can be found in the chapters on each individual motion.

Common sense and logic indicate that objects affect and are affected by a given MTM motion in definite physical ways. This is the essence of the third requirement stated above. It is obviously a physical impossibility to perform a Move with a hand that has just released the only object previously held. The only logical motion that could follow would be a Reach. Likewise, as noted in the Turn data, the empty hand cannot perform a T90L, nor can the motion following a T90L be a Reach. At least an RL1 must intervene before a Reach is physically possible. Many such examples of the third requirement above could be stated, and reasoning alone should convince the reader that limitations such as these on individual motions may be imposed automatically by the objects and the workplace.

Subject to the limitations mentioned, almost any series of consecutive motions can be executed. The symbols and method of assigning times is shown in the examples below. The motion is recorded in the proper left- or right-hand column on the analysis form, following the rules given for each motion in earlier chapters. The times for each motion are then inserted in the TMU column and these are totaled for the time required by the entire consecutive combination. Note that each motion is written on a separate line of the motion pattern.

Example 1: An adult using both the left and the right hand in a consecutive motion-sequence to give a child a cinnamon ball.

Description	LH	TMU	RH	Description
Wait		12.9	R10C	To mixed candy in dish
		9.1	G4B	Select cinnamon ball
↓		12.2	M10B	Candy from dish
To child's hand	R6B	8.6		Hold candy
Grip child's palm	G1A	2.0		↓
Hold child's palm		10.6	M8B	Bring candy to palm
		2.0	RL1	Drop candy into child's
Total TMU		57.4		palm

Example 2: Dyeing Easter eggs using consecutive motions.

Description	LH	TMU	RH	Description
To hardboiled egg	R16B	15.8		Hold dipper
Pick up egg	G1A	2.0		
Egg toward dipper	M16C	18.7		
Place in dipper (with care)	P1SD	11.2		
Let go of egg	RL1	2.0		↓
Wait		11.8	M8C	Carefully place egg in
Total TMU		61.5		Easter dye

It is obvious that this annual operation can be done with consecutive motions, although other methods are also possible.

Combined Motions

*COMBINED motions are two or more complete motions **by the same body member** performed during the time required by the limiting motion.*

1. *The body as a whole, excluding the limbs, may be considered as a single member for such action.*
2. *Motion assistance often is a factor in combined manual motions.*

Note that this definition specifically excludes consecutive motions. A simple illustration of a combined motion combination is a Turn or Regrasp combined with a Move. This clearly shows that combined motions are non-consecutive.

Combined Turns, as previously discussed, brought out the many variables present with motion combinations of this type—such as assistance, measurement of the variables integral with each motion and as affected by being combined, and the method or convention of showing the limited and limiting motion(s) and time. The reader can best learn to distinguish combined motions by context or by sight.

The normal possibilities of dual combined motions can be listed as at right. While other combinations can and do sometimes occur, these are found with greatest frequency. The possibilities of triple combined motions are, however, harder to list because they occur more rarely and often require a special

M with G2

R with T or RL

M with T or RL

dexterity based on repetition of handling certain objects in certain ways. One common triple, however, is a Turn and Regrasp combined with a Move.

The time required by a set of combined motions is that for the individual motion in the set that consumes the greatest time and limits out the others. While there is no absolute rule by which one can, upon examination only, decide which of the motions is limiting, knowledge of the relative time values of the individual motions is useful and will often save the analyst the necessity of referring to the data card to aid his decision, even though the data card is the only authority to consider for a sure answer.

The symbol conventions and times for combined motions are recorded on the work form as follows:

One line of the form is used to record each motion in the set, although the order in which they are listed is not important. The reason the order means nothing is because all motions in the set occur within the same time interval and also that a bracket is used to set off the combination from preceding or following motions of the pattern.   Without this bracket, it might at times be difficult to tell at a glance whether the motion(s) limited out (identified with a slanted line) was combined with the motion above or below it; the bracket removes this doubt.

The time for the limiting motion is shown in the TMU column, with a horizontal dash being shown in this column to indicate a time that is limited out.   The right- or left-hand column will, of course, help identify the body member which performed the combined motions.   The slanted line through the limited out motion(s) also helps identification.

While combined motions can be illustrated as a set apart from preceding or following motions, the place they have in a work pattern and their purpose is more obvious if the examples include the combined motion set in a sequence of other motions.   The descriptive columns also help promote clarity.   The examples which follow are of this type.

Example 1: The familiar toss aside motions shown in the Move chapter are used here to cull dead blooms from plants in a flower pot.

Description	LH	TMU	RH	Description
Hold stem of flower		14.4	R14B	To dead bloom
		2.0	G1A	Grip dead bloom
		4.0	D1E	Pull off stem
		8.6	M10Bm	Toss bloom aside
		—	RL1	Open fingers
		4.7	mR4E	Hand overtravel
Let go of stem	RL1	2.0		Wait
Total TMU		35.7		

Example 2: A triple combined motion employed by many people in returning pencils to their shirt pockets.   Note the G3 properly shown in the receiving hand column.

Description	LH	TMU	RH	Description
Obtain pencil from RH	G3	5.6		Let go of pencil
Pencil toward pocket	M12C	15.2		Wait
Shift grasp	G2	—		
Preorient tip	T60S	—		
Locate to shirt pocket	P1SE	5.6		
Total TMU		26.4		

Simultaneous Motions

As generally used in work measurement, simultaneous motions are any combination of motions done at the same time by more than one body member. As will be seen later in this section and the following one on compound motions, this general idea can include combinations ranging from the extremely simplified situation covered in the restricted definition next given to very complex combinations which include multiple motions with overlap and involve many body members. For reasons which will shortly be apparent, however, the MTM system approaches the interpretation of simultaneous motions on a categorical basis to permit the development of practical tables to aid the analyst in describing and timing such combinations. The categories are of the simpler variety but the reasoning may be extended to more complex cases after the simpler kinds of simultaneous motions become familiar to the reader.

SIMPLE SIMULTANEOUS motions are two complete motions by different body members performed during the time required by the limiting motion.
1. *This definition covers a restricted case of general simultaneous motion.*
2. *The commonly accepted abbreviation is simo, taken from the first two letters each of simultaneous motion.*
3. *Balancing tendency and interaction time may be factors during simple simo motions.*

The concepts in the third sub-definition will be discussed at a later point in this chapter, after all the control factors have been presented.

For convenience in discussion, the terms "limiting member" and "limited member" have been coined to describe, respectively, the body members controlling and not controlling the total simo time. The time for the limiting motion, as found from the individual motion data already given, is the performance time allowed for a simo combination. Understanding of simultaneous motions, however, requires that the actions of the *limiting and limited members* be clearly delineated from both the motion and time standpoints.

The *limiting member,* which by the Limiting Principle controls the time, always performs a basic or elemental motion (as previously defined in the MTM procedure) that completely fulfills all the conditions and requirements of any given Type I, II, or III motion and consumes the MTM data time predetermined for it. In this sense, the analyst may conveniently regard the limiting motion in an isolated manner. The time for a pattern would be correct as long as the limiting motion was included in the proper place.

It is in *describing* the action of the *limited member* that the explanation of simo motions becomes more involved. The reason is' that present application rules are based on the idea that the limited member will ordinarily be assigned only one motion which requires the same or less time than does the limiting member.

In order for the analyst to gain some insight into the simo variables, however, it is necessary that he know at least the rudiments of the limited action of the combination. Clarification of this type will be given here; but, for readers who desire further data and have a theoretical bent, the discussion of motion overlap and symbolic plotting of it is well covered in R.R. No. 105 and *A Research Methods Manual*[4] available from the MTM Association. Basically, the application simplification given above avoids the motion and time complication actually present in simo motions due to overlapping and interaction phenomena. While recognition and evaluation of these factors are essential for research evaluation, the practical analyst has no way by which to account for them in arriving at assignable times for simo occurrences. Perhaps progress in this deficiency will be a future advance of motion analysis that will come from further research by the MTM Association and others.

In truth, however, the researcher or statistician investigating simultaneous motion must recognize this to be a practical arbitrary approach that enables the analyst to make concise choices from the data which has been developed regarding motion times. With this reservation in mind, the theoretician can conduct research that may be codified into "simo tables" which give practical answers when followed, even though they do not reflect fully the precise nature and interaction of the two motions being combined in a simultaneous manner. The regular MTM analyst will not ordinarily concern himself beyond this level of theory, being content to leave the complexities to the theorists and researchers. Analysts apply the working rules represented by the "simo tables" with an understanding of the variables—to be discussed—and leave the question of statistical accuracy in the hands of the people who have considered the true factors and reflected the practical interpretation in the "simo tables" developed.

According to the restricted simo definition, the limited member, the same as for the limiting member, performs a single, complete motion which requires the same or less time for its performance. Actually, however, the true performance of the limited member may include complete or partial motions and/or motion sequences. It could consist of a single, complete motion as noted by the restricted definition. It could also consist of partial increments of two motions, both of which overlap the limiting motion at the beginning and end of the limiting action. One or more complete motions, together with a partial motion comprises another possibility. An additional possibility would be one complete motion with two

[4]R.R. 107. (Available from the U.S./Canada MTM Association.)

partial motions overlapping the limiting motion. The present simo data and rules cover only the first case mentioned, which is the reason for the restricted simo definition. An approach to handling the other cases will be given under "compound" motions presented later. In any event, the time for the limited member's action will not be allowed, since it is the limited time of the combination. This is the same idea meant when MTM analysts speak of the limited motion being "limited out" of time inclusion.

With this introduction to simo motions, further discussion will be deferred to more complete elaboration on the variables in a later section of this chapter. It is worthwhile, however, to present the symbol conventions here. Both motions in the simo combination are recorded on the same line of the analysis form, with the left and right columns reflecting the handedness involved. The limited motion is encircled, unless it is identical to the limiting motion, and the time for the limiting motion is written in the TMU column. Omitting the circle for identical motions is merely a saving of clerical time in pattern writing, since the simo time is not affected by adding it. Use of this data is obvious in the following example which involves seasoning eggs with salt and pepper.

Seasoning the breakfast staple by this method should improve its tastiness in a matter of little more than five seconds. Note that the simo Moves limited are either of shorter distance than the limiting Move or else require less time than the Turn employed. Circle omission is also well illustrated. The convention for writing multiple frequencies of individual motions is also exemplified in the shaking action, where four short Moves in each direction require the time for a single Move to be multiplied by eight as indicated by the numeral 8 in the No. column.

Description	No.	LH	TMU	RH	No.	Description
To pepper shaker		R10B	11.5	R10B		To salt shaker
Obtain shaker		G1A	2.0	G1A		Obtain shaker
Shaker toward plate		(M6B)	12.2	M10B		Shaker to plate
Hold			7.4	T135S		Turn top down
↓			36.8	M2B	(8)	Salt egg 4 shakes
Shaker to plate		(M4B)	7.4	T135S		Turn top up
Turn top down		T135S	7.4	(M4B)		Shaker out of way
Pepper egg 4 shakes	(8)	M2B	36.8			Hold
Turn top up		T135S	7.4			↓
Shaker to table		M10B	12.2	(M6B)		Shaker to table
Let go of shaker		RL1	2.0	RL1		Let go of shaker
Total TMU			143.1			

Compound Motions

*COMPOUND motions are the simultaneous performance **by different body members** of complete single or combined motions during the time required by one limiting motion in the set.*

1. ***Single*** *compound sets include a single motion by one body member and a combined motion by the other body member.*
2. ***Dual*** *compound sets include combined motions by both body members.*
3. ***Higher order*** *compound sets may include motion overlap or entire sequences by both body members which mesh only at start and finish points.*

Either of the body members may be the limiting member, as may be determined by sequential application of the Limiting Principle to each member individually and then a reapplication of the Limiting Principle *between* the limiting motion(s) of the individual members.

The concept just defined has often been referred to in other MTM writings as "simultaneous and combined," which tacitly recognized that another motion combination besides consecutive, combined, or simultaneous exists.

For ease of explanation, it is well to assume that the body members in question are the left and right hands and to base the discussion on the right hand being the limiting member in all examples. Logical deduction may then be used to isolate these compound possibilities:

(1) The RH does a single, complete motion while the LH performs a combined motion involving two or more individual motions.

(2) The RH does a single, complete motion while the LH performs a consecutive series of two or more individual motions which together take equal or less time than the right-hand motion.

(3) The RH does a combined motion that includes two or more individual motions while the LH does a combined motion comprised of two or more individual motions.

(4) The motions of the RH and LH may include any of these three compound actions with the additional feature of overlap of partially completed motions.

To see where this concept fits in, it is helpful to continue the listing to show that all other possible combinations have been previously defined;

(5) The RH may perform consecutive motions while the LH is idle or holding. The ordinary definition of consecutive motion confined to one body member fully describes this case.

(6) The RH does a combined motion while the LH holds or waits.

This instance fulfills all the restrictions of the definition for combined motions.

(7) The RH does a single, complete motion while the LH also performs a single, complete motion. This accords with the restricted, practical definition of simultaneous motions described earlier. The reason for restricting simo motions to this case now becomes clear.

Case four (4) introduces the complication of motion overlap. For practical purposes, overlap must be evaluated in terms of the RH performing a sequence which may include both single and combined motions while the LH can perform a like sequence. (Not strictly confined to the limits of the compound motion definition.) In other words, a compound motion in the general sense may be considered as the combination of entire sequences in both hands, during the same time interval, that cannot be separated into other motion combinations due to overlap—the only requirement to be considered as a compound set being that the sequence in each hand begins and ends at the same instant of time. No limiting or time allowance rules have been predetermined for this most general of all motion combination cases; in a real sense either hand may be considered limiting so long as the entire sequence is included in the total time allowed.

To show the application of the Limiting Principle to compound motions, the first three cases listed will be discussed in detail since they are the ones of practical value which have not already been discussed or where the solution is not obvious. The Limiting Principle says that the analyst must first examine each body member separately to determine which of the motions is limiting and apply the conventional symbols accordingly, and he then should examine the resulting limiting motions in each member in terms of simultaneous rules. Or:

(1) The lone RH motion is limiting for that hand. One of the motions comprising the combined set in the LH is limiting. The limiting RH and limiting LH motions are then considered as a simo combination, with the time resulting from this step being the total compound time.

Example: LH   TMU   RH

M8C	?	M10B
T60S		
G2		

The finished result will appear like this:

M8C	12.2	M10B
T60S	—	
G2	—	

RH analysis: The lone M10B requiring 12.2 TMU is obviously limiting.
LH analysis: The M8C is 11.8 TMU, the T60S is 4.1 TMU, and the G2 is 5.6 TMU. The M8C is thus limiting.
Simo analysis: The M10B in the RH limits the M8C in the LH, so a circle is used to limit out the LH motion.

In the finished analysis, the steps taken above are obvious by the conventions used to show limited motions. The dashes indicate that the T60S and G2 are not allowed. The encircling of the limiting motions of the LH analysis shows the result of the simo analysis.

(2) The lone RH motion is limiting. The consecutive motions in the LH require less total time than the RH motion. Considering the simultaneity, the RH by definition will always limit in this case of compound motion.

Example:

LH	TMU	RH
G1A	?	R22B
M4B		
RL1		

The analysis result will be:

G1A	20.1	R22B
M4B	—	
RL1	—	

RH analysis: The single R22B is limiting with a time of 20.1 TMU. *LH analysis:* Adding the time for the three motions gives 2.0 + 6.9 + 2.0 or 10.9 TMU total. *Simo analysis:* The motion in the RH limits the consecutive sequence in the LH, which is encircled to show its limited status.

Note here that the absence of slanted lines or a bracket makes it clear that the LH motions were not combined. (They are actually physically impossible to do combined in this example.) The dashes show that no time was allowed for the components of the simo LH sequence written on succeeding lines of the form.

(3) While care must be exercised that this case includes individual motions that are really possible to perform in the manner written, such cases do appear rather frequently in actual motion patterns. The combined motions in the RH and LH will have their respective limiting motions, which then may be analyzed for the simo effect.

Example:

LH	TMU	RH
T90S	?	M8B
M3A		G2

The end result is:

T90S	10.6	M8B
M3A	—	G2

RH analysis: The 10.6 TMU for the M8B limits out the G2 of 5.6 TMU. *LH analysis:* The T90S of 5.4 TMU is limiting over the 4.9 TMU for the M3A. *Simo analysis:* The RH limits.

The use of the conventions should be obvious by now, and they clearly show the limiting process used in getting the answer.

Novices to MTM sometimes ask why the limited motions are shown in a methods analysis, since they do not determine the time for the motion combination.　It is because the limited motions may be essential in providing a complete and understandable methods record on which the time was based.　The superiority of methods description afforded by MTM in contrast to time study is partially due to this inclusion of limited motions in the pattern.　However, it is equally unnecessary to be extreme in the use of symbolism by including every small motion that might conceivably be a part of a given pattern.　For instance, a short R–E to balance the motion rhythm or provide working space for the other hand need not be shown; contrarily, safety requires that an R–E to remove the hand from the operating area of a punch press ram must be shown.　Also, minor Turns (less than 30 degrees) made combined with Reaches and Moves— especially where their occurrence is either obvious by the work conditions or they are non-essential—can be omitted.　It is only important to show limited motions that definitely affect the method or impart essential actions to the objects where these factors would otherwise not be obvious from the context.

General Definition of Simultaneous Motion Combinations

As noted after item (7) under Compound Motions, the complications of overlap and interaction time have not all been evaluated from the standpoint of limiting or time allowance rules.　No approved research results or practical application procedures presently exist to handle overlapping combinations, so MTM patterns cannot adequately recognize or indicate overlap at this date.　While it is therefore impossible now to give full data for the general case of motion combinations, it is nevertheless possible to attempt a full definition.　By way of summarizing the reader's thoughts on types of simultaneous motion combinations, then, the authors offer this definition of the most general type:

*GENERAL SIMULTANEOUS motions are the performance, during the **same time interval**, by one or more body members of a complete or partial single motion, motion sequence, or combined motion while one or more **other** body members simultaneously perform a complete or partial single motion, motion sequence, or combined motion.　The word "partial" recognizes the behavior known as motion overlap.*

This definition covers all of the possible types of combinations previously discussed.　It also describes such complex situations as where one hand reaching toward a fixed object is assisted by a Sidestep and partial Turn Body while the other hand Reaches, Contact Grasps, and partially Moves a tool into a position for subsequent usage.　The generalized definition, in a more particular way, presents a possible avenue of future thought and analysis that may lead to more practical means of interconnecting manual and body motions in motion patterns; the data on this concept is rather sparse now, as noted later in this chapter.

The present data concerning motion combinations given in the remainder of this chapter can be applied by the reader if he fully appreciates the meaning of the more detailed descriptions of the restricted definitions for consecutive, combined, simultaneous, and compound motions presented above.

CONTROL—THE KEY TO COMBINATION ANALYSIS

In the performance of work, two principal agencies are present—the operator and the objects being handled. The key to what can, may, or will happen when these agencies interact is the *control* involved. Control, for this purpose, has a dual meaning. First, from the standpoint of the operator—Of what kinds and amounts of control is the human mechanism capable? Secondly, from the viewpoint of the objects—What kinds and quantities of control do the various characteristics of the objects permit and/or require? These questions, when answered separately, provide clues which can be wedded by synthesis and categorization to yield answers to the total problem of work control.

To evaluate the question of *object control,* the MTM analyst must rely upon his powers of observation, his engineering or other technical background, and the use of measuring devices where appropriate—seasoned with practical common sense and experience factors. Definitive rules and guides are difficult, if not impossible, to establish because the objects can range in size and characteristics from a pin head to a two hundred pound sack of flexible material.

The question of *human control* is better defined because it has been the subject of very exhaustive investigation by psychologists, scientists, physiologists, and many engineers dealing in branches of the engineering art that include the human problems of operation and direction. The mass of MTM research on motions and their combination is basically grounded to this orientation. The subtle differentiations in both the basic motions themselves and in the simo data all depend on control characteristics. In this respect, industrial engineering is certainly attuned to other fields of engineering that concern themselves so greatly with differing kinds of control and their characteristics.

General Control Factors

In general, the higher the degree of control in a motion, the more difficult it is to perform either alone or in combination. Conversely, motions requiring little control are readily accomplished either alone or with other motions.   Research has also established a most important fact related to simultaneous motions—*The amount of control required tends to increase when a given motion is performed in combination with other motions.*

The control capability of an operator depends most directly on:

1. APTITUDE AND/OR SKILL.   The muscular, nervous, and mental equipment of a worker obviously influence his initial control capability. When he has acquired general or specialized skill, however, this has the effect of improving his basic ability to control motions. While natural aptitude for coordination is basically an inherited trait, it is possible to train people to the point where additional coordinating ability is gained.   This idea is obvious when considering the differences between amateurs and professionals in the athletic world; the striking disparity between them can be reduced by a regimen of training and supervision of the neophyte.   A sandlotter is unlikely to have the basic ability of Babe Ruth or Henry Aaron, but he can easily be taught to bat more effectively and attain greater hitting power.   A similar situation obtains in regard to the aptitudes and skills of industrial workers as reflected in their control capabilities.

2. PRACTICE.   The discussion of output variables in Chapter 2 made clear the role of training and practice in increasing operator skill. The repetition, or practice, of motions has an influence on control even beyond what was noted above in reference to skill.   Practice has such a marked effect on the question of simultaneity that the choices of simo performance are directly based, in part, on three categories that express the degree of practice the operator will likely have with the combination being decided.   More discussion of this appears later.

3. WORKING CONDITIONS.   Peculiarities or other properties of the workpiece and arrangement of the workplace may aid or inhibit the use of certain motion combinations.   One of the major gages of this, discussed later, is the field of vision.   Some motion combinations obviously require that placement of the objects concerned occurs within the field of vision.   In the MTM data, two categories for the field of vision aid choices of simultaneity.   Other factors can also affect the methods or combination of motions used.   Sharp projections that endanger the skin is an example.   Such factors will be more fully explained in a later chapter dealing with the Laws of Motion Economy.

Kinds of Control

The kinds of control involved in motions by an operator are not new to the reader, since they were listed briefly in the discussion of selection time in the Grasp chapter. At this stage, however, it is possible to be more precise and achieve greater understanding than was true at that point.

All of the three kinds of control listed here have been considered in the simo tables. Although lengthy discussion of them is possible, a practical condensation of each is more useful to the MTM analyst. Basically, the worker does his tasks by employing the muscles of his body in a mechanical manner, in much the same way as the components of a mechanism function to cause a whole machine to accomplish its assigned role in production. But even the machine must be fed input information by dial settings, electric signals, pneumatic action, and hosts of other ways. In this regard, the human performer utilizes the sensory information gathered—mainly by the eyes—and filtered through the "master control"—the brain—to direct and limit the action of the muscles in a most effective manner.

The kinds of control, accordingly, depend on the components of this action:

1. MUSCULAR. This includes the effort or muscular strain involved in starting and stopping the motion, keeping it on the proper path, and any accuracy or precision demanded by the kind and classification of the motion. These all influence simultaneity.

2. VISUAL. Eye motions are often used in a minor way when the only control demand is normal orientation of the operator. On the other hand, when exact discrimination is the type of control required, the exertion of visual powers to control the motion is very high. The frequency with which the eyes transmit information to the brain obviously must be greater when the need for more precise direction of the muscles is present. In passing, it is also well to note that the muscles have the special faculty of kinesthetic sense which can take over the control functions initially provided by visual means when the control demands are of such nature that they permit it; sufficient leeway in the operating characteristics and much practice of certain motion sequences tend to favor this possibility. The sense of touch, in such cases, must provide enough data about the surrounding conditions that the higher sensing ability of the eyes is not essential to control. For most instances where high control is needed, however, the visual avenues must be used and relied upon.

3. MENTAL. All work requires at least a modicum of mental control. Certain motions, such as Case C Moves or Positions, obviously require more mental control than others, such as Case E Reaches. This control factor covers simple decisions affecting the performance that must be made consciously, but in no way approaches that involved in the work elements known as setup or plan. The mental control being described might be categorized easiest as the instan-

taneous coordination of the eye and muscles within the brain. Work that requires relatively more coordination is commonly called high control work. Conversely, mental coordination is minimal when the control demands of the work elements are low. Obviously, it is harder to combine two motions requiring high control than two involving low levels of control.

Variables in Control

A host of factors were taken into account in development of the simo tables. Many are of such detailed nature that they are not discussed at length in this chapter. However, the following list of questions reveal the nature of the variables that have influenced the simo answers to some degree:

1. At which portion of a motion is the control exercised? Normally, the higher degrees of control are exerted toward the end of a motion, the lower degrees near the beginning and the middle portions.

2. Do the degrees and/or types of control needed conflict? Remember that the left side of the brain controls the right side of the body, and vice versa, so that dissimilar commands tend to produce difficult motion sequences. This touches on the Law of Motion Economy regarding symmetrical motion paths. It is easier to perform motions whose directions tend to complement one another than motions having odd path configurations or paths of travel in opposite directions in relation to the body.

3. What effect do differences in the predominant purposes of the two simo motions have on their ease of performance? Obviously, one would expect similar purposes to cause easier motions, while divergent aims would tend to complicate matters.

4. Does the use of certain muscle groups involve special difficulties or does coordination suffer when certain combinations of muscle groups are used together? Much study of motor ability by medical and psychological people has been devoted to this question. Their answers tend to be complex and cause one of the major difficulties of simo analysis.

5. Does the speed and force of one of the motions affect the other of the pair? A partial answer to this is mentioned later under balancing delay, but there evidently is an interconnection. One reason why more data for this problem does not exist is the experimental difficulty of measuring either the speed or force of motions directly.

6. What is the effect of preceding or succeeding motions on the control demanded of the pair of motions being considered? The many rules and definitions which provide answers to this question in relation to

individual motions contrast with the paucity of information concerning the same problem under simo conditions.

It should be apparent from this list that many variables have been considered and evaluated to some extent in arriving at the simo data; but the need for much more detailed research to permit even better answers in future data on motion combinations is also obvious. Other variables not yet isolated may well be as pertinent as these.

Degree of Control

In much of the preceding material, reference has been made to the amount of control inherent in motions. It is necessary to know the degree of control for individual motions in order to intelligently combine them. For this reason, an attempt has been made to list the degree of control in the general order of difficulty. This data can be seen in Table 1, where suggestive categories are used. These should not be considered as absolute or restrictive, but they do indicate where the motions usually fall in the over-all scale of control.

Low control implies that they may be easily done with little or no practice, and accordingly would tend to be easily performed simo with other motions. *Medium control* suggests that at least a moderate amount of practice is needed, so that their presence in simo combinations could either indicate an easy pair or one requiring practice—depending on the performance conditions. *High control* infers maximum utilization of all three kinds of control, such that the motion can be done simo only with difficulty unless high skill and much practice indicate otherwise.

The simo tables, to assure uniform interpretation by analysts of the many combination possibilities, have generally been based on the degree of control needed by each of the individual motions. They can be roughly checked by using the listing in Table 1 with these generalized rules:

A. Two low control motions can be done simo under almost any conditions

B. One low control motion can be normally done simo with a medium control motion under almost any conditions

C. One low control motion can usually be done simo with a high control motion under almost any conditions

D. Two medium control motions can be done simo if they occur within the Area of Normal Vision, but they require practice (as later defined) if they occur outside the Area of Normal Vision

E. A motion of medium control can be done simo with one of high control within the Area of Normal Vision if practice has been had; it is

Table I. Degree of Control for MTM Motions
(Suggestive only)

Motions Made	Control Category		
	Low	Medium	High
Reaches	R–A R–E	R–B	R–C R–D
Moves	M–A	M–B	M–C
Turns	Usually	—	—
Apply Pressures	—	Usually	—
Grasps	G1 A G2 G3 G5 (Always)	G1B G1C— (Depends on Case of Reach preceding)	G4–
Positions	None	P1SE	All others
Disengages	D1E	D1D D2E	All others
Releases	Always	—	—
Eye Motions	—	ET EF	—
Body Motions	Some	Most	Some

    difficult to combine these simultaneously outside the Area of Normal Vision

F. Two high control motions are difficult to perform simo under almost any conditions, normally they must be considered as uncombined.

Exceptions to these rules obviously will occur for two reasons: First, because the rules themselves are only approximations; and secondly, because the control listings for the individual motions are also only suggestive categories. In instances where extreme control is required, the performance of motions in a simo manner would obviously be very unlikely. At the other end of the scale, however, it would seldom be that relatively uncontrolled motions could not be combined simultaneously. Remember that control varies on a continuum, and assigning breakpoints at all is an arbitrary procedure at best. Application of these rules will, however, permit practical answers in most instances of need.

SIMULTANEOUS MANUAL MOTIONS

    Recall from all the background information previously given that control is the key to simultaneous coordination of motions. More specific factors will now be discussed that lead to an explanation of how the simo tables disclose whether a given pair of motions can be done simultaneously.

In actual usage, the analyst must choose, with the aid of the tables, whether the higher or both motion times must be allowed for two motions being considered as possibly simo. This is regardless of the number of categories by which to judge which class of combination—as next discussed —is correct for the pair of motions being considered. In effect, this limitation leaves *two choices;* either the pair can be done simo and the longer of the two motion times is allowed, or else it cannot be done simo and both times must be allowed because the motions are really consecutive. As for notation, the first choice requires that the motions be listed on the same line of the analysis form, while the second will necessitate writing them on successive lines.

The simultaneous motion tables existing at present cover only the manual motions. The research necessary to arrive at this state of knowledge was very complex and took much time; even so, these tables in no way represent final, unchangeable answers. To extend the knowledge of simo motions, and to further inquire into the even more complex compound motions in greater detail, will require a major effort in future researches. The benefits to be derived from proper usage of the present tables for manual motions, however, are sufficiently large so that when used with prudent judgment they can help solve a host of work measurement problems beyond the immediate concern of writing simo patterns.

General Classes

Simo manual motions can readily be classed in three ways—identical, similar, and dissimilar. The latter two can be subdivided respectively into three and four sub-classes each. All of this data is given in Table 2 and is self evident upon critical examination. The three classes, their features, and examples are all included in the table and are discussed in detail in a later section.

Table Categories

Interpretation of the manual simo tables shown later in this chapter depends on knowledge of the two means or factors by which the combinations are listed to show which motion pairs are possible, probable, or the converse. These two factors are the field of vision and the degree of practice. In addition, there are several special variables which influence the observation of motions and the choice of pairs as possibly simo rather than the manner in which the tables are used directly.

    A. *Field of Vision.* The principal means of control—visual, muscular, and mental—must all be exerted in space; but only the visual control is limited in respect to the location in space of the motions being

Table 2.　Classes of Motion Combinations

Class of Combination	Features			Examples		
	Kind of Motions	Class or Distance	Motion Case Code	LH Motion	TMU Allowed	RH Motion
IDENTICAL	Same	Same	Same	(1) R5A	6.5	R5A
				(2) G1A	2.0	G1A
				(3) M6B	8.9	M6B
SIMILAR	Same	May differ	May differ			
Subclass A	Same	Same	Different	(R10A)	12.9	R10C
Subclass B	Same	Different	Same	(M10B)	13.4	M12B
Subclass C	Same	Different	Different	(1)(R10A)	12.9	R12B
				(2)(G1A)	9.1	G4B
DISSIMILAR	May differ	May differ	May differ			
Subclass A	Different	Same	Same	M4A	6.1	R4A
Subclass B	Different	Same	Different	(M4A)	6.4	R4B
Subclass C	Different	Different	Same	R14B	14.4	(M6B)
Subclass D	Different	Different	Different	P2SE	16.2	(G2)

DEFINITIONS OF THE CLASSES

IDENTICAL —Identical simultaneous motions are the same as to kind, class, and case code for both members
SIMILAR　　　—Similar simultaneous motions are the same as to kind, but may differ as to class and/or distance and case code between the two members
DISSIMILAR—Dissimilar simultaneous motions are always different as to kind, but may be the same as to class and case code between the two members acting.
　Note in the examples that the simplification of the limiting convention for identical—omitting the circle around whichever of the two motions is limited in the theoretical sense—makes no difficulty in assigning times or describing the motion occurrence.
　Note also that the case code identity for different kinds of motion has little meaning other than that the case code "letter" is the same or different, since the "A" in R__A does not mean the same thing as the "A" in M__A or G1A.

controlled.　This is due to the inability of the eyes to cover an area effectively that much exceeds the Area of Normal Vision.　This area, as was defined in the Eye Motion chapter, is that area within a cone from the eye to a 4-inch-diameter circle located 16 inches from the eye.　This provides a guide to the visual span of an operator during simo performance, for within this area objects can be seen without shifting the line of sight.

Thus if two motions being performed are terminated within 4 inches of each other at around 16 inches from the eyes, maximum visual control is possible; therefore, the likelihood of simo performance is favorable.　If the endpoints are not within this distance proportion, and visual control is needed, it is less likely that the two motions can be simo.   To distinguish this factor in the simo data, the tables include separate columns for motions ending Within the Area of Normal Vision (denoted by W) and Outside the Area of Normal Vision (denoted by O) in all cases where visual control is an important variable.   In choosing the proper column for field of vision, the MTM analyst automatically accounts for visual control in simo motions.

B. *Degree of Practice.*   The basic assumption is made in the simo tables that the worker will have ample practice on long production runs and highly repetitive cycles to develop the best degree of skill; whereas under jobbing conditions or short production runs, sufficient practice to achieve simo performance will not be gained.   The tables are constructed to permit a choice of simultaneity, therefore, that depends on the degree of practice the analyst anticipates will be applicable to the motion pattern in which the motions being examined occur.   The three categories are:

E—*Easy* to perform with little or no practice.   Such motions can be done simo, so only the longer of the two motion times should be allowed in the motion pattern.

P—*Practice* is required in at least a reasonable degree before the two motions can be simo.   If the analyst decides there is enough opportunity for practice, he will show the motions simo and allow only the longer of the two times.   If he decides, according to the basic rule given above, that practice cannot be enough to effect simo performance, he will show the motions consecutively and allow both motion times.

D—*Difficult* combinations, even after long practice, indicate that the analyst should almost always write the motion pair in a consecutive manner and allow both motion times.   Of course, he might justify such combinations as simo when above average operators are observed or where exceptional practice and training are found on unusual jobs under highly repetitive conditions.   Since the MTM standard is for average operators under average conditions, however, he will almost never decide for simo when the tables show a difficult combination.

It is obvious that the simo *tables must be used with judgment* when deciding whether the degree of practice permits simultaneous performance.

Included in the judgment of degree of practice should be the effects of lot size, cycle length, and pattern complexity.   A good operator can learn with practice to do progressively more simo combinations; he can also reduce the case, type, or class of a given kind of motion or even eliminate certain motions entirely.

Experience with similar motion sequences on other jobs will tend to carry over to newer tasks. For these reasons, the total lot size alone will not always correctly show the degree of practice to assign. Also, standards resulting from analysis of large lots or high production may be too "tight" for subsequent lots small in size, or smaller lots produced after a notable time lapse since the larger lot was run. Similarly, a long cycle will require much more practice to learn than a short one because there is less opportunity to repeat each motion sequence in the cycle. When the motion combinations or the individual motions are of the more difficult or complex varieties, a greater amount of practice will be needed to achieve equal skill and simo ability compared to a pattern involving only simple motions and combinations.

Additional insights regarding the degree of practice required to achieve MTM-1 motion standards may be found in MTM research reports[5] covering the Learning Curve project. Specific MTM research data on how simo motions affect learning[6] also is available from the MTM Association. These sources show that inexperienced operators may require thousands of cycles to attain the MTM standard times.

C. *Other Factors:* Personal differences in the ability of individual workers or groups of operators, particularly in aptitudes, will affect the decision of whether to allow simo or consecutive performance. The motions possible are directly affected by the distances and arrangement of the workplace, as well as the equipment and tools provided. Good facilities favor the assigning of simo combinations, so the work analyst should either improve the facilities or else realistically be fair to the operator by not assigning simo motions when the facilities are unfavorable. Besides these rather general decision factors, there are two special variables which the analyst must sometimes evaluate during simo analysis.

One of these is called *balancing tendency.* For the simultaneous similar class of motion combination, of the type shown in Subclass B in Table 2, the MTM research showed that when the motions in the two hands differ only in distance, the motions tend to balance each other as to time needed. Actually, the hand traveling the shorter distance will tend to delay slightly, while the hand going the longer distance will tend to speed up slightly. Both hands therefore tend to arrive at their destination together. This means that the

---

[5] R.R. 112 and R.R. 113A. (Available from the U.S./Canada MTM Association.)

[6] Stanley H. Caplan and Walton M. Hancock, "The Effect of Simultaneous Motions on Learning," *Journal of Methods-Time Measurement,* Vol. IX, No. 1, MTM Association for Standards and Research, Ann Arbor, Michigan, September-October, 1963.

actual time taken for the combination will lie between the limiting and limited times, although the difference between the actual time and the limiting time due to balancing tendency is too small in most cases to change the time allowance in practical usage. This does indicate, however, that the decision of Easy—where permitted by appearance of a Practice designation in the simo tables—will be favored if this type of simo similar combination is encountered.

Interaction time is the other special variable that affects the simo decision from the table categories. This concept covers the observed cases where several motions appear to be done simo, but which are difficult to perform simo according to the data tables. What is actually occurring is similar to the action of Moves against a stop discussed and illustrated in the Move chapter, where it was shown that double Moves sometimes occur even though the second Move is too short for the eye to observe directly. Reliance on a "rule of thumb" developed by the researcher viewing each motion picture frame was shown to provide proper analysis for such cases. For example, an operator may appear to be doing two M7C simo outside the Area of Normal Vision, which is difficult according to the tables. Without knowing about interaction time, the analyst would incorrectly assign both times, or 22.2 TMU. The research showed this combination to require 12.1 TMU. Simo performance would require the time for only one M7C of 11.1 TMU, which is low. Knowing about interaction time would cause the analyst to record the proper motions and assign times as shown below. This obviously gives a better answer that is on the safe side of the actual time. The M6B can be done (with practice) out of vision with the M7C, so that the simo rules are not disregarded by this analysis.

LH	TMU	RH
(M6B)	11.1	M7C
MfC	2.0	
Total	13.1 TMU	

SIMULTANEOUS MOTION TABLES

Two sets of simultaneous motion data exist in the MTM system. One is the very detailed information that resulted from the MTM researches. In order to permit handy reference in a condensed form to fit the MTM data card, the other set of data was developed. Actually, these tables differ mainly in degree of preciseness and the detail of coverage of the simultaneous manual motions. The data card table, because it is condensed, includes more cases in one listing and is thereby more conservative. It will sometimes indicate a higher degree of simo difficulty for a pair of motions than will the detailed tables. In addition, some simo combinations cannot be found separately listed in the condensed table; but they are readily found in the detailed table, which covers almost all possible combinations between the manual motions.

Detailed Tables

The detailed tables are Tables 3A, 3B, and 3C, which include an explanation of the practice coding. Remember, when consulting the tables, the analyst decides either that the motions can be simo or that they will not be simo under the visual and practice conditions pertaining to the motion pattern in which the two motions concerned appear. If they can be simo, the motions will be recorded on the same line of the analysis form and only the longer time allowed; if not, the motions will be written on successive lines and both motion times allowed.

Regarding the visual conditions, the "W" columns are used for motions occurring within or ending within the Area of Normal Vision. The "O" columns are used for motions ending or occurring outside the Area of Normal Vision. If the field of vision has no bearing on simultaneity, the columns are either labeled "ALL" or the heading is omitted.

Note that the Case B Moves are distinguished both for Type of Motion (m) and definitional case. The basis for the latter subdivision appears in the definition for M–B as follows: "Move object to *approximate* or *indefinite* location." Apparently, some differences in the ability of operators to perform B Moves simultaneously depend on a distinction between the precision with which the end point of the motion path of an M–B can be associated. Separate rows and columns for each class of Case B Move therefore appear in the detailed simo tables. In symbols, the approximate case is often written M–Ba while the indefinite distance is shown as M–Bi. This listing should involve no difficulty of interpretation so long as the analyst is careful to decide which kind of Case B Move he has to evaluate.

Other special distinctions in the tables can be seen. Turns are considered as either approximate or exact in most of the tables; a further subdivision is made for Turn with Disengage depending on the degrees of Turn being greater or less than 90 degrees. Release has a general rule that shows it is easy to perform simo with any other motion. A note for Apply Pressure indicates that special analysis is needed when deciding whether it can be simo. There is a difference for Disengage that depends on the presence or absence of care. And finally, Grasp-Grasp combinations may require reference to several notes that make the simo decision depend on the type of Reaches preceding the Grasps.

Table 3. Detailed Simultaneous Motions*

☐ an EASY combination with little or no practice. Allow only the longest of the two motion times.

☒ a combination requiring PRACTICE. Under highly repetitive conditions, allow only the longest of the two motion times. For job-shop or other non-repetitive conditions, allow both motion times (not simo).

■ a DIFFICULT combination even after long practice. Unless exceptional training and practice prevail under highly repetitive conditions on unusual jobs, allow both motion times (not simo).

W = *within* the Area of Normal Vision ⎫ not over 4″ apart
O = *outside* the Area of Normal Vision ⎭ at 16″ from the eyes

REACH simultaneous with ⟶

R_A		R_B		R_C		R_D		R_E		Area of Normal Vision	
W	O	W	O	W	O	W	O	W	O		
☐	☐	☐	☐	☐	☐	☐	☐	☐	☐	R_A, R_E	REACH
☐	☐	☐	☐	☐	☒	☐	☒	☐	☐	R_B	
☐	☐	☐	☒	☒	☐	■	☐	☐	☐	R_C, R_D	
☐	☐	☐	☒	☒	☐	■	☒	☐	☐	M_A	MOVE
☐	☒	☐	☒	☒	■	■	☒	☐	☐	M_B (approximate)	
☐	☐	☐	☐	☐	☒	☒	☒	☐	☐	M_B (indefinite)	
☐	☐	☐	☐	☐	☐	☒	☒	☒	☐	M_Bm	
☒	☒	■	■	■	☐	☐	☐	☐	☒	M_C	
☐	☐	☐	☒	☒	☒	☒	☒	☐	☐	G1A, G2, G5	GRASP
☐	☐	☐	☒	■	■	☒	☐	☐	☐	G1B	
☐	☐	☐	■	■	☐	☐	☐	☐	☐	G1C	
☐	☐	☒	■	■	☐	☐	☐	☐	☐	G4	
☐	☐	☐	☐	☐	☐	☐	☐	☐	☐	Approximate	TURN
☐	☒	☐	☒	☒	☐	■	☐	☐	☐	Exact	
☐	☐	☐	☒	■	■	☐	☐	☐	☐	P1SE, P1SD, P1SSE, P2SE	POSITION
☒	☐	■	■	■	☐	☐	☒	☒	☐	P1SSD, P1NSE, P1NSD	
☒	☐	■	■	☐	☐	☐	☐	☐	☐	P2SD, P2SSE, P2NSE	
☐	☐	☐	☐	☐	☐	☐	☐	☐	☐	P2SSD, P2NSD, all P3	
☐	☐	☐	☐	☐	☒	☒	☐	☐	☐	D1E	DISENGAGE
☐	☐	☐	☐	■	■	☐	☐	☐	☐	D1D, D2E	
☐	☐	☒	■	■	☐	☐	☐	☐	☐	D2D	
■	■	☐	☐	☐	☐	☐	☐	☐	☐	D3E, D3D	

* The Table 3 pages differ from the condensed table (Page 620, Table X on the MTM-1 Data Card) mainly in their greater preciseness and degree of coverage as stated on page 615. The basis is R.R. 105. (See Ch. 10.)

Table 3 (Cont.). Detailed Simultaneous Motions

MOVE simultaneous with

M_A		M_Ba		M_Bi M_Bm	M_C		a = approximate i = indefinite	
W	O	W	O	All	W	O	Area of Normal Vision	
			X				M_A, M_Ba	MOVE
							M_Bi, M_Bm	
			X		X	█	M_C	
			X		X	X	G1A, G2, G5	GRASP
			X			█	G1B	
		X					G1C	
			█				G4	
							Approximate	TURN
	X		X		X		Exact	
			X		█		P1SE, P1SD, P1SSE, P2SE	POSITION
X	█		X		█		P1SSD, P1NSE, P1NSD	
					█		P2SD, P2SSE, P2NSE	
█	█		█		█		P2SSD, P2NSD, all P3	
						X	D1E, D1D	DISENGAGE
					█		D2E	
	X				█		D2D	
█	█				█		D3E, D3D	

GRASP simultaneous with

G1A, G2, G5	G1B		G1C		G4		NOTE: G3 is simo by its nature	
All	W	O	W	O	W	O	Area of Normal Vision	
							G1A, G2, G5	GRASP
	*I	*II	X			X	G1B	
			*I	*II	X		G1C	
	X	X			█		G4	
							Approximate	TURN
	X	X	X		█		Exact	
█							P1SE, P1SD, P1SSE, P2SE	POSITION
	█						All other Positions	
	█						D1E, D1D	DISENGAGE
█	█						D2E, D2D, D3E, D3D	

*I. EASY after R_A or R_B, otherwise DIFFICULT.
*II. PRACTICE after R_A only, otherwise DIFFICULT.

Table 3 (Cont.).   Detailed Simultaneous Motions

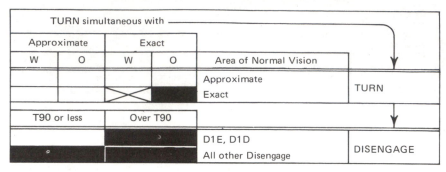

TURN simultaneous with					
Approximate		Exact			
W	O	W	O	Area of Normal Vision	
				Approximate	TURN
				Exact	
T90 or less		Over T90			
				D1E, D1D	DISENGAGE
				All other Disengage	

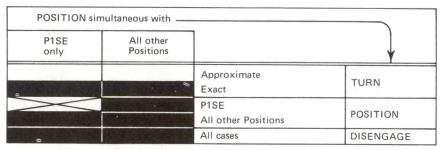

POSITION simultaneous with			
P1SE only		All other Positions	
			Approximate	TURN
		Exact	
		P1SE	POSITION
		All other Positions	
		All cases	DISENGAGE

DISENGAGE simultaneous with			
No care		Care required	
		No care	DISENGAGE
		Care required	

APPLY PRESSURE	simultaneous with any motion may be EASY, PRACTICE, or DIFFICULT. It must be observed or tried by the analyst, for no definitive rule can be stated.

RELEASE	simultaneous with any motion is always EASY.

Examples:
1. R–B with R–C. This combination is easy in vision, but will require practice out of vision.
2. R–D in one hand with an exact Turn in the other hand. To do this within the Area of Normal Vision requires practice, and it is difficult to do it out of vision.
3. M–Bi and G4. This can be done easily regardless of visual conditions.

4. G1C in combination with D2. This is always difficult to do simo.

5. P3SE with RL. Release can always be easily done simo with any other motion.

The answers given provide information with which the analyst can apply his knowledge of the variables in simo performance to reach a decision.

Data Card Tables

Table X of the MTM-1 data card is reproduced in Table 4.

Note that in this table, the intersections for Difficult are in heavy black as they are in the detailed tables. Generally, it is easiest when using this table to enter the right-hand column with the motion nearest the top of the column and then read the row headings to the left until the intersection desired is found. For instance, the quickest way to find an R–B and G1C combination is to find Reach B in the right-hand column, follow that row to the left until it intersects with the row heading for G1B and G1C, and discover there that the two motions are easy in vision but require practice out of vision. When attempting to enter the table from the top headings, time will be lost because of the manner in which the listings are made.

Table 4. Condensed Simultaneous Motions (MTM-1 Data Card Table X)

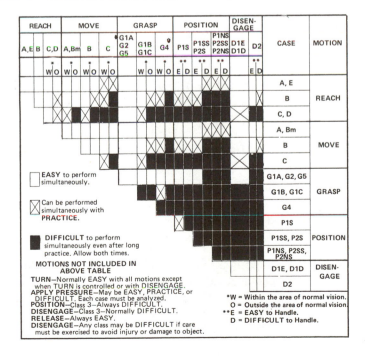

The column headings labeled "W" and "O" refer, as do the detailed tables, to the field of vision. Those headed "E" and "D" refer to easy and difficult handling categories as they normally pertain to Positions and Disengages, that is, they are the ending letter for the symbol. For instance, in the P1S column, there are two sub-columns so marked; the left one is used for P1SE and the right one for P1SD.

At the bottom of Table 4 (Table X of the data card) are listed some guide rules that help decide simo for combinations not listed in the table. It is obviously much easier to use the detailed tables when instances such as this are found, but the data card (usually carried) may often be the only information source the MTM analyst has available when he must make a simo decision. Remember that the condensed tables have the utility of being conveniently available with the rest of the MTM motion data; they are not intended to provide final answers in every case, but they do offer a convenient form and location of data. To make sure that the answers are generally on the safe side, check any doubtful combinations by reference to the detailed tables:

Examples: 1. R–B with R–C. Easy in vision, practice out of vision; this answer is the same as found in the detailed tables.

2. R–D in one hand with an exact Turn in the other. The special note says this is not easy because the Turn is controlled, but whether it requires practice only or is difficult cannot be found on the condensed table. This is a case where the detailed table gives an answer more enlightening than the Data Card table.

3. M–B and G4. The card shows this to require practice in vision and to be difficult out of vision. Because the "indefinite" could be identified on the detailed tables, the answer there was more liberal—it said this combination was easy. Notice, however, that the M–Ba answer for the detailed tables agrees with the condensed tables.

4. G1C in combination with D2. The data card agrees with the detailed tables that this is a difficult combination to perform.

5. P3SE with RL. The answer here also agrees—this is easy to do.

This comparison should make obvious both the advantages and limitations of the data card simo tables in relation to the detailed tables. It is better and more correct to use the detailed tables when they are readily available, but the data card must often be relied upon in practical situations where the detailed tables cannot be carried or located easily.

Simo Analysis by Class of Combination

If the MTM engineer has a firm grasp of the general rules on which the answers are based, he can often save time by not having to look at either set of simo tables. These are given here grouped by classes as they were defined earlier in Table 2.

Identical and Similar Combinations:

REACH-REACH. In most cases, identical Reaches are easily done simo, but some require practice out of vision. The main exception is that identical Case C and Case D Reaches require practice out of vision. For the similar class cases, the same general rules hold; thus, only the B, C, and D Reaches need to be looked up in the tables.

MOVE-MOVE. In vision, similar or identical combinations are all easily simo with the exception that the identical Case C requires practice. Out of vision, some combinations require practice and C Moves of the identical class are difficult to do simo. Recall in this particular case, however, the analysis given earlier for interaction time.

TURN-TURN. Approximate turns (relatively uncontrolled) are easily simo regardless of the field of vision. Identically exact Turns, however, require practice in vision and are difficult out of vision.

GRASP-GRASP. Generally, the tables should be consulted for Grasp combinations, because of the variety of answers. Note that G1A is always easily performed simo with other Grasps, and this is also true for G2 and G5. Identical G4 cannot be done simultaneously; both Grasp times must be allowed. Similar Grasp combinations including G4 result in various answers. The reason is that search and select control is needed for each hand that makes a G4, so the operator must concentrate on one hand at a time. G3 is by definition a kind of Grasp that always is simo in nature. Identical G1B and G1C in vision are easy when preceded by R–A or R–B, but are otherwise difficult. Identical G1B and G1C out of vision require practice when an R–A precedes, but are otherwise difficult. For other similar combinations, no rule can substitute for looking into the simo tables.

POSITION-POSITION. For average operators, identical P1SE require practice to do simo. All other identical or similar Position combinations are difficult to do.

DISENGAGE-DISENGAGE. All identical or similar combinations not requiring care can be done simo, but all other possibilities are difficult.

RELEASE-RELEASE. Release is easily done simo with any other motion.

APPLY PRESSURE-APPLY PRESSURE. Although each case must be examined, these combinations tend to be easier if the pressures are applied in the same direction by like body members. All other cases either require practice or are difficult, the latter being especially true when unlike body members are used in different directions.

Dissimilar Combinations:

REACH-MOVE. Case A and B Reaches may be easily simo with any case of Move except Case C within the normal vision area; out of vision, they may require practice. All R—E can be easily performed with any case of Move except Case C Move out of vision (needing practice), since the eye will follow the Move rather than the Reach. Case C Moves usually require practice to combine with low control Reaches; it is difficult to combine them with high control Reaches.

REACH-TURN OR MOVE-TURN. If the Turn is approximate, it will be easy to combine with any case of Reach or Move. When it is controlled, however, the need for practice or the difficulty of combining depend mainly on the field of vision and the relative control in the Reach or Move.

REACH-GRASP. Simultaneous performance is easy for any case of Reach with G1A, G2, and G5. Practice is needed to combine G1B with controlled Reaches, and either practice or difficult choices are possible for G1C with controlled Reaches; either of these Grasps are easily done with low control Reaches. Case A and E Reaches are easy with G4, Case B in vision requires practice, and all other G4-Reach combinations are difficult.

MOVE-GRASP. With G1A, G2, and G5 any Move can be easily combined. Practice is needed for M–Ba out of vision with G1B and G1C, most M–C require practice with either of these Grasps, but G1B and G1C can otherwise be easily done together with Moves. G4 can be easily performed with M–A and M–Bi, but it is difficult to Select Grasp with most other Moves.

TURN-OTHER MOTIONS. Approximate Turns are easily done with all other motions except Disengage. Exact Turns, however, are usually difficult to combine.

APPLY PRESSURE-OTHER MOTIONS. The remarks given for identical and similar simo combinations hold here, except that dissimilar combinations involving AP are less likely to be easily done unless special conditions are true.

RELEASE-OTHER MOTIONS. Again, Release is easy to do with any other motion.

POSITION-OTHER MOTIONS. Generally, it is difficult to combine Positions with most other motions, although some cases can be done with practice. The only frequent exception to this occurs with P1SE, but even these are usually difficult to combine with other motions involving higher control.

DISENGAGE-OTHER MOTIONS.   Some Disengages of the D1 and D2 classes can be readily performed with certain Reaches and Moves of lower control categories.   In almost all other combinations, Disengage is difficult to do simo.   It is never simo with Positions, and seldom with any kind of Grasp.

Special Notes:

Sometimes people uninitiated in work measurement think that by using both hands at all times together production will be doubled.   With knowledge of the extensive simo data, it is extremely obvious why this is ordinarily impossible.   Not all of the right- and left-hand motions can be simo.   Because of this, it is considered good if 20 to 35 per cent productive increase results from changing an operation to two-handed, other things remaining the same.

Remember that all of the simo data given is predicated on the capability and dexterity of average operators.   It is readily understandable, therefore, that highly skilled workers with much practice can perform motion combinations that apparently belie the simo data.   For example, an average worker may perform two Case B Reaches simo and follow these with non-simo G1B; whereas a highly skilled operator may change this to two simo R–A followed by simo G1B.   Positions that require practice, or are difficult, to achieve simultaneously by the average worker may actually be readily done by experts; the expert may even find ways to eliminate entirely the Positions considered.   The MTM standard, to repeat for emphasis, is applicable to *average* workers exerting *average* skill and effort under *average* conditions.

Occasionally, an MTM analyst observes motions being performed partially simo.   If the appropriate simo table indicates the motions can be done simultaneously, this is the method which should be specified.   However, if the chart does not indicate this, the partial simo should be considered as due to above average operator skill and non-simo motions should be assigned and allowed.

SIMULTANEOUS BODY MOTIONS

Some remarks concerning the likelihood of combining Body Motions with each other and with manual motions were made in the last chapter.   The comments here provide additional information.

The fact that Body Motions generally are time consuming, and therefore limit out manual motions concurrently performed, is the main point to remember.   It is not usual that simo motions can be done with body actions involved.   It is fairly common, however, to find combined motions involving the body, and compound motions with body action included are also encountered. Since no tables have been developed to provide ready information—this is another area which research should cover in the future—these general concepts can be applied with judgment in the analysis of Body, Leg, and Foot Motions where simultaneity appears worth investigating:

1. During Body Motions, it is essential to maintain a certain degree of balance. Because the manual members are frequently used to provide stabilizing and restraining action, they are often unavailable to perform productive motions themselves.

2. Heavier objects are usually handled by manual motions to bring them close to the body before Body Motions are made; additional manual motions are then needed to move them away from the body prior to performance of succeeding manual motions. If all such motions are recognized and analyzed, rules previously given should be sufficient to arrive at reasonable performance times.

3. Grasps to pick up objects in conjunction with Body Motions are usually performed only after the body action has been completed.

4. The motions of the right side of the body are controlled by the left half of the brain, and vice versa. Dissimilar commands in regard to kind of motion or direction of motion path to each side of the body will tend to result in confusion, destroying effective action. Also, different body members controlled by the same half of the brain must be given harmonious commands if successful motions are to ensue.

These ideas indicate that much is yet unknown about the human body in spite of the large amount of data on human motions available from existing research. When an MTM analyst must record Body Motions, therefore, it is wise practice to observe more than one operator if possible so that he will not be misled by special ability of the first operator he was watching. In addition, when Body Motions are being visualized, an experienced analyst will usually try out the various motions himself to see whether motion combinations he has in mind are possible.

DECISION MODELS FOR MOTION COMBINATIONS

The two sheets of Fig. 1 provide both the required Decision Model for practical application of the simo data and a summary of the graphical symbol conventions.

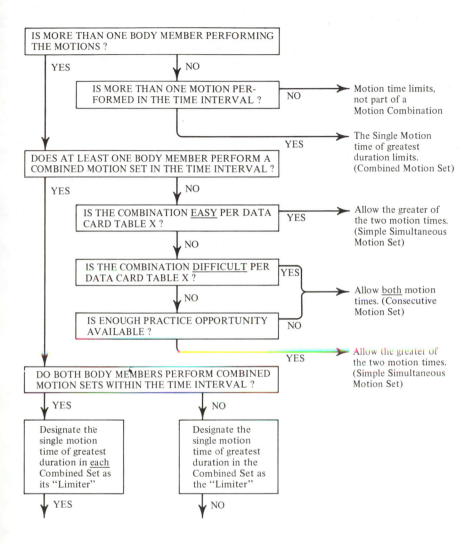

Fig. 1A.   Decision Model for Motion Combinations.

(both lines continued from last sheet)

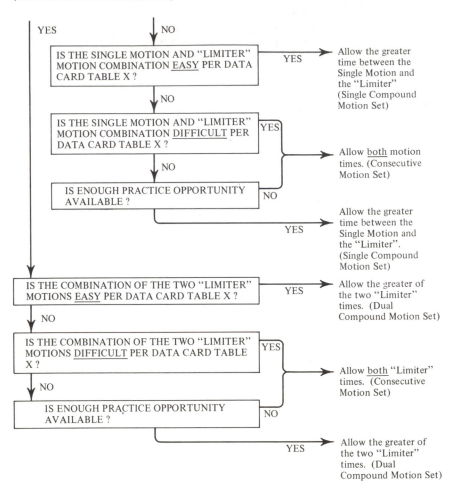

NOTE: The graphic symbol conventions for Motion Combinations are:
1. <u>Consecutive</u> Sets. Write each single motion on successive lines in the correct hand columns.
2. <u>Combined</u> Sets. Write the single motions on two or more successive lines in the same column, and enclose the set with a "(" for the right-hand column or a ")" for the left-hand column. Use a slanted line "/" through <u>each</u> "limited out" motion, leaving only the "limiter" motion unmarked.
3. <u>Simultaneous</u> Sets. Write motions found to be simultaneous on the same line, left- and right-hand columns. Encircle " ◯ " the "limited out" motion unless both motions have equal time values. When the "limited out" motion is itself a "limiter" of a Combined Set, encircle the whole Combined Set.
4. <u>Compound</u> Sets. Use appropriate mixtures of the above markings.

Fig. 1B.   Decision Model for Motion Combinations.

SUMMARY

The four kinds of motions combinations are:	Consecutive Combined Simultaneous Compound
The three classes of simo motions are:	Identical Similar Dissimilar
The main basis to judge motion combinations is:	Control
The main practical factors in simo data are:	Degree of Practice Field of Vision

There is no substitute for practical judgment when analyzing motion combinations.

CHAPTER 17

Motion Patterns—Methods Evidence

The visible evidence of Methods-Time Measurement analysis is the recorded motion pattern. Without the written record preserved to substantiate the work method and/or resultant time standard, one cannot use the analysis for methods specification and/or improvement, training, development of standard data, an incentive system, or the many other applications (see Chap. 28) of MTM analysis. Readers unfamiliar with the elements of a complete motion pattern are referred to Figs. 1 and 2 of this chapter, which are described in detail later.

The preceding chapters have given all the motion information and data the reader needs to write motion patterns. However, little detail has been given previously concerning the procedures, cautions, or other factors incident to the writing of motion patterns. Therefore, this chapter shows how the MTM motion data and motion combination knowledge can be coordinated into motion analyses that comprise the practical, useful aspect of MTM in its final form. Although sometimes seeming laborious, the "paperwork" stage of MTM analysis—writing motion patterns—is the key to maximum benefits from usage of the techniques. For this reason, the astute reader will wish to know every possible aid to quick, accurate writing of motion patterns that will reduce the effort of recording without leaving doubt in the minds of those reading the patterns as to any intentions or decisions regarding the work method.

Most of the remarks above and following in this chapter apply equally well, generally, to motion patterns written with other predetermined time systems. Many of them have all the inherent advantages stressed in this chapter.

WHAT ARE MOTION PATTERNS?

DEFINITION: *A motion pattern is the recorded evidence of the motion analysis of a work method at the element, sequence, or task level that provides lingual and/or symbolic information in a condensed form which fully and clearly delineates the means of performing the work in a manner permitting precise* **reproduction of the method.**

To be consistently successful in solving the many problems of work measurement, an MTM analyst must fully understand the characteristics of motion patterns. He can do this only if he is able to practice his art with full awareness of *what* he is doing, *why* he is doing it, *how* it may be done, *where* and *when* it is best done, and *who* is affected or a valid party to what is being done. Note how the six basic questions used by the industrial engineer are as applicable to evaluating the engineer himself as they are to evaluating the work he is studying.

A motion pattern is a useful evaluation device solely because it answers for anyone reading it, in compact form, the six basic work study queries. It is adequate only to the degree it faithfully and completely mirrors the method employed (or to be used) in performing the work analyzed. This concept does not deny that time determination is a necessary and useful part of the motion pattern, but only emphasizes the motions and descriptions—with the time values being an added feature. It is of little worth to be overly concerned about time determination if the analyst lacks ability or confidence in writing the motions on which, when using a predetermined time system, the time must necessarily be based. Since most of the true methods engineering has been achieved prior to this point, assigning and computing the time values on a motion pattern are primarily clerical in nature. This latter work can be done by a trained clerk or a computer—they can even check for such errors as incorrect simo analysis. In fact, if the basic motions for each work element are determined by the analyst and interrelated or keyed as to time and frequency of occurrence, all other pattern work can be accomplished with relative ease by a computer.

However, to appreciate the place and importance which motion patterns can hold in the chain of events that create production, remember that one of the *prime* functions of industrial engineers—perhaps their most utilitarian one—is to develop and write processes. Industrial engineers, in cooperation with any other concerned specialist, staff, or executive personnel in the plant, must perform the vital step of processing. Starting with the design blueprints and the bills of materials (often called the "Bible of Production") as basic data, they must play a vital role in translating these data into a finished product ready to be shipped to the customer.

To help illustrate the importance of this essential step in the manufacturing enterprise, the following three definitions[1] are quoted:

Process Engineer:

An individual qualified by education, training, and/or experience to prescribe efficient production processes to produce a product as designed and who specializes in this work. This work includes specifying all the equipment, tools, fixtures, and the like that are to be used and, often, the estimated cost of producing the product by the prescribed process.

Processing:

1. The act of prescribing the production process to produce a product as designed. This may include specifying the equipment, tools, fixtures, machines, and the like required; the methods to be used; the workmen necessary; and the estimated or allowed time.
2. The carrying out of a production process.

Process:

1. A planned series of actions or operations which advances a material or procedure from one stage of completion to another.
2. A planned and controlled treatment that subjects materials to the influence of one or more types of energy for the time required to bring about the desired reactions or results. Examples include the curing of rubber, mixing of compounds, heat treating of metals, machining of metals, and the like.

Inherent in many of the acts and functions thus defined is the necessity and desirability of evaluating human performance. *This implies that work measurement is essential to adequate processing and will both aid and measure it for the many industrial parties who are interested in profitable manufacture.* The implication, indeed, is closer each day to being an absolute fact as products and production technology become increasingly complex. The day when persons other than qualified industrial engineers can successfully design processes that will meet the competition of other vendors is rapidly passing; the growing demand for all types of engineering effort and skills is ample evidence of this truth.

To continue the reasoning, any valid work measurement procedure must be directly related to all the major factors in production. It must aid their attainment and give practical clues to better ways of doing things. The efforts of all technical, staff, and executive departments in the plant impinge on the worker as he makes the product, so that correct and complete analysis of his work requires attention to how the aims of these departments are served and contribute to the success of the business. Such analysis must reflect all the factors the industrial engineer must face and choose from, both in developing and in improving productive processes. Most of these factors can definitely be shown to be an integral part of the working methods. *The motion pattern is the most evident proof and pic-*

[1] "Industrial Engineering Terminology," *ASME Standard 106,* American Society of Mechanical Engineers, New York, New York, 1955. These definitions are used because the newer ANSI Z-94 publications do not define these terms with the exact connotation and emphasis desired here.

ture of the work method contained on any piece of paper in the entire factory. Motion patterns, therefore, are important to the degree with which they properly mirror productive factors. As practiced in the techniques associated with Methods-Time Measurement, this kind of reflection in motion patterns is extremely well done.

Having regarded the general picture presented above, it is easier to look more closely at the connection of motion patterns to and with processes. The industrial engineer's work is aided by subdividing processes into increasingly finer components, as follows: Operations, Sequences, Elements, and Motions. For example, the *process* may be that of producing a typical hand-held electric mixer. The *operation* in question might be the final assembly of the mechanism into the outer housing. This would include the *element* of inserting the assembled mechanism into the outer housing and fastening these two items securely. All of the *motions* needed to perform this task element would be analyzed by MTM to provide a record of the method of inserting and securing the mechanism. From such a motion pattern would spring all the information and data with which the work analyst would proceed to institute, guide, and improve production. MTM motion patterns of correct order and frequency *specify* the process and associated tooling and/or fixtures. Actually, the work instruction sheets given the supervisor and operator—often called Process Sheets—are easily written from motion patterns merely by transposing the pattern data into common shop language and descriptive words the operator can readily understand.

A motion pattern is basically built from the MTM motions that account for what, how, and where the operator performs the part of the process under human control; to be complete, it must also include pertinent information covering "process time" that is under the control of other agencies and factors. When the rate or time for performing the element covered by the pattern is controlled by, say, the cycle time of a machine, this must be obtained by time study or calculated by some alternate method —the MTM data covers only the manual time. It is, therefore, obvious that predetermined times do not supplant time study. In fact, certain essential features must appear in analyses by either technique. For example, both must include necessary sketches of parts and/or workplace layouts, identification of tooling and equipment, and similar data which aids in providing a complete picture of the work being done. A technique like MTM, instead of supplanting time study, requires the integration of time study into the total analysis task.

What Patterns Show

Basically, the only thing a pattern of motions can show is a *method*. True, the attendant time is also a part of the written pattern, but such time is meaningful only to the extent and completeness with which the pattern accurately and truly describes a given method. However, writing the correct motions does not guarantee that the times will be correctly applied, or that there is complete lack of error in the predetermined times; nevertheless, the time could not be correct unless the proper motions have been written. This emphasis on the method employed to do a task is basic to good MTM analysis, and from it springs most of the benefits of MTM usage. It is doubtful that anyone can properly and completely analyze a method without writing motion patterns in striving to improve or simplify it. Recording also is essential to later verification, checking, and possible further modification. For instance, whenever an improved part, material, tool, or fixture change affects only a portion of a previously correct motion pattern, it is very easy to determine a new, more accurate time.

Because methods are the principal feature shown by motion patterns, it is essential that the concept of method be crystal clear to an analyst. A method is a *way* of performing work—how the operator uses his muscles, eyes, and other faculties—along with any machinery, tools, etc.—to perform a useful or required task. An MTM pattern permits very fine distinctions along this line. An element might consist of twenty motions that include a P3SSD. Redesign of a part or a fixture might change this *one motion* to a P1SE. Although only this motion might be changed in the pattern, technically and for time purposes the pattern represents a new method. In actual use of MTM, such single motion changes in highly repetitive, short cycle jobs are frequently productive of significant cost savings. In the example given, 46.5 TMU would be saved each time the element was performed; this means that .028 minutes multiplied by the frequency of that element in a total task would be the saving of labor resulting from merely reducing the positioning time involved. As cited here, this example therefore shows that calling such a change in a pattern a new method is highly logical and justified.

While the MTM procedure is supported by more publicly documented research than any predetermined time system, its practical application is assured only by suitably trained analysts. This does not imply that such analysts when observing the same operator together will write identical motion patterns. Some minor differences in the allowed motions may well appear. However, unless the analysts have made gross errors, their total allowed time will be practically equal. Perfectly valid reasons explain this situation. No two people can physically occupy the same vantage point for observation, and therefore logically they cannot see things exactly alike. Also, average operators normally employ minutely differing methods and motions during successive cycles of a given task, since they are not automatons. In addition, certain of their motions may at times be extremely difficult to classify, which clearly indicates possible honest differences between analysts when they exercise their judgment as to the true motion classification for an observation. In spite of all these factors, proper application training including guided practice will enable analysts to write valid motion patterns with minimal differences. Finally, the practical effect of such differences upon the allowed time usually will be of such small magnitude as to be well within accepted time study limits.

The MTM neophyte will learn from practice that a well-written motion pattern provides precise answers to the basic industrial engineering questions that should be answered when any work is analyzed. How the patterns show these answers will be described in the next section, but the questions regarding the worker and his actions are: *What?, Why?, When?, How?, Where?*, and *Who?*. These six basic questions were discussed in Chapter 3, and could be reviewed profitably at this point.

1. *What* is the worker doing? *What* happens to the workpiece and/or parts?
2. *When* does he do it? *When* is the exact time in the cycle to do a certain act that will achieve the purpose of the work?
3. *How* does it get done? *How* does the worker use himself, the tools, the equipment, and the parts to get the job done?
4. *Where* on the workplace, the assembly, the parts, or the tools does the worker direct his actions? *Where* do the workpieces go during the task?
5. *Why* are certain actions taken or certain motions used? *Why* is one motion, rather than another closely similar, used in the working method?

It is obvious that if a motion pattern can answer these questions, there will be little doubt in the mind of anyone reading it concerning the analyst's interpretation of the method, and this can readily be checked against actual performance for verification, detection of changes, and clues to operator skill.

Insofar as machine time or process time are concerned, the motion pattern, by times and word descriptions, can show only what function they serve or add to the progress of the work being done. Whether such time was properly determined can be checked only by retiming that portion of the cycle by any of the numerous means and devices described in Chapter 3. It is also sometimes possible to calculate such times based on known factors and data.

Finally, a motion pattern with its required and associated data provides for future reference a concise compendium of facts regarding the task analyzed and comprises the best known type of methods record at the

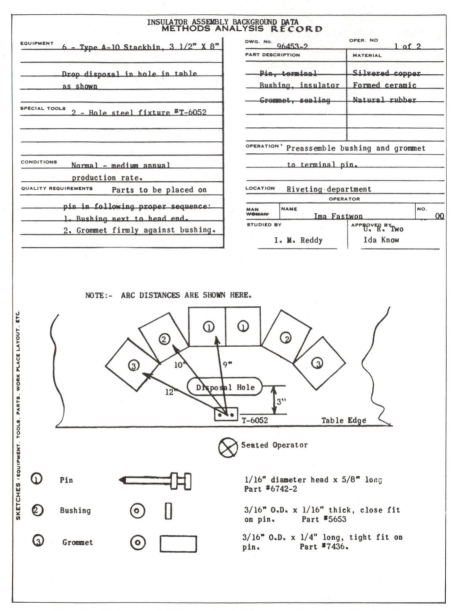

Fig. 1.   Background data for insulator assembly analyzed in Fig. 2.

present time.   Evidence, legal proof, time changes easily verifiable, detection of errors and difficulties in the task, the need for tools, and many other features are recorded in a form permitting easy confirmation or reconstruction of the method for any desired purpose.   Any task de-

METHODS ANALYSIS CHART						PART No. 96453-2
PART INSULATOR ASSEMBLY				DATE 7/26/56		OPER.No. 1 of 2
OPERATION Assemble bushing and grommet				ANALYST C. I. Kno		SHEET No. 1 of 1

DESCRIPTION — LEFT HAND	No.	L H	TMU	R H	No.	DESCRIPTION — RIGHT HAND
A. OBTAIN PINS, PLACE IN FIXTURE						
to pin supply		mR6C	7.2	mR6C		to pin supply
			9.1	G4B		select one pin
select one pin		G4B	9.1			
pin to fixture hole		M9C	12.7	M9C		pin to fixture hole
orient head down	2	G2	--	G2	2	orient head down
			21.8	P2SD		push into hole
push into hole		P2SD	21.8			
let go of pin		RL1	2.0	RL1		let go of pin
			83.7	TMU		TOTAL ELEMENT A
B OBTAIN, ASSEMBLE BUSHINGS						
to bushing supply		R10C	12.9	R10C		to bushing supply
			12.9	G4C		select one bushing
select one bushing		G4C	12.9			
bring to pin shaft		M10C	13.5	M10C		bring to pin shaft
preorient in fingers	2	G2	--	G2	2	preorient in fingers
push onto pin		P2SD	21.8			
			21.8	P2SD		push onto pin
let go of bushing		RL1	2.0	RL1		let go of bushing
			97.8	TMU		TOTAL ELEMENT B
C. OBTAIN, ASSEMBLE GROMMETS						
to grommet supply		R12C	14.2	R12C		to grommet supply
			10.9	G4C		select one grommet
select one grommet		G4C	12.9			
bring to pin shaft		M12C	15.2	M12C		bring to pin shaft
preorient in fingers	2	G2	--	G2	2	preorient in fingers
			48.6	P3SD		push onto pin
push onto pin		P3SD	48.6			
			152.4	TMU		TOTAL ELEMENT C
D. ASIDE ASSEMBLY IN HOLE						
get new hold – all parts		G2	5.6	G2		get new hold – all parts
pull assembly from hole		D2D	11.8	D2D		pull assembly from hole
assembly to dispose hole		M3Bm	3.6	M3Bm		Assembly to dispose hole
drop into hole in table		RL1	------	RL1		drop into hole in table
			21.0	TMU		TOTAL ELEMENT D

Fig. 2. Motion analysis of insulator assembly.

scribed in a motion pattern can be resumed easily and/or reinstated following even long lapses of time since the task was discontinued. The value of such detailed methods records is unquestionably great for practicing work analysts and others.

All of these facts comprise a powerful justification for using MTM instead of time study.

How Motion Patterns Show Facts

Perhaps the easiest way to portray the means by which a motion pattern shows the various facts cited in the last section is by analysis of an actual pattern. Figure 1 gives all the background data for an Insulator Assembly that is made by the motions analyzed in Fig. 2. This data is a necessary part of the motion pattern. A summary for the operation is shown in Fig. 3. Before proceeding, it is suggested that the reader familiarize himself with the two figures in detail. Bear in mind the previous remarks while doing this, since the following discussion will thereby be more meaningful.

Before proceeding with a discussion of the motion pattern itself, it is important that the data given in Fig. 1 be explained to show how vital it is to the meaning of the method and resulting time standard based upon it. Why this is so also is evident from the definition[2] for method:

Method:
1. The procedure or sequence of motions used by one or more individuals to accomplish a given operation or work task.
2. The sequence of operations and/or processes used to produce a given product or accomplish a given job.
3. A specific combination of layout and working conditions; materials, equipment, and tools; and motion pattern, involved in accomplishing a given operation or task.

Logically, if a method subsumes a *specific* layout together with the materials, tools, and equipment pertinent to the method, any change in these factors indicates, at least in some degree, a different method or slightly different motions needed. From the MTM viewpoint, *any* motion or time change in a pattern, however slight, constitutes a different method.

To substantiate a given method, therefore, requires the use of factual data, sketches, photographs, blueprints, and even models that clearly depict the essential features of the layout and job conditions which affect the method. The means employed will naturally be suited to the degree of precision required in the standard and the expected difficulty of supporting the standard; background information could even be recorded to the extent of making motion pictures which critics would find extremely difficult to refute. With these remarks, the significance of the data in Fig. 1 should be apparent upon examination.

A word of warning—it is far too easy and all too frequent that work analysts bypass or neglect this stage of analysis, and find later that it is the key to unanswered questions which embarrass them. The few extra minutes of analysis and recording time are well invested. In fact, the thinking and effort to clearly discern these factors in a job even lighten the work. of motion recording and analysis. It is likewise obvious that

---

[2] *Ibid.*

				OPERATION SUMMARY		PART No. 96453-2 1 of 2			
PART INSULATOR ASSEMBLY					DATE		OPER No.		
OPERATION Assemble bushing & grommet					ANALYST C. I. Kno		SHEET No. 1 OF 1		
No.	ELEMENT DESCRIPTION	ANALYSIS CHART REF.	ELEMENT TIME TMU	CONVERSION FACTOR .0006 LEVELED TIME	15 % ALLOWANCE	ELEMENT TIME ALLOWED minutes	OCCURRENCES PER PIECE OR CYCLE	TOTAL TIME ALLOWED min.	
A	Obtain pins, place in fixture		83.7	.05022	.00753	.05775	1	.05775	
B	Obtain & assemble bushings		97.8	.05868	.00880	.06748	1	.06748	
C	Obtain & assemble grommets		152.4	.09144	.01372	.10516	1	.10516	
D	Aside assembly in hole		21.0	.01260	.00189	.01449	1	.01449	
	TOTALS		354.9	.21294		.24488			
	ALLOWED RATE			490.0 pieces per hour or .12244 min. per piece					

TOTAL TIME ALLOWED PER____

Fig. 3. Operation Summary of elements for insulator assembly.

changes in any of these factors that might justifiably affect the method can
easily be checked and evaluated at any later date. The value of such data
can therefore be great.

Referring now to Fig. 2, there are natural divisions of the motion pattern
which answer the basic questions raised in the preceding section. These

will be pointed out and discussed in the same order previously used:

1. *What?*   The description columns—left hand and right hand—when properly written show with distinct, specific words what is happening during the elements.   There are a few good rules to follow that help achieve this aim.   Origins and destinations of Reaches and Moves, since they represent the major manual displacements, should be noted by "from" and "to" designations.   The object, part, tool, or other material on which the action is exerted should be identified by words that leave no doubt in the reader's mind.   It is wise to use shop terms and nomenclature[3], as well as to differentiate between similar or successive parts by finer terms or number references such as "first," "second," etc.   As soon as parts lose their identity in units produced, it is necessary to refer to them as assemblies.   Naming the assembly is also required when more than one assembly is included in the same element.   It is unnecessary to repeat in the description columns the words such as "Reach, Move, etc." describing what the operator does in the way of motions, since the symbol columns tell this precisely and in much better defined manner.   In summary, the description columns tell *what* happens to the objects, *where* it happens on the workplace layout, and *how* the tools and equipment are used.

2. *Why?*   Answering this query amounts to defining the purpose of any given motion and differentiating it from other motions closely similar. The purpose of a motion is partly defined by the symbol itself (because of the MTM definitions which the symbol represents) and partly by key words in the descriptive columns.   The differentiation feature is revealed by the class, case, and/or distance assigned to any given kind of motion indicated in the symbol column.

3. *When?*   This question is answered by the order in which the elements are displayed, the succession between the motions, and the frequencies that show repetition of either motions or elements.   As long as a pattern shows the complete method, there is no doubt as to when things occur.   Sometimes, when a pattern merely contains a collection of necessary elements, not necessarily in their proper order—as when used to develop standard data, for instance—this feature is less obvious until recourse is made to the summary of elements such as appears in Fig. 3.   In actual practice, it is usual to find this summary on a separate form rather than at the bottom of the motion pattern.

---

[3] Where they exist, in codes and standards and even legal sources, standardized terms and definitions should be used.   At times these might generate the need for educational activities among shop and staff personnel.   When such aids or commonly accepted shop terms do not exist, creating and establishing new terms may be a necessary and fruitful pursuit—especially when new technology is introduced into daily practices.

4. *How?* The columns labeled LH and RH containing the MTM motion symbols are used to show precisely how the operator does the work. The condensed, shorthand system which the symbols represent are already quite familiar to the reader from earlier study of previous chapters. To write the same information in words and numbers would require prohibitive space, which emphasizes the necessity and economy of the MTM system of symbols. The same statement applies for virtually all predetermined time systems—as mentioned earlier in this chapter. The symbol conventions also show the application of the Limiting Principle as discussed in Chapter 23. A properly written motion pattern must be carefully edited from the standpoint of which lines are used to record the symbols, since this convention and uniform practice becomes a great aid to interpretation when different readers consider the patterns written by different analysts.

5. *Where?* Part of this answer is provided by the pattern itself, as was noted in the remarks concerning the descriptive columns. But it is even more important to this answer to refer to the workplace layout and the parts prints or sample parts, since a visual aid is worth many words.

   In spite of this general fact, however, the finer distinctions of location are often inherently found in the motion symbols and carefully worded descriptive columns. Developing skill in designation of locations is one of the more difficult features of motion patterns for the neophyte to learn; his best success in analysis requires, however, that he attend to this matter at the very beginning of his MTM practice. There are two general ways to show workplace layout distances. Blueprint type dimensions from reference lines usually indicate that the distances shown are actual measurements on the projection given. A more frequently used strategy, however, is to employ arrowed lines radiating from the operator symbol and between parts that show, instead, the allowed Reach and/or Move distance (including arc distance) in any motions made between the locations. With this practice, it is easy to reconcile the motion symbols and layout. Where the method record warrants it, however, it might be desirable to show both types of layout for future reference.

6. *Who?* Of course, the motion pattern does not directly show or list which person, labor code, or job title will do the work described. However, the detailed information and data revealed by the pattern can be a powerful aid for those charged with personnel assignment responsibilities. Because personnel policy, supervisory, and even contractual questions also affect such decisions, the pattern can be only one criterium; nevertheless, routine means should be found to

assure that pertinent facts reflected in the motion patterns do actually come to the attention of those responsible for assigning personnel.

That the MTM motion pattern fully answers all the basic work study questions in a concise and clear manner is apparent from the discussion above. Remember, though, that these remarks apply only to well-written patterns; patterns lack methods value to the degree in which they depart from the criteria of good patterns just enumerated.

MOTION PATTERN MECHANICS

With the nature and desirable characteristics of motion patterns clearly in mind, the next problem is how to reduce motion analysis to a systematic approach. This section will give both general and specific ideas toward solving this problem.

MTM Working Forms

Paperwork of any kind can be systemized most readily when prepared forms adequate for the task and flexible enough to handle the range of situations are utilized. Accordingly, MTM practitioners make use of several general types of forms; the illustrations are typical of the forms adopted by many firms and industries applying MTM. A variety of such forms is available from the MTM Association at nominal cost, as mentioned in Chapter 10.

1. Since the method always depends on the conditions and surroundings during its performance, a Methods Analysis Record such as shown in Fig. 1 should be made a part of each pattern written. The principal exception to this suggestion occurs when standardized workplace layouts, tool designs, and other factors in the process have been developed in detail and assigned coded symbols. In this case, reference to the symbols on one of the other forms might lessen the need for the Methods Analysis Record. The divisions of this form are obvious from the printed designations.

2. The form which actually displays the motion pattern, commonly known as a Methods Analysis Chart, is seen in Fig. 2. Further discussion in this chapter revolves principally around this form. It is used both for studies of the visualization type and studies based on direct or indirect observation. Indirect observation consists of viewing the actual job to note key facts and complete the Methods Analysis Record, with the actual motion analysis and recording being done away from the job. It has the advantage of minimizing

interference with the work and operator, while permitting writing of the motion pattern for standards purposes.   With the facts compiled on the Methods Analysis Record, the descriptive portion on the Methods Analysis Chart completes the method details.   If the purpose is to document what an operator *is* doing, the motions will be written accordingly.   However, for standards purposes, the chart should show what the operator *should* be doing; that is, one must write the *average* motion pattern rather than the motions actually used by one certain operator.   The Methods Analysis Chart is, therefore, in all respects, the key MTM form.

3. After the MTM analyst completes a number of Methods Analysis Charts, he normally wishes to summarize the resulting data and develop the time standard for a complete operation or process.   The Operation Summary form generally designed as in Fig. 3 fulfills this need.   Essentially, it compactly collects data from many motion patterns and represents the finished data derived from the total analysis of an operation.   This form is such as to permit its usage to summarize time studies also, as sometimes is done by practicing engineers.

Familiarity with the format and content of such basic analysis forms will assist the MTM analyst in his work, and permits the paperwork to be systemized.

Elemental Division

When considering the mechanism by which industrial engineers are able to analyze even highly complex work routines, one striking feature stands out.   This is the old and obvious truism, sometimes lost sight of in these modern days of high pressure and exciting variety: *An observer can see only one thing at a time and understand it.*   A corollary statement might well be: *A person can think only one thought at a time clearly.*   Any reasonable or logical use of human powers in analysis requires recognition of these concepts and adaptation of them to the problem at hand.

The difficulty of divided attention, for example, is found in conventional time study because the analyst must almost simultaneously (1) observe the work, (2) observe and operate the timing device, and (3) record the details of timing.   This is a basic reason why such study is often inadequate, poorly done, or arrives at unwarranted conclusions unless employed by an analyst well versed in the intricacies of time study and experienced in the art of close observation.   The MTM procedure differs from this in that attention need be, indeed should be, directed to only one factor of the total analysis task at any given instant.   Among the features of the procedure that permits this desirable state is the means or method of subdividing tasks into elements for MTM motion study.

Having determined the general orientation to a process and divided it into operations and sequences, the analyst must next decide what elements should be chosen and define their starting and stopping points in the work cycle. Skill in this part of his effort will greatly simplify his work, increase the value of his finished analysis for many uses of methods and times, and result in a truer picture of the work being analyzed. There are a number of good principles which, when known and properly understood, shorten the time needed to develop such skill.

To begin the exposition of these principles, consider three additional definitions[4] as follows:

Element—a subdivision of the work cycle, composed of one or a sequence of several basic motions and/or machine or process activities which is distinct, describable, and measurable.

Element Breakdown—1) the separation of a work cycle into two or more elements.

2) a listing of work elements with individual descriptions and/or calculations for each.

Element Time—the time to perform a given element. May refer to the observed, average, selected, normal, or standard time.

The fineness of breakdown of operations into elements is thus not limited by the commonly accepted definitions. Furthermore, practicing work analysts know well that the current and future demands of work study indicate the need for both more finely divided and large elements—depending on the application requirements. In this respect, MTM has an advantage over time study. It permits complete freedom as to the point at which the element is split and revision of that point later without affecting the method in any way. Further, existing micro elements can easily be combined in various ways to construct needed macro elements. It also gives keys—in the right- and left-hand descriptive columns on the analysis form—as to the most appropriate name for the element, because the principal action or effect during the element is pinpointed with one or more of the motions included.

Length of element, while optional, should generally be suited to the particular work studied; one widely accepted opinion is that an MTM element should be limited to 12-20 motions or less for maximum ease of analysis. Even if elements of such length include many motions of high time value, they will be unlikely to total more than 100–200 TMU and are usually less than 100 TMU. Note that even 100 TMU is only .060 minutes. Most time study men agree that timing with a decimal minute watch cannot evaluate accurately elements requiring less than .02–.03 minutes; elements are commonly selected to require few instances where

[4]From ANSI Z94.1-12 IE Term. (Footnote 3, Chap. 1).

less than .10 to .15 minutes are recorded. Thus, MTM elements certainly permit finer element breakdown than do time study elements.

Another sometimes vexing problem with work elements concerns their identification as constant or variable, together with the cause of any variation. A constant element requires the same performance time on each and every occasion it may be repeated in the task of which it is a part. The time for such elements is constant chiefly because the method for the element does not change with added occurrences and/or the motions comprising the method do not contain any variable factors, such as differing distances. Variable elements, on the other hand, may require unlike performance times on each occurrence in the task. The time variation may be due to employing a different method to accomplish the purpose of the element, or else the length, case, and class of motions used may include variable factors that are not the same in time influence at each repetition of the element. Especially in time standards and time formula work it is important that the analyst be able to determine constancy or variability and detect reasons for any variation.

If constancy or variability are not recognized when the length and/or defined content of an element is initially determined, much wasted effort and difficult manipulation can result in later usage of the element. With MTM, these problems are minimized because the breakpoints between elements can be revised at any stage of the analysis without much effort; rewriting of the known motions into more convenient groupings is easily done. In addition, with each MTM motion and its frequency shown, one cause of element variability is readily pinpointed. Also, the motion symbols and verbal descriptions in the pattern yield obvious clues as to any cause of variation; the length, case, and class of the motion can all be examined for variability conveniently.

Time differences in time study elements are less readily explained and justified. The time variation might be that which could normally be expected, although the degree of variation—especially when slight but nonetheless significant in the standard—is often difficult to justify or suspect. The analyst would most likely scrutinize distances for readily detected causes of time variation, but he would find it difficult to isolate the cause in a time study element if this hypothesis did not prove fruitful. Accordingly, it is not as easy to initially establish or later revise the content of time study elements as it is for MTM elements; nor can the classification or analysis of time study elements be as thorough and defensible in regard to constancy or variability. This would be especially true if the element in question had the same name as another element covering an entirely different method of doing the task, since the time difference shown by the watch yields no real clue to the reasons why the method differs.

The MTM convention for separating the elements and how and where to write the name of the element is well-illustrated by Fig. 2.   Space of at least one line is left at the end of the preceding element—often used for totaling the time for that element—and the next element name is written in capital and underlined to make it easily distinguished from the rest of the pattern.   Of course, many alternate ways of doing this are possible and used, but conformity to such a convention would aid transmission of information between MTM analysts.

The Frequency Problem

In the experience of many MTM instructors, one of the more vexing problems for people learning MTM is determining and assigning frequencies to motions, motion series, and repeated elements.   It would be wasteful of analysis time, analysis forms, and the time of any party reading an analysis if each and every motion, series of motions, or job element were chronologically listed singly in an analysis.   Such repetitious writing could well destroy much of the effectiveness of the MTM technique simply due to the increased laboriousness of the analyst's work.   Motion economy principles should certainly be used by the analyst who hopes to get them adopted in the job he is studying!

The frequency problem can be greatly minimized by a number of intelligent attacks upon it, which will be described briefly in this section.   But first, it is well to show the mechanics by which frequencies are indicated on a study.

When single motions or motion combinations are immediately repeated in the work, a numeral to show the frequency is written in the "No." column (or columns) on the analysis form.   Note, for example, the double frequencies of G2 in the first three elements shown in Fig. 2.   This indicates that, on the average, several Regrasp motions will be performed while bringing objects to the assembly nest for insertion.   The same columns are equally usable for cases where motion series appear; such a series would be common in running down a bolt with the fingers.   The series of Reach, Grasp, Move and/or Turn, and Release could be repeated any number of times within an element, depending mainly on the number of threads on the bolt and the distance it must be advanced into the threaded hole.   It could vary also with the possible need to repeat the series more than once for a complete revolution of the bolt.   In any event, whatever frequency is decided as appropriate would be shown by numerals in the "No." columns on the analysis forms.

When whole elements are repeated in a given task, several means can be used to account for the extra time involved.   A new element could be written, with a note of its identity to a preceding element, and the total element time copied from the earlier analysis.   Or, the frequency could merely be multiplied at the point where the element is totaled and reference to the proper place of the repetition inserted in later stages of the analysis. The most usual place, however, in which to account for element frequency is in the column on the Operation Summary form headed "Occurrences per Piece or Cycle." In Fig. 3, for instance, the unitary frequency of each element in a total cycle is shown by "1" numerals in this column.

In any of the cases cited, of course, the motion, motion series, or element time values would be multiplied by the frequencies to obtain total task time.

Many MTM neophytes, while knowing where and how to indicate frequencies on their analysis, still experience difficulty in deciding the proper frequency to assign. Logic, common sense, and experienced judgment must be used in this decision; and it is quite common to find that the greatest source of error in task time by newcomers to MTM analysis is caused by their lack of these attributes at that stage of their experience. One good reason for this difficulty is that often frequencies are not constant, but vary with each cycle of the task; the need for reaching an assignable figure requires practice and knowledge of the task being studied, as well as familiarity with the various ways in which frequencies can be ascertained and/or what may usually be expected. The fact that an observed instance may differ from what would be proper to allow for the average operator in the finished analysis is a further complication.

The most obvious way to determine frequency is by direct counting as the work is done, with or without a counting device to aid when the frequency is high. Averaging of a sufficient number of cycles would then provide a reasonable expectancy of the correct count. Sometimes physical factors suggest directly the proper frequency, as would be true for the case of the bolt cited earlier. In many cases, application of the principles of applied mechanics, particularly where the hand action is on mechanical devices involving gear or similar ratios, will benefit the analyst if he is familiar with them. Statistical and ratio-delay analysis (described in Chapter 7) can be employed to derive mathematical approaches to critical frequencies. The practical analyst often will find that estimation will yield a quick and usually satisfactory answer, particularly if this is checked with the foreman, worker, or other analyst. It is evident that practice with all of these techniques will help the analyst in his problem.

Another factor with series of motions is that the frequency might not be the same for each and every motion in the series. To illustrate, if the worker had hold of the bolt before he started tightening it, the Reach and Grasp in a series like that earlier noted would be assigned a frequency one less than that allowed for the Move and Release motions. A further case in which the analyst without experience could err occurs when a tool is used that requires alternate approaches and departures from the workpiece. For instance, to hammer four blows could require—depending on conditions preceding and following the Moves written for the hammer blows—three, four, or five upstrokes; in any event there would be four downstrokes. Close observation and discernment of changes in the direction of motion must be relied upon to catch such frequency variations.

It is noteworthy that many of the frequency problems here discussed would never be encountered in time study simply because single motions are not evaluated in that procedure. Much of the frequency problem is buried in an element which is merely variable by the watch reading. On the other hand, the knowledge of frequency resulting from MTM analysis often provides one of the very best keys to method improvement; reducing the frequencies associated with motions, series of motions, and elements by various strategies is often one of the most time-saving tactics that can be utilized in work measurement.

A word of caution. Accounting for frequencies can be overdone in terms of analysis cost. This is especially true for "occasional" elements, which are often best handled by allowances, variance factors, or standard element times that apply to all infrequent occurrences of a given type and description. An "occasional" element could comprise minor setup, minor tool care, tagging, tabulating of pieces produced, etc., that covers necessary work which does not occur each and every time the complete work cycle is performed. When such work should be written into a pattern in detail, or when it should be handled by an alternate scheme is another instance where judgment and experience help the analyst greatly.

Synthetic Elements

Many times in the work of engineers, the use of synthesis helps achieve answers when direct facts or the limitations of their definitions do not permit them to give unequivocal answers to problems. In a sense, the employment of synthesis is a major difference between the successful engineer and one who becomes mired in a morass of uncertainty, indecision, and inability to reconcile facts with the necessity for a practical, workable answer. Synthesis, as used here, implies that the result or answer given is a reasonable approximation, although strict interpretation of the procedures and rules under which the answer was obtained would not warrant the given result. The use of such synthesis is a highly legitimate procedure provided the analyst does not fool himself into unbounded faith in the result or refuse to revise his decision in the light of new facts or better technology on the problem at hand; successful engineers have no qualms at using synthesis properly when it is appropriate and provides answers that apparently check against performance.

In work measurement, synthetic elements, motions, and motion equivalents are all basically a means of reducing analysis time and achieving answers which are required but do not merit extensive study or precision. For example, it is possible to argue endlessly whether a hesitation or shifting of the fingers in handling an object is one kind of MTM motion or another.   The real problem, oftentimes, is to allow sufficient time in the motion pattern if the shifting is legitimately necessary to the job.   If an analyst finds that, for given occurrences of this type, assigning of several Regrasps covers the time, he is using appropriately a synthetic equivalent motion.   An equivalent motion is one that does not precisely delineate by definition or actual occurrence the action taking place, but which does have a predetermined time that seems correct for the action.   It sometimes happens that minor planning pauses can be covered in this way, particularly when the time is too short to catch on a stopwatch as normally used for process time.   Hosts of cases where such synthesis can be applied advantageously could be cited.   It should be sufficient to refer, however, to the usage of synthesis in the theories of Grasp, Position, and Disengage given in those respective chapters.   Even the definition for an APB includes the idea of its differing from an APA by "a Regrasp or its equivalent."

The analyst must always remember, when employing synthesis, that his analysis is open to question and must sometimes be defended (such as in a grievance hearing or even an arbitration case resulting from an unsettled grievance) and the allowed time demonstrated to be correct.   While disputants to an analysis will seldom question the assignment of a motion which complies fully with basic MTM definitions, they will often provide much argument on synthetic motions which the analyst should be prepared to meet.

Most firms using MTM have compiled series of equivalent motions commonly accepted in their own usage as being time savers.   Many of such equivalent motions and/or typical motion sequences are peculiar to certain trades, lines of activity, and the practice of certain crafts.   An instance is the set of sewing motions widely used in the needle trades employing MTM.   Lists of equivalent motions and typical motion sequences of a more general applicability are available from many MTM consultants and firms using MTM.   All such synthetic devices are a strategy by which to reduce MTM analysis time, particularly where experience and validation by stopwatch check have shown the device to give reliable answers.   However, no claims to such devices being official MTM data are valid unless the MTM Association has sanctioned the particular synthesis through validated research.

Another use of synthesis consists in using special code schemes to indicate in motion patterns those motion series that are extremely common in a given type of work.   This device is merely a logical extension of the idea of predetermined times.   It must be used with caution, however, because there is a danger of assigning such coded equivalents when they do not apply or of bypassing good methods study due to the habit factor in usage of coded elements.

Extensive experience with synthetic coded elements actually "set the scene" for the development of the MTM-based combined and simplified systems—such as: MTM-GPD, MTM-2, MTM-3, etc.,—described in Chapter 11. The danger mentioned above is avoided with such systems by their integrated design for compatibility.

Clerical Aspects

Another saving in MTM analysis time results from judicious use of less skilled and lower pay personnel to relieve the MTM engineer of some of the clerical detail in pattern writing. The whole problem here is what can safely be delegated without sacrificing the effective methods work of the MTM analyst. Perhaps the most important factors are the degree of training, intelligence, and amount of dedication to perfection of the clerks. The answer will depend somewhat on whether the study is of the observation or visualized type. It will also vary depending on what the constants and variables in the task and the motions might be. There are several items on an MTM analysis, however, which can safely be delegated to a properly trained clerk or technician.

One of these is the insertion of captions on the forms, particularly when a large number of analysis sheets for a given job must be labeled identically. Another is the insertion, totaling, and extending of time values on the analysis and summary sheets. Posting of standard elements to master sheets is another possibility. This means of reducing the demands on the MTM analyst is worthy of wide acceptance, as it has been successfully used in many places. The whole concept is based on the idea that the method record sheet and actual motion analysis and description requires the talent and attention of an MTM engineer, while the remainder of the total job of finishing the work analysis is primarily clerical in nature. The completed study returned to the engineer can always be checked and revised, if necessary, to assure the efficiency of the clerical help utilized.

SUMMARY

The crux of good MTM usage consists in the writing and recording of good motion patterns that mirror the working method accurately. There are many skills which are a part of this task and which are essentially available only from an adequately trained MTM analyst. This is one good reason why application experience is absolutely basic to make knowledge of MTM theory and data meaningful for the MTM student. On the other hand, parts of the analysis task that can be removed from the shoulders of the MTM analyst should properly be assigned to technical or clerical personnel capable of performing them.

PART III

Reducing Labor Costs

CHAPTER 18

Labor Cost Control

ALLOWANCES AND VARIANCES

Before starting the discussion of our main subject, a few closely related facts and general work measurement standard's concepts need to be disclosed for the benefit of the less knowledgeable (about work measurement procedures) reader.

MTM-1 produces a time standard for the work to be performed that is known as Select or Normal Time—the same holds true for a properly performed and correctly performance rated Time Study. this time can be considered a Base Standard. It should contain *all* the readily measurable and/ or frequently occurring *work* elements (work events as well as basic motions) involved in the daily routine of performing the work. To meet such a standard the operator would have to work continuously with average skill and effort under average conditions using the method specified in the base standard.

The expectation that the worker can or should be expected to perform the specified operation for the entire work day is not realistic. He/she will almost always need to go to the restroom. An allowance to provide time to do this is called Personal (P) Allowance.

Second, the worker will almost certainly encounter situations that prevent him or her from working as specified. Reference here is not being made to the occasional built-in production system delays such as tool changing, set-up, accommodating delivery of a skid load of parts, the finding of an occasional part that will not work because of a defect, etc. Such delays should be noted on the work study, the events timed, their frequency determined, and a proper time allowance should be built into the Base Standard. The delays we are here discussing are those that are different from those recorded in the study, such as a power failure for a significant period of time, a delivery of mostly defective part(s) required in the operation, a stoppage of work by management to conduct a training session, a major machine breakdown, etc. These latter delays are unavoidable by the operator and are sometimes classed as Major and Minor. Usually there is a small Delay (D) Allowance built into the Base Standard for minor delays. The delays that are not built into the standard are Unavoidable Delays for the operator, and management must create a special allowance when such an event occurs. Therefore, such delays are often called Management Delays.

A third kind of allowance built into the Base Standard as an allowance is for company approved and sanctioned Rest (R) Periods and (where present) a tool, machine, workplace end of the day Cleanup Allowance. The duration, number, and clock time of the rest periods are specified as well as the duration of the cleanup time. Such events are built into the Base Standard as an allowance. These are known as Rest Period and Cleanup Allowances.

A fourth allowance built into the standard is a Fatigue (F) Allowance. We cannot discuss it here in detail and the reader is referred to *Advanced Work Measurement,*[1] Chapter 4. There is a strong historical precedence to have such an allowance, and one is usually found in the standard whether or not it is needed—usually in the range of 5–10 per cent. This allowance to compensate for the effects of fatigue is also a integral part of the Base Standard.

Once the so-called P F & D Allowances (including rest periods, cleanup, etc.) are included in the labor standard, the work performance time specified is called Allowed Time. There is very considerable body of additional related information about such subjects as operator capabilities, operator selection and training, learning and learning curves, decision time, etc. Again the reader is referred to *Advanced Work Measurement*[2] for more information. Also, a fairly extensive discussion of the kinds of special allowances encountered in industry is found in the third edition of *Engineered Work Measurement.*

INTRODUCTION TO LABOR COST CONTROL

As explained fully in Chapter 2, the control function of a manager can *not* be performed adequately without the capability and data to compare performance against a standard. This means, among other things, that one needs a standard of performance against which to base evaluative judgment. The control process is very similar to the sequence by which a

---

[1]Delmar W. Karger and Walton M. Hancock, *Advanced Work Measurement,* Industrial Press Inc., New York, N.Y. 1982.
[2]*Ibid.*

wall thermostat compares the existing space temperature to the preset standard and then initiates corrective action by actuating heating or cooling equipment to revise the temperature.   In the case of labor control, the more explicit the labor standards can be, the easier it is to perform the control function successfully.

This chapter, therefore, explores the control of labor costs based on standards, which may be established with the techniques described in this text.   While most examples in this text are of manual work, or at least work with a largely manual content requiring no significant mental activity or "thinking" time, the control principles are equally adaptable in a greater or lesser degree to the full range of productive human effort.

Substantial sums of money can be saved by determining proper labor standards for the work to be performed and then using them intelligently to gain better cost control.   Resulting savings then can be used to expand productive enterprises, which in turn will benefit both the workers and their management, as well as the ultimate consumer.

To illustrate this fact, one of the authors participated in the direction and installation, within a large industrial concern, of a dual program of methods correction and labor standards (including a standard cost accounting system) accomplished in less than a year.   The combined program resulted in *yearly* savings of well over one million dollars, with about half of this total resulting from each phase of the program; the half-million dollars saved just from establishing labor standards dramatically emphasizes that statement.   However, it was not merely the establishment of labor standards nor the standard cost system which produced the savings, but rather that the firm took advantage of both factors to attain a better control of labor costs.

Labor standards are normally used to best advantage in conjunction with either an incentive pay system and/or a standard cost-accounting system. The standards program mentioned did not involve wage incentives, but relied instead upon measuring performance against the standards in terms of individual operator efficiency so that supervision could implement any indicated corrective action. According to one source,[3] the introduction of labor standards and a measured daywork system (described later) over the "no standards" condition usually improves productivity about 14.6 per cent, as based on a survey of more than 400 companies. If an incentive system is associated with the introduction of the labor standards, the productivity improvement over "no standards" will be about 63.8 per cent. However, the associated cost savings would be smaller since they stem not only

[3]Mitchell Fein, "Rational Approaches to Raising Productivity," *Monograph Series No. 5,* Work Measurement and Methods Engineering Division, American Institute of Industrial Engineers, Atlanta, Ga., 1974.

from the increase in productivity from the "no standards" condition to the point where incentive pay is received, but also include the benefits of greater production from the involved capital investment. The approximate money savings for the measured daywork case were about 12.8 per cent and for the incentive case about 24.6 per cent. It should be obvious that not all the increase in productivity is a labor performance savings.

BASIC CONCEPTS OF INCENTIVE PLANS

Before discussing the details of incentive plans and their associated standards, a quick overview of these subjects is advisable.

First, consider standards and their use. One can have standards merely for management to make comparisons between actual performance and that required by standard. This sometimes is done, rather obviously, where the worker is paid an hourly or day rate; for it generally represents the first case described above, where the supervision is expected to implement required corrective action. The firm may or may not have different levels of pay for the same kind of work. The latter situation is more commonly found for such crafts or "trades" as electricians, carpenters, welders, etc., than for the more standardized operations performed by most ordinary production workers. Whether or not different pay levels exist, if certain workers obviously are very poor performers management will discuss individual performances against standard with such workers. This essentially describes a straight daywork with standards pay scheme.

In subsequent discussion, two basic methods or plans for using labor standards must be recognized. One is a basic *incentive pay* plan. As shown later, this has many variants which all establish pay rates based on short-term performance. For example, either a worker is paid a fixed amount for each 100 pieces produced or else the result of working for an hour at 125 per cent of standard merits pay of 125 per cent of the basic hourly rate. If the individual workers and the plant management agree on a long-term performance at 125 per cent of standard for 125 per cent pay, then the pay scheme is effectively converted to the second basic approach of a *measured daywork* plan of payment. Here the essential feature is the fixing of the pay level on condition that a specified level of performance is maintained—in the long term. Pay does not fluctuate with short-term variations in performance. Such measured daywork plans may or may not be modified to permit periodic bonuses that, in effect, provide a long-term straight incentive adjustment.

A recent in-depth study of the increasing usage of measured daywork

[4]"Measured Daywork," Office of Manpower Economics, Great Britain Department of Employment, Her Majesty's Stationery Office, London, 1973.

schemes in England[4] found that they covered 9 per cent of United Kingdom workers. The three main variants found all pay the worker on a *time* basis. One plan requires performance to match that specified by established labor standards. Another pays a bonus for longer-term maintenance of the standard performance level. A third arrangement allows the worker to choose among a series of performance levels commanding differing pay rates and agree to maintain the chosen level. Obviously, each of these schemes can have sub-variants.

Although agreement is less than unanimous among industrial engineers, one finds three performance levels associated with the three pay plans already discussed in this chapter. The lowest exists with the straight day-work pay scheme. The next higher level occurs with the measured day-work system, especially where any higher-than-average performance by workers is rewarded by bonuses or pay raises. Rather obviously, the payment of direct incentives results in the highest level of average performance.

The closer worker pay is tied to labor standards, the more they will be interested in them. Accordingly, both industrial engineers and managers must recognize certain facts. One is that the standards must be as correct as it is economically possible to make them and they *must* be updated each time any product, process, or technology change affects the workers' ability to produce. Another fact is the likelihood of more complaints filed by the workers which might lead to grievances, especially in an incentive installation. With proper and expeditious handling, this situation need not be intolerable to either the workers or their management. It has been handled successfully in numerous and highly varied experiences within the full range of companies.

Finally, most of the existing labor standards installations causing difficulty either do not properly recognize the factors listed below or else are faulty as to their implementation or maintenance. Work standards need to be multidimensional, in the sense that they account for:

1. All the factors that the operator is expected to control—such as product quality, protection of and responsibility for equipment, etc., as well as the speed of performance in relation to expectations.
2. The variability of operator performance times.
3. The variation in learning time, especially that due to differing lengths of work cycle.
4. The continuous reduction in the unit cost of labor due to continued experience (a form of learning) by everyone, not the production worker alone.
5. Continuous refinements in work methods, product design, and production technology.

6. Machine interference within the work cycle affecting the attainable operator productivity.
7. The effect of lead-time compression in the competitive introduction of new products and new technology.

Each of these items is treated in this book as indicated parenthetically.

Labor Standards with Direct Incentive Pay

Most of the direct incentive plans pay a bonus in direct proportion to performance above the standard base. Although early incentive installations often used some sort of sliding scale approach, very few of these still remain in modern industrialized nations.

As mentioned previously, the method of computing an operator's pay under an incentive plan is based either on a price for each 100 pieces produced or on a base rate for a standard hour's work. The piece rate systems have virtually disappeared because one must revise *all* rates every time pay rates are changed; a monumental task. No such problem exists with the other schemes. The basic procedure for pay calculations under the piece rate system is obvious and needs no elaboration. The procedure under the base rate plan is almost as obvious.

One method is to calculate the operator's efficiency in terms of hourly, daily, or weekly efficiency by dividing actual production by standard production for the period involved. This efficiency is then multiplied by the base rate; of course, adjustment must be made for the number of hours worked at the calculated efficiency on the particular incentive job.

A second method divides the actual total number of parts produced by the hourly standard (pieces per hour) for the operation to obtain the standard hours of work. These standard hours multiplied by the base rate will determine the amount of pay due the operator. The actual detailed steps followed by any given plant varies with the individual situation.

Labor efficiency is officially[5] defined as:

Efficiency, Labor—1) the ratio of standard performance time to actual performance time, usually expressed as a percentage.
2) the ratio of actual performance numbers (e.g., the number of pieces) to standard performance numbers, usually expressed as a percentage.

Note that the methods indicated for efficiency calculation are in accord with these definitions.

It is quite common for incentive installations to begin payment of the incentive 10 to 20 per cent below what is actually considered as normal performance. This policy utilizes a so-called incentive allowance which actually has the effect of adding time to the base or select time for the

[5]From ANZI Z94.1-12 IE Term. (Footnote 3, Chap. 1).

operation. Beginning payment of an incentive at a point below actual normal performance is based on the theory that an operator who begins to earn an incentive premium will be further motivated to attain the best possible performance. Actual results in the field indicate some factual support of this theory and some benefits can be derived from this approach. Such policy is, of course, often reflected in the base rates—often they are lower than would be the case if payment of incentive began at normal or 100 per cent performance.

Many situations exist in industry where an operator works part time on an incentive job and part time on a non-incentive job. This type of situation can be handled by reporting practices connected with the daily time card system. Also often involved in such a case is a problem of two different base rates. The incentive job may have one base rate and the non-incentive job another base rate. However, it is normally undesirable to require operators to work under such conditions. An operator placed on an incentive job should be provided, wherever possible, such tasks that all assigned work will be on an incentive rate, even though two or more different operations on different parts are involved.

Also encountered are situations where during a given pay period operators work on two or more incentive jobs that do not carry the same base rate or price per hundred pieces. Again, the time card system can take care of such situations. Operators are often paid average earnings when machine breakdowns occur, when taken off incentive jobs to instruct new operators or try out new machines, when their normal work is not available, or when they report for work at the beginning of a shift and for some reason cannot perform incentive tasks (parts are not available, etc.). The payment policy for such situations varies from plant to plant. While it may be questionable to pay average earnings for machine breakdowns, lack of parts, etc., the practice is unquestionably proper for temporary absence from regular work to instruct new operators or to try out a new machine. These latter two occasions obviously occur at management's discretion, so an operator should not be penalized for possessing acquired skills valuable to management.

It is generally conceded that an incentive pay system should never be started unless the majority of operators can be covered by the system. Sixty per cent would constitute a bare minimum, a much more practical and desirable coverage being 75 to 90 per cent. Once an incentive system is started, those not covered by incentives will normally begin demanding coverage. If it is obvious that the system cannot be extended to cover most of the direct labor operators, then it should never be started at all, since it will only cause labor difficulties within the plant when attempted under such conditions. An obvious solution under these circumstances is to set labor standards where possible and measure individual operator and/or department performance against them, which data can then be used as an aid in securing adequate performance.

Individual Versus Group Incentives

On individual operations, it is natural and easy to provide for an individual incentive. It is generally agreed that individual incentives are also the most effective and, therefore, desirable type; however, individual incentives cannot always be established. This is particularly true for progressive assembly operations where any particular operator's productivity depends upon the productivity of all team mates. Group incentives are logical in this type of situation. Group incentives are not as effective as individual incentives; however, some increased performance is usually obtained.

Under group incentives, particularly those involving progressive assembly lines, the maximum production is limited by the slowest operator. It is usual in these situations to find the more experienced operators aiding inexperienced persons, thereby reducing the earnings loss due to training of new operators introduced to the line. It is also true that they encourage laggard operators to strive for their best performance. This action by the laggard operator's team mates is often much more effective in obtaining maximum production than is pressure from the supervisory group.

Another beneficial aspect of such group incentives is the fact that when an operator obviously cannot keep pace with normal and/or superior operators, they will approach the foreman and actually request replacement. In non-incentive installations, on the other hand, a laggard operator's team mates often keep quiet or even insist that the laggard be retained, since their pay is not affected and the amount of work expected of them is reduced.

Process-Controlled Work and Incentives

Whether the operator is rewarded in direct proportion to actual production or on a sliding scale, one of the most controversial elements in an incentive plan is the correct procedure when all or a large portion of the work cycle is machine- or process-controlled. The greater the proportion of the work cycle controlled by machines or other process requirements not directly controlled by the operator's pace, the less is the opportunity and hence reward for maximum operator effort unless a special incentive plan is developed and used to compensate for this situation.

With industry constantly increasing its use of mechanization and/or automation, the problem of incentive payment on such jobs is ever more critical in relation to maintaining satisfactory labor relations and maximum output. Part of this problem are the facts that a fair incentive plan must permit a reward for operator efficiency and that performance above standard has two effects on work including machine- or process-controlled elements. By maintaining a faster pace in the manually controlled portion of the cycle, the operator obviously accomplishes both the external and internal manual work in less time; and *this also increases the machine utilization,* because the essentially constant machine cycle then is performed more often per unit of time than if the worker had merely sustained an average or normal performance pace. In addition, if the operator continues the faster pace during internal work, extra idle time then makes possible either enough rest so that the higher pace can easily be continued without undue fatigue or a capacity to absorb extra machine assignments that will enhance the value of the operator to all concerned.

With an understanding of the facts just presented, the design of an incentive plan for wage payment on tasks not fully under operator control resolves itself into a decision on the rewards due the operator for superior performance and then a proper application of fundamental mathematics so as to yield the desired result. This is easier said than done as evidenced by the wide variety of apparent solutions found to this problem in industsry. A survey study[6] which evaluated the practices with respect to a number of important criteria is rephrased by the authors as follows.

1. Is the plan simple and understandable? That is, are the calculations and ease of explanation such that the operator will be confident of it?
2. Does the plan afford adequate incentive earnings considering consistency, fair differentials, comparative attainment, and maintenance of proper attitudes by all parties to the plan?
3. Will the plan protect management's interests in regard to equipment performance, manual labor output, ease of developing and explaining labor standards, and simplicity and accuracy of standards calculation?

The approaches or solutions made by industry, from the source cited, can be classified as follows:

A. Normal piecework or standard hour (previously discussed)
B. Constant machine incentive allowance combined with standard hour
C. Constant work incentive combined with standard hour
D. Empirical determination of work performance
E. Empirical determination of both work and equipment performance.

The first approach has already been discussed. It is unsatisfactory for process-controlled work because it does not provide an adequate incentive opportunity on such work. Although the second solution can provide financial incentive, its inconsistencies undoubtedly tend to result in dissatisfaction by the workers. That is, although it might yield usable and reasonably correct results for a given task, it contains no adjustive feature for task variation.

Approach C appears to be the best over-all procedure, except for one factor—the pace adjustment used is a constant which yields exactly correct

[6]Dr. John C. Scheib, Jr., "Incentive Wage Practice for Restricted Work," *Journal of Industrial Engineering,* Vol. VIII, No. 3, May-June, 1957.

results only when the operator works at the forecast pace. It provides a generally satisfactory financial incentive with relative simplicity of mathematical computation for development of standards. The earnings computation is simple and straightforward, with the difficulty of explanation being slight. By using pre-calculated tables, etc. the slightly complex computations required for standards development can be greatly simplified also. The standard is developed by summing *all* manual work elements (both external and internal) and *adding to this the idle time which would be incurred if the operator would work at a certain pre-established level of performance.* Of course, this standard will be *exactly* correct only if the operator works at that pace, since too much idle time will be included in the standard if he works faster and too little when he works below the pre-established pace. Actually, this error normally tends to be small. The approach does allow or adjust for task differences, as well as providing the other benefits previously indicated. The authors feel it is the best practical solution to the problem for most situations at the present time.

The pay plans involved in Approach D are based on empirical determination of earnings by the use of tables, etc. *which depend primarily on the manual work time involved.* Since such tables can usually be constructed to provide a correct result only for a given task (in terms of percentages of external and internal manual work and idle man time), the standards and earnings calculations both become extremely complex if the plan attempts to adequately and properly cover a wide range and variety of tasks. This leads to a further major detriment—the operator will not understand it.

Although the last solution is the most complex, it is technically the best in that it can yield a precise answer to every task evaluated. However, its complexity tends to negate the fact that it provides satisfactory incentive earnings and adjusts well to differing tasks. This procedure attempts the admittedly desirable task of giving consideration to the maintenance of machine capabilities as well as to the manual work performance. While neither of the resulting dual standards—one covering the machine capacity and the other the performance of the worker—is too complex in its development or application, the time standard has little meaning to the average operator and the earnings calculations are of necessity very complex. This complexity results from the necessity to relate each of the two standards to actual time and then relating these two results to each other through the use of formulae or prepared tables. This plan therefore does not appear as practical as Approach C.

It cannot be too strongly emphasized here, however, that the success or failure of any wage system does not depend solely on its merits, but more importantly on how wisely and fairly it is administered. While undesirable features can be overcome, it is best to take full advantage of all possible benefits and/or features during the original design of the system. Hence,

Approach C appears to offer the best solution to incentive payment for operations involving process-controlled elements.

A more recent solution not included in the source cited but which is very close to Approach D is known as Multi-Factor incentives. Here, the incentive is based on combinations of job-related factors, some of which have no connection with the magnitude of the output from the work cycle. Rather, some of the factors concern product quality, composition, etc. All or only some of the factors may be controllable by the operator; therefore, others may be machine- or process-controlled. Multi-Factor incentives can be developed for most man-machine operations and for operations which are only partially process-controlled. Developing Multi-Factor incentives involves essentially the following steps:

(1) Determine those factors other than mere quantity of output which the operator can control, e.g. thickness of coatings, composition of mixtures, percentage of impurities, fineness of finish, brightness of polish, etc.

(2) Establish for each such factor in Step 1 its relative importance to the finished production

(3) Assign numerical "weights" by which each of the controllable factors may be allocated, i.e., the degree of effect controlling of the given factor should have on the amount of an operator's pay

(4) Isolate and measure the amount of machine- or process-controlled time in the work cycle

(5) Establish the time required by the manual portions of the operation

(6) Combine the data of Steps (4) and (5) to establish the basic work cycle time.

(7) Synthesize the results of Steps (3) and (6) into a wage incentive formula for the given operation.

Such Multi-Factor incentives are the most sophisticated approach possible. They normally are used only when operator-controllable factors have a sufficient degree of economic importance to justify the development of this kind of incentive. This approach also can be used in a general way to set performance goals for such tasks as engineering and accounting if the work measurement engineer is creative and imaginative in applying this procedure to such work.

Job Evaluation with Incentives

Before initiating a program of labor standards, it is usually wise to couple it with a job evaluation program. Basically, job evaluation is a systematic procedure for determining proper differentials in pay between the various tasks in a plant. Note that it essentially determines *differentials* in pay and not the exact base rate for each job. A good industrial engineering course will cover the details of a job evaluation program and it is not the intent here to cover this procedure. The industrial engineer's attention is merely called to the procedure and the fact that it is wise to assure the establishment of proper pay differentials between the various jobs.

Job pay-rate differentials sometimes become quite distorted in plants where certain groups of employees performing certain key operations have been very aggressive in the management and direction of their labor union for many years. These groups then often achieve special benefits for themselves which makes their rate of pay out of proportion to that of others in the plant.

Similar disproportionate pay often results from changed conditions. For instance, special premiums are often introduced for performing operations involving exposure to heat or fumes which at some later date following installation of modern ventilation systems and/or cooling systems may be entirely unjustified. Yet the pay differentials originally established because of the poor conditions usually continue. Similar situations often exist when originally highly developed skills were required and later mechanization eliminated the need for such skills.

Summary of Incentive Pay

To summarize, maximum increase in productivity through the use of labor standards is attained when the standards are coupled with an incentive pay system. The increase in productivity above normal is approximately 20 to 30 per cent. A 20 per cent productivity increase does not mean that management has gained by 20 per cent, but rather by some factor less than the 20 per cent, since extra payment is made for production exceeding the normal. However, management gains in addition by somewhat more production at base cost because less floor space per unit of production is required; fewer machines are needed; fewer facilities in general are involved; and also the total cost of personnel services, insurance, and other similar practices are reduced. Incentives also provide management better control of production so that they can more accurately forecast completion dates of orders, etc. These benefits to management are quite tangible and substantial.

It is sometimes argued that production under an incentive pay system does not yield a high quality product. However, quality is not necessarily sacrificed because of an incentive installation; but it is quite important to maintain strong inspection and quality control organizations. Actually, this is important under either an incentive or non-incentive pay system; but it is somewhat more important to have such organizations under incentive conditions. Just how much more important is somewhat difficult to evaluate, since there are many instances of poor quality work in non-incentive plants.   There is a greater *tendency* in incentive plants for the workers to neglect quality *if* they are not watched and/or know they are not being checked.

LABOR STANDARDS WITHOUT INCENTIVES

Labor standards systems without incentives are usually called "measured daywork" systems. This is defined[5] as follows:

Measured Daywork—1) work performed for a set hourly nonincentive wage on which production standards have been established. (Most frequent use.)

2) an incentive plan wherein the hourly wage is adjusted up or down and is guaranteed for a fixed future period (usually a quarter) according to the average performance in the hourly period.   (Infrequently used.)

The reader will recognize the latter sense as the one discussed earlier in this chapter; now the main usage of the term will be discussed.   The problems involved in establishing standards are essentially the same as for incentive systems.   The standards must be equally good for a measured daywork system to be worth the effort required to establish and administer the plan.

To achieve satisfactory production, that is production approaching (but practically never equalling) that in an incentive plant, means that every known management technique must be utilized to the fullest extent.   Having standards makes it relatively easy to detect below-average performance. The standards also provide a definitive goal for the operator.   This is usually all they will do under a measured daywork system, and it is up to the foreman and other supervisors to get adequate performance.   This means that they must be well trained and motivated to demand adequate operator performance.   This is easier to say than to achieve; however, it is not beyond the realm of possibility—if management recognizes the problem and gears itself accordingly for the effort required.   This applies to management at all levels from the factory manager down to the lowest supervisor.

Fortunately, most measured daywork systems are coupled with a standard cost system that automatically generates reports showing variance from standard performance.   The factory manager, supervisors, and industrial engineers must analyze these reports and determine reasons for any substantial variance from standard or normal performance.   Industrial engineers, especially, will find that daily or weekly efficiency reports showing variance from standard performance are a necessity.   In some cases, such reports can be in terms of departmental performance and in others they

[5]From ANSI Z94.1-12 IE Term. (Footnote 3, Chap. 1).

must be based on piece-part and/or individual operator performance if usable data is to be developed.

When a new job is started, operational performance reports are very desirable and in many cases a necessity. If over-all departmental type reports are utilized and there are many different kinds of operations in the department, it will be virtually impossible to detect which operators and/or which operations are causing any over-all low performance. It is only when the engineers and supervision are able to detect those operations most in need of corrective action that the necessary concentration on the problem can result.

These so-called efficiency reports are ordinarily produced by the factory cost accounting department and/or timekeeping department, but in some cases the industrial engineering department produces them. Under certain conditions, especially where a particular kind of job has been in operation for a long period of time, it is possible to resort to efficiency reports developed by foremen. This normally is not a good procedure because it makes it possible for the foremen to hide poor performance. However, it does have one beneficial effect in that it requires the foreman to take direct cognizance of the efficiency of his department. The usual approach, when having the foreman calculate his department's efficiency, is to have the industrial engineering department give him the number of standard hours credit per week that he should receive based upon the scheduled production or in terms of standard hours per unit of production. The foreman then needs only add up the number of actual hours expended by the personnel in his department and divide this sum into the number of standard hours to determine departmental efficiency.

TROUBLE SIGNALS

Low efficiency is always a "red flag" signal indicating a need for corrective action.

Low efficiency could indicate poor or insufficient training, low operator performance, incorrect methods, improper material, poor dimensional tolerance and/or specifications, improper parts, machines in poor condition, insufficient material, etc. This list could be continued ad infinitum. The industrial engineer's tasks, in cooperation with the other operating departments, are to locate the cause and to secure correction when and where required. Neither of these is an easy task.

Locating the cause of the trouble will often call into use all of the analysis skills and procedures at the command of the engineer. Generally, the problem should initially be tackled as follows if the probable cause is not already fairly well established:

1. Interview the production supervision involved
2. Interview the operator
3. Check with inspection and/or quality control personnel.

By this time, the probable reasons for the low efficiency will normally be fairly well established.

If the engineer dealing with departmental efficiency does not have operation efficiency reports available, he will be working under a great handicap and likely will not be able to isolate the cause of the trouble and secure correction. The reasons for this statement are multitudinous. One operator may be experiencing material delays, another is using incorrect methods, still another has a poorly operating machine, etc. Even if some bad conditions are corrected, new ones can develop without the engineer being aware of them, much less knowing their nature and location.

Once the general nature of the difficulty has been isolated, the engineer needs to select his analysis tools and proceed to finalize the exact cause of the difficulty. The various techniques mentioned in Chapter 3 will often help. Other hints involving work measurement analysis techniques have been given in other chapters. Common sense application of this knowledge usually yields a satisfactory answer as to what is wrong.

As soon as the *cause* of low efficiency is located, a major portion of the corrective task has been accomplished. Many hints and suggestions regarding the correction of work measurement difficulties have already been given. Others will be found in succeeding chapters.

Low efficiency is not the only indicator of trouble! Abnormally high efficiency is also a "red flag" indicator; in fact, it does not even need to be abnormally high. For example, assume that a new department has been established in a measured daywork plant to manufacture a new product and that after two or three weeks some operations show efficiency of 40 to 50 per cent, the majority show 70 to 90 per cent, and about one-fourth of the operations show performances of 110 to 130 per cent. Obviously, the 40 to 50 per cent efficiency is too low and needs attention. In like manner, the 110 to 130 per cent efficiency operations also need attention. The high performance, while not out of line for operating an incentive installation, is not normal in a measured daywork installation—especially not under the circumstances enumerated. The causes could be use of improved methods, poor standards, reduced quality requirements, above-average performance of operators, widened dimensional tolerances, better tools than those specified, etc. All of these may be desirable; however, all but one of the reasons indicate the need for a new labor standard. If such conditions are not corrected when and as they occur, the work measurement program soon deteriorates to being worse than useless; it really becomes a hindrance to operating personnel because no one believes in it and both labor and management try to use it to their individual benefit.   Chaos is the final result.   Work measurement programs must not only be carefully engineered in the beginning, but they need *expert* and *continuous* attention throughout their life.   This concept should always be remembered, not only by the industrial engineer but also by management.

THE STANDARDS EROSION PROBLEM

All the previous indicators of trouble with an incentive system, or even with a non-incentive standards program, are detectable because of "signals" which can be rather obvious to a knowledgeable observer.   Not so obvious is the creeping deterioration of the quality or correctness of standards called "standards erosion."   It is not detectable on a day-to-day, week-to-week, or even on a month-to-month basis; and often it escapes detection even when comparing one year's experience with another.   It probably is also one of the main reasons why only about 26 per cent of industrial workers are on incentive.

This situation is as serious as it sounds.   Earlier in this text and again in this chapter, mention has been made that people learn and their organizations accumulate experience.   A worker or his supervisor often will conceive or learn how to make very minor improvements—not that they fail to conceive of large-scale improvements, but these are obvious and are usually handled through a suggestion system.   Also, in time, incoming materials and the tooling used are improved.   These events result in a sort of creeping improvement of perhaps 1 to 3 per cent per year.   Unless these cumulative changes are picked up by a sensitive reporting system or some other strategy so that the standards can be kept adjusted to reality, their value as a basis of control is steadily eroded.   No matter what the industrial engineers tell them, the supervisors figure that such minor improvements will not hurt anyone—so why rile the employees by telling management, assuming the workers do know about them.   Often the supervisors themselves, however, do not know about the improvements.   One year's effect may not really matter, but the compound effect of 5 to 10 years of such gradual improvement is to raise the *average possible* bonus payment from the range of 20 to 30 per cent to an inordinate level of 40 to 60 per cent.   If the system has been permitted to slide into this state due to inaction toward correction, about the only recourse is to scrap it and start over.   Little imagination is required to see what can happen to labor tranquility in the concerned plant.

Actually, what usually happens to avoid such dire results is that the workers sense they can "get by" with raising bonus earnings to about 30 per cent, maybe to 35 per cent—and they usually can.   Individuals raising incentive earnings above this range risk a good chance of having their jobs restudied, and this probably will happen because most industrial engineering departments do look for such obvious signals of trouble with standards. Rather than take this chance, most workers hold or "peg" their production to keep their incentive bonus at about the 30 to 35 per cent range and work more slowly or loaf part of their working time. Having most of the plant in this condition is almost as bad as having the average worker earning 40 to 60 per cent bonus; it is costly to everyone concerned in either case.

If management and the industrial engineering department *both* understand the above process and realize that permitting it to happen is why most incentive systems are abandoned, they *can* act to prevent it. How? Simply by auditing—restudying the job—it does not take long to merely check each standard at least once each year and then make required minor corrections. Incentive programs can and do succeed over the long term because of such action.[8] An even more reliable audit strategy is the continuous random checking of standards on a sampling basis that assures complete annual coverage, especially if the auditing target is known in advance by *no* one except the analyst. Many firms use these and other, often sophisticated and complex, auditing strategies to prevent standards erosion.

COMBINING MTM AND TIME STUDY

It would be inappropriate to leave the subject of labor cost control without discussing MTM versus time study. The work measurement engineer is often required to make a choice between these two techniques.

Some of the more prevalent work study situations encountered in industry will be explored by first delineating the problem and then giving the more important reasons why time study and/or MTM should or can be used. In some cases there is little technical choice between them. Occasionally, entirely non-technical considerations—such as union attitudes—may force the choice of one or the other.

Operator Not Meeting Standards

The existing condition is that the operation has been processed, the standard set, etc., but the operator claims inability to meet the standards. The problem facing the engineer is to determine whether the standard is

---

[8]Mitchell Fein, previously referenced (Footnote 3 in this chapter), has proposed what he admits is a more radical solution—buy back the minor improvements. The reference describes how his hypothesis *might* be reduced to practice.

wrong for the process or if the operator is failing because of inherent poor ability, not following the process, deliberate malingering, or insufficient training and/or practice. Time study is the most commonly used technique in such situations, regardless of whether MTM or time study was used to set the standards. A time study will first be made to check the actual versus standard time. MTM training will then help the engineer most in reconciling any lack of agreement in the two times. The real questions to be answered in such cases are:

1. Is the specified method being followed? Are ineffective motions being substituted?
2. Have conditions changed? Are the method motions different to compensate for this?
3. Is the operator reasonably well trained? Was the method otherwise well instituted?
4. Is the operator working with average skill and effort? Does his rating correct the actual time back to standard?
5. Is the operator deliberately trying to avoid meeting the standard? Does notable lack of consistency in the stopwatch readings point to malingering?
6. Is the standard satisfactory? Does obvious confirmation of the standard time show the problem to be non-technical instead?

Items 1, 2, and 3 can probably be answered best by MTM; however, time study could be used. Time study yields better answers in less time to the other questions. The clue as to which technique to use first is the fact that the watch readings provide a quick indication of whether further technical study is justified. Items 4 and 5 especially almost demand usage of the time study technique first. Lack of consistency in performance for each element would raise the suspicion that the operator is deliberately trying to fail. Consistency of poor performance, on the other hand, would indicate a lack of ability or insufficient practice; the supervisor could help in determining which of these is involved. If the definite cause still cannot be found a detailed MTM study should be made.

Both time study and MTM can serve to check the standard in this manner. However, MTM will provide the most accurate check for the manual parts of the process, while time study is essential to checking the accuracy of process times and it also provides quickest the first clues as to the proper course of action in resolving the operator's complaint.

If the task must be reprocessed and/or a new standard established, all of the previously mentioned factors regarding MTM and time study must be considered in making the choice between MTM or time study.

Low Volume, Non-Repetitive, Low Manual Labor Content Operations

Examples of this situation would be the operation of mills, lathes, etc., in a toolroom, maintenance shop, etc. What technique should be used to set the standard? How can measurement of such work be economical?

The technique to use is not clear. Time study is the usual approach; however, this means the job must be running. This, in itself, is not a major hindrance in many such cases, since time study of the machine time would be required even if MTM were used. The engineer would then, of necessity, be present during the performance of the manual operations anyway and could just as well use time study.

As the manual work content increases, the choice of MTM becomes more important. Also, a more important consideration then is the possibility and need for standard data, where MTM gains the advantage for reasons stated elsewhere in this text. Some firms have solved this problem by using MTM almost exclusively to build standard data and/or time formulas.

Actually, for the situation originally stated, the combined approach of MTM and time study will often prove most fruitful. Standard data, either by element or by task, can be developed to provide time standards for later recurrences of operations of this type. It must be recognized, because of the non-repetitive factor, that the expense of obtaining extreme accuracy with either technique is not warranted since a good approximation is sufficient. The best approach, therefore, seems to be estimation of the manual portions by simplified versions of MTM data and rough indications of the process times by stopwatch checks of the few cyclical occurrences in which they appear. After such analysis on several occasions, reasonably useful standard data could be compiled.

High Volume, High Manual Labor Content Operations

Examples of this situation are: assembly operations, certain test operations, riveting operations, etc. Which technique should be used? How can they be combined?

Normally, MTM with time study of process times is the best over-all choice because:

1. The correct method can be established before starting the job when MTM is used. The process times can be recognized and determined by time study after the job is running.
2. The best technique for improving and/or establishing the method is MTM, while time study can provide over-all time checks to assure that no pertinent factors were overlooked.

3. MTM provides a sound labor standard, provided time study is integrated with it where appropriate.

Exceptions can always be found. For example, if engineering time is limited, an immediate standard is needed, and training time is virtually non-existent, then time study of a few cycles would generally provide a quite satisfactory temporary answer. This answer could then be improved later, when analysis time is available, by utilizing the approach indicated above that relies mostly on MTM procedure. This combined approach provides better processes, more accurate labor standards, helps in training the operator, and yields standard data with least over-all effort. Normally, more analysis time becomes available after an operation has run sufficiently long to be "debugged." Using the latter approach, the standard will also be more accurate.

High Volume, Low Manual Labor Content Operations

Examples of this situation are: rolling mill operations, automated machine tending, etc. Again, the question involves how to combine the two techniques in setting a labor standard, although main reliance here should be on time study.

Time study would definitely be needed to determine the machine time, which would be a high proportion of the total time. In the rather extreme examples mentioned, the manual labor content probably would have minimal effect on the standard. Time study would probably be combined with a ratio-delay study on the two examples in order to determine delay time, adjustment time, setup frequency, etc.

However, MTM often can be very useful in answering such questions as how many stations one operator can adequately tend if the final answer, based on the time study data, is not certain. After time study discloses the over-all cycle time and/or the limiting time at each station, the motions needed by the operator at that station can be analyzed with MTM. An attempt to reconcile and balance this operator time against the limiting time at one or more stations can lead to greatly increased efficiency and better labor utilization. It is even possible to justify extensive mechanization of operations by this combined procedure.

High Volume, Intermediate Manual Labor Content Operations

When to use each technique in these situations cannot be clearly delineated. MTM loses some of its advantage as the manual work content of an operation decreases. However, any manual work in a high volume operation is worth evaluating; and if the manual work is at all complex, then MTM probably should be used to help set the method and standard on that portion of the work.

Related to this problem are such factors as whether the job is already running with an apparently satisfactory method. If so, then time study would appear to be a logical choice. If it were a new job, then it would appear to sway the decision to MTM since an MTM motion pattern can be developed without necessitating an *operator* to perform the task.

The possibility of standard data development also enters. Standard data on manual work can normally be obtained more quickly and precisely with MTM, hence the amount and degree of complexity of the manual work guides the choice. Each situation merits individual evaluation.

Low Volume, Non-Repetitive, High Manual Labor Content Operations

Examples of this problem would be the assembly of small lots of special machines, jigs, equipment, etc. The following are possible considerations when deciding which technique to use:

1. If the work has not been started, and similar assemblies have not previously been produced, MTM is the only analytically sound procedure by which to establish any desired detailed methods and work standards prior to production.
2. If work has been started, and the task is relatively simple, it will obviously be faster and less expensive to use time study to establish a fairly accurate standard. It will also provide generally adequate information in these situations to aid sound methods improvement.
3. If the task is such that many elements will repeat on future assembly operations (such as fit 1-inch face, 20-tooth gear on shaft; position ¼–20 screw in hole, etc.), then MTM is by far the best technique since a simple analysis of one such element will provide accurate standard data for future occurrences of the element.

Summary of MTM Versus Time Study

It is evident that almost every work measurement situation encountered merits individual consideration in selecting the way to use MTM and time study. The advantages and disadvantages of each must be weighed in the decision.

MTM can claim the following major advantages:

1. Yields good labor standards with no necessity to view the actual operation.
2. Makes possible the establishment of good methods prior to the start of an operation and thereby cuts training time, yields minimum manufacturing cost, uses less over-all engineering time, etc.

3. Provides a scientific and practical tool for methods improvement of manual operations.
4. Provides valid data for standard data and/or time formulas from even one properly written motion pattern.

Time study has the following advantages:

1. When the job is running, it will provide the quickest labor standard. If standard data is not being developed, it will produce the standard at the lowest over-all cost.
2. It is a technique more universally recognized and understood by all concerned parties.
3. It provides the best data for analyzing operator performance levels.

Using these points as guides, it is possible to analyze most work study situations to arrive at a solution as to which technique to use. Remember, the real problem is not whether to use MTM *or* time study, but rather *when* and *how* to use *both*.

CHAPTER 19

Computers and the Work Measurement/Industrial Engineer

The objective of this chapter is to inform the industrial and/or work measurement engineer what kinds of work measurement/industrial engineering computer programs (computer software) are available to them in their work and give a very brief description of what these programs can accomplish. Most of this chapter deals with microcomputer software. However, a few of the programs were developed for and initially used on a mainframe or minicomputer. A small portion of the material even discusses programs for "handheld" computers.

It should also be mentioned that an attempt was made at the end of the chapter to suggest things that needed more attention from industrial engineers, managers, and even politicians.

If you are a work measurement/industrial engineer successfully using a personal computer (PC) in the various facets of your work, skip ahead to the first subtitle. You can do the same if you are using a mainframe and/or minicomputer in your work. The next paragraphs are for those who are not yet using a personal computer and are delaying their start because of perceived complexities and difficulties. Personal computers are also known in the trade as microcomputers.

As to learning how to run one, relax. If you can read and have average intelligence, you should have little trouble becoming adept in using a PC, since virtually all work measurement or engineering professionals exceed the stated requirements. You do not have to take a course.

First you will likely be given a learning program magnetic recording disk and an associated booklet to help introduce you to your new computer. They are not complicated and you should go through at least part of it if computers are new to you.

Also, when you purchase a computer you get an operating manual, really a Disk Operating System Manual. First read it and then "format" (A computer process to prepare a floppy magnetic disk for use on your computer) a couple of disks, and try out a few commands such as Diskcopy, Check Disk, and Directory (of a disk). These are always quite well explained.

When you buy a PC, you need a program or two to make it intelligent enough for you to get some value from the machine. Everyone with a PC needs word processing software. This assumes that you or your secretary type. If you do not type, obtain a software program that teaches one how to type. Let your computer teach you.

Learning how to use word processing software is normally not too difficult unless you try to start with a software program such as Word Star or Symphony—these are good but very large and complex programs. Almost any standalone word processing software program is appropriate for your first effort. Examples are Superscripsit for the TRS-80 series of computers, Apple Writer for the Apple computers, and Multimate for the IBM PC. IBM also offers such software. Enable is a comparatively simple piece of integrated software that includes a good set of word processing software. There are many other satisfactory word processing programs. Read the software manual that comes with your program and start to use it. Many programs contain a teaching disk to aid your learning; if so, use it.

Once you have learned how to use one program, learning another will be much easier, simply because there is a great deal of similarity in the command structure of software systems.

Nothing will be explained in this chapter as to the details of using a computer; you can learn this on your own. In fact, that is the best way to make the procedures stick in your memory. The only thing the beginner needs to be told is that he/she must communicate with the computer in the *exact* manner described in the program manual—nothing else will be accepted by the computer. Generally, the computer says "bad command or file name" if you goofed.

No, you do not need to know a programming language; but if you do want to learn one, we would suggest starting with BASIC. You likely will be given a BASIC Operating Manual, really a command structure manual, with your computer. There are a number of texts that explain the rudiments of programming in BASIC[1] (a relatively simple and much used programming language). However, one of the authors (D.W.K.) uses a PC (mostly a IBM—PC, but also he has a TRS-80, Model III) 25–30 hours per week and in the last three years he has never needed to use BASIC or Fortran or any other computer language. Buy your programs—it is faster and more economical than creating your own if your time is at all valuable.

All of the above has been very positive and indicates that there are few barriers to the relatively easy use of microcomputers. Actually there is one

[1]James S. Coan, *Basic Basic, An Introduction to Computer Programming in Basic Lanuage,* 2nd edition, Hayden Book Company, Inc., Rochelle Park, NJ. There probably are as many as two dozen or more such books on the subject.

handicap that users have to overcome, it is that software program manuals often are rather poorly written.

The last comment brings up a related matter. All engineers, computer users, managers, etc., must do a lot of reading—and most people read at 300–400 words per minute, some even slower. Let your computer *at least* double your reading speed. Secure a speed reading software program and use it!

PREDETERMINED TIME SYSTEMS

This portion of the chapter deals with the main predetermined time systems for which there is a mainframe, minicomputer, and/or microcomputer system program. Essentially all of these will run on an IBM PC or PC compatible computer; however, it is usually required that these have a hard disk recording system.

METHODS-TIME MEASUREMENT (MTM)

The MTM Association can provide computer programs for all of the heavily used MTM systems, including MTM-1, MTM-2, MTM-C, MTM-M, MTM-V, MTM-MEK, MTM-UAS—and these will first be described. In addition there are other computer programs available to help you computerize one or more of the MTM systems or one of the other predetermined time systems. Most of these will also computerize other work measurement procedures.

4M (MMMM), Micro Matic Methods Measurement

This is the largest and most sophisticated program offered by the U.S./Canada MTM Association and is therefore first described. It is designed to computerize the use of MTM-1. While some information on what the program does is given here, its details, including the manner of application and the general structure of the system, are not fully explained. For more information see *Advanced Work Measurement* (AWM).[2]

4M is at least five times faster to use in setting standards when compared to the normal method of applying MTM-1. In fact, it is faster to use than MTM-2 and "in the ballpark" when compared to using MTM-3. Figure 19-1 is taken from AWM and shows the average time to set standards using the various MTM systems.

[2]D. W. Karger and W. M. Hancock, *Advanced Work Measurement,* Industrial Press Inc., 200 Madison Ave., New York, NY 10016. This text also provides full information on MTM-2, MTM-3, MTM-C, MTM-M, and MTM-V as well as such topics as learning curves, worker selection tests, system selection, and work standards for professionals.

Speed of Application

System	Average Time to Set Standard
MTM-1	250 X
MTM-2	100 X
MTM-3	35 X
MTM-V	10 X
MTM-C1	125 X(tentative)
MTM-C2	75 X(tentative)
4M DATA	(15–60)X

Figure 19-1. The average time to set standards using the various MTM systems where X is the nonrepetitive cycle time and all time units are the same.

When the increased speed *and* accuracy of 4M is taken into account with the cost of settling labor disputes and the productivity loss caused by less accurate or less correct labor standards, the relative monthly costs for setting the same example sized average standard and the same example total TMU of set standards is shown in Figure 19-2, also taken from AWM.[3] Note that when all these factors are taken into account, it is far preferable to use 4M than MTM-2 or MTM-3.

System Application Costs

System	MTM Applicator	Complaints	Productivity Loss	Total
MTM-1	$4200	$1200	$ 840	$6240
MTM-2	1680	2500	2099	6279
MTM-3	588	3600	4200	8388
4M DATA	504[a]	1000	420	1924

[a]4M DATA applicator time assumed as 30X.

Figure 19-2

Figure 19-2 shows the relative system cost per month based on a complaint threshold of 105 percent and when taking into account MTM applicator costs, costs of handling complaints, and cost of lost productivity due to inaccuracy of standards. The cost of computer time is not shown since in many cases it is almost negligible and in others more substantial. Actual total covered costs will vary with each such installation. Moreover, the effects of inflation, time value of money, etc., were not taken into account—but the relative costs are essentially as shown.

The description of the 4M program in the introduction to 4M in AWM reads as follows:

[3]*Ibid.*
[4]*Ibid.*

One of these firms, a member of the U.S./Canada MTM Association, was successful in achieving significant overall benefits—Westinghouse. They turned over the copyright of the system to the Association, who not only made it publicly available, but also are continuing its further development and improvement. It is known as 4M DATA, an acronym adopted in 1973.

The original system has now been significantly revised and is titled 4M MOD II. MOD II is a "structured program" of *modular* construction. This makes it relatively easy to add additional future program functions.

The 4M DATA system is so structured that its files may be "loaded" with the elements of functional MTM systems such as MTM-V and MTM-C. Users of these functional systems may utilize the 4M DATA System for standards application and operator instruction generation. Computerized MTM-V and MTM-C are not considered to be "modules" since no additional programming is necessary to use them.

The 4M program deck is written in ANSI (American National Standards Institute) COBOL in either DOS or OS environment. A 50K storage partition is required, and auxiliary storage on a disk pack or a similar medium is also needed.

Users of 4M MOD I usually used punched cards as the input medium. However, a CRT data terminal with video can also be used, in conjunction with an interim storage medium such as a "floppy disc" when the program is run in the batch mode. This is and was the usual case, since it reduces computer requirements and costs in most cases. However, if desired, the 4M program can be run in the time-sharing mode.

Naturally, an output printer is needed. Moreover, a magnetic tape drive or other suitable device for back-up storage is very desirable.

The MOD-IIB program is similar to the original version with respect to the above computer requirements.

MOD IIA contains interactive programming which makes it suitable for "online" as well as "batch" operation. MOD IIB is the "batch only" version of 4M. The same *User's Manual* describes both systems, the principal difference between the two being the input procedures.

In the interactive mode inputting is by remote terminal (keyboard and CRT display). The analyst is "prompted" for the necessary input data. As soon as the data are entered the files are updated, which means the inputting can be done in one or more input sessions. Output is available immediately or later.

The batch mode is possible with both versions. Inputting may also be accomplished by Remote Job Entry (RJE) or by keypunching.

Conversion of any original system installation to MOD II is a minor problem and can be accomplished rather easily.

Through proper user programming a computerized set of standard data can be built as the program is used. This approach not only retains all of the advantages of MTM-1, but it also can make its application more efficient and effective. Last, but not least, as previously mentioned, the 4M program enhances the methods analysis aspect of MTM-1. The economic considerations are advantageous both for the firm with a computer of the proper size and characteristics or through the use of leased or rental services (time-sharing) if there is a sufficient work measurement work load to justify the expense of acquisition, installation, training, and operation. For the larger firm the advantages are overwhelming.

Since the preceding quotation was written, personal computers with much larger RAM memories, faster operating systems, and larger magnetic storage systems became available. This has made it possible to offer 4M for use on IBM PCs and many of the "compatibles." It requires a hard disk system (provides a greatly expanded and faster magnetic storage system than the standard floppy disk system used on the standard IBM PC). The IBM AT or XT are such hard disk systems—however, a hard disk system can be added to the standard IBM PC and to many of the compatibles.

As of the date of this writing the 4M program is being updated to include such functions as formula capability and the production of routing sheets. Also, the clarity and ease of usage of the program manual, which was already reasonably "user friendly," are being improved.

The 4M system, in addition to MTM-1, provides for incorporating any or all of the following data sets: MTM-UAS, MTM-MEK, MTM-V, MTM-C, and MTM-M. Moreover, the 4M system will accept and apply data developed from any other source such as time study or standard data. Additionally, the 4M System is designed for structured coding of operations and has the ability to easily build and maintain standard data.

The 4M system has a number of excellent features such as the Methods Improvement indexes. Through these indexes, the 4M user is supplied automatically with a guide to work simplification and methods improvement. Five indexes are calculated by the computer to allow the practitioner to judge the relative effectiveness of the method in regards to simultaneous motions, workplace layout, tooling, fixturing, and the use of internal process time. No other system offers such a powerful tool for methods improvement, and these indexes are fully explained in AWM.[5] Moreover, the 4M system will accept and apply data developed from any other source such as time study or standard data.

The system uses the logic of the computer to reduce the number of application rules, thereby reducing the potential variation between analysts and significantly decreasing the application time—partially explained in the material which follows.

Features of the 4M system include a powerful Mass Update routine utilizing a single command to update all standards affected by a given element change. Also included is the ability to combine MTM and user-defined elements, a quick change feature for similar operations, identification of manual elemental times internal to process times, and simulation capabilities.

Many 4M systems (over 100 at the time of this writing) have been purchased and installed by government and industry, and in other business areas. More specifically, with respect to the government contractors and subcontractors, the 4M system has enabled these organizations to meet and

[5]*Ibid.*

exceed government requirements, including compliance with MIL-STD 1567A. Audit trail and documentation provided by the 4M system is excellent.

4M, The Derivation or Source of Its Name

The acronym 4M (or MMMM) was derived from the following four major attributes of the system:

Micro. The system retains all of the elements of MTM-1 in their complete detail and thereby is able to provide significant advantages and features.

Matic. Really automatic, since the analyst no longer needs to refer to the data card or to tables. This is possible because of the simple and easily remembered coding used by the analyst to describe the work. The coding is automatically translated by the computer, which determines a near-optimum MTM-1 method. Any compromises in optimization are analyst imposed through the degree of completeness of descriptions.

Methods. The 4M system develops and defines sound manual methods to the same degree as analyst-used MTM-1. In fact, it not only retains this advantage, but it also extends it by automatically providing improvement indexes (ratios) that indicate the approximate degree of optimization—or, conversely, additional methods opportunity.

Measurement. While GET and PLACE codes are used, the computer system interprets these and their special numerical descriptors so that it recognizes and applies the proper MTM-1 components for both hands. In fact, when necessary, it even automatically reduces C MOVE into two increments—a Class B MOVE and a short Class C MOVE. Furthermore, MTM-1 simo rules and the case of motions are all properly applied automatically by the computer.

Analysis of frequency of motion occurrence indicated that about 90 percent of the motions performed in industrial motions are those of REACH, GRASP, MOVE, POSITION, and RELEASE—essentially the design basis of MTM-2. The 4M GET classification is used to cover REACH, GRASP, and RELEASE elements. PLACE is used to cover the MOVE and POSITION elements. The remainder of the MTM-1 elements are applied using virtually unchanged MTM-1 notations.[6]

While Get and Place categories have been made an integral part of 4M in order to simplify MTM-1, an examination of Figure 19-3, which is taken from AWM,[7] outlines the coding of Get—G. This clearly indicates how the basic or underlying Case of Reach, Distance of Reach, and Class of Grasp are *all* specified by the coding entered by the analyst. The same coding *scheme* is used for Place—P.

[6]*Ibid.*
[7]*Ibid.*

GET

GXXX xxx

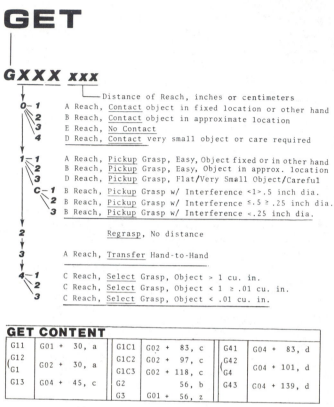

```
                    └──Distance of Reach, inches or centimeters
0─1   A Reach, Contact object in fixed location or other hand
  \2  B Reach, Contact object in approximate location
   \3 E Reach, No Contact
    4 D Reach, Contact very small object or care required

1─1   A Reach, Pickup Grasp, Easy, Object fixed or in other hand
  \2  B Reach, Pickup Grasp, Easy, Object in approx. location
   \3 D Reach, Pickup Grasp, Flat/Very Small Object/Careful
   C─1 B Reach, Pickup Grasp w/ Interference <1>.5 inch dia.
    \2 B Reach, Pickup Grasp w/ Interference ≤.5 ≥ .25 inch dia.
     3 B Reach, Pickup Grasp w/ Interference <.25 inch dia.

2           Regrasp, No distance

3     A Reach, Transfer Hand-to-Hand

4─1   C Reach, Select Grasp, Object > 1 cu. in.
  \2  C Reach, Select Grasp, Object < 1 ≥ .01 cu. in.
   3  C Reach, Select Grasp, Object < .01 cu. in.
```

GET CONTENT

G11	G01 + 30, a	G1C1	G02 + 83, c	G41	G04 + 83, d
G12		G1C2	G02 + 97, c	G42	
G1	G02 + 30, a	G1C3	G02 + 118, c	G4	G04 + 101, d
G13	G04 + 45, c	G2	56, b	G43	G04 + 139, d
		G3	G01 + 56, z		

Figure 19-3. The essentials of GET coding. © MTM Association for Standards & Research, Inc., Fair Lawn, NJ.

ADAM SERIES OF PROGRAM

The ADAM series of MTM-based programs were developed under the direction of the U.S./Canada MTM Association for utilizing some of the generic and functional MTM systems in the development of standards and standard data with full storage, retrieval, and mass updating capabilities. It was developed for the microcomputer, and while emphasizing MTM systems, it also has the ability to handle data from any source. This program is also available from the U.S./Canada MTM Association.

The program obviously works to a reasonable degree based on the fact that a fair number of sets of ADAM software have been sold, but a review of the program manual indicated to this writer that it is not "user friendly" and anyone without programing and considerable microcomputer experience may find using it a substantial task.

Automatic Data Application and Maintenance (ADAM) is the full name of the system which allows the user to operate a work measurement system with the aid of a computer to increase the speed of application while maintaining the accuracy of the specific technique being used. The ADAM system is available from the MTM Association for Standards and Research, Fair Lawn, New Jersey. It was introduced in 1980 as the first predetermined time system developed for the microcomputer designed to create and maintain labor standards derived from standard data.

ADAM is a fully integrated work measurement system. Just as a word processing system reduces effort and errors in composing and editing, ADAM does the clerical and computational work involved in creating and applying labor standards. Basic elements may be derived from any measurement technique, whether it be an MTM-based predetermined time system, time study, historical data, etc., through the use of appropriate modules. The user is able to combine the basic elements to form intermediate levels from which the standards themselves are ultimately set.[8]

The system was designed to incorporate, in any combination, the following MTM systems: MTM-2, MTM-UAS, MTM-V, MTM-C, and MTM-MEK. Adam will also accept and apply data developed from any other source such as time study or standard data.

Future MTM Association systems will be adapted to run on either the ADAM and/or 4M systems.

Features include provisions for user-customized keyboard controls and options regarding printout format. The user can also assign both manual and process allowances as well as choose the precision and type of summary calculations to be displayed. Formula capability is also fully supported by the ADAM system.

The basic ADAM system was written in the PASCAL programming language, and can be run on any microcomputer that has the capability of compiling PASCAL. It can be purchased without data sets as self-contained software for the application and maintenance of company-developed data bases, whether MTM-based or not. The data sets mentioned above can be purchased in addition to the basic system as desired in any single or multiple combination.

ADAM-2

The use of the MTM-2 data set with the ADAM system will be the only such data set described. More information can be obtained from the U.S./ Canada MTM Association.

ADAM-2 can be used wherever MTM-2 can be applied manually. The advantages of easy input, high speed of application, consistency, and accuracy are retained. All of the nine elements and two weight factors of MTM-2 are present in the system.

ADAM-2 analyzes simultaneous motion combinations and applies the proper overlap values when required. It also analyzes combined motions

[8]*Ibid.*

(those done by the same body member at the same time) and "limits out" the shorter motion(s).

Elements developed with ADAM-2 may be stored as standard data and retrieved for use within studies or other higher-level elements.

Inputting of the data to the ADAM System is easy to do. Inputs are of "free format," releasing the analyst from the restrictions of the 80-column card with designated field sizes and locations. Figure 19-4 illustrates an ADAM-2 input. The "," notation separates left-hand motions from right-hand motions.

/FILE-HAND TOOL-CIRCULAR MOTIONS

GB2 , GB6
GB6 , PC6
, PB6 × 18
, A × 18
R × 9
PA6 × 9, PB6 × 18
, PA6

Figure 19-4 ADAM-2 input.

Figure 19-5 is the output prepared by ADAM-2 from the input of Fig. 19-4. For each MTM-2 notation, ADAM-2 supplies an appropriate "action" word, while the analyst need only input object information. Of course, the automatic action word can be suppressed, allowing the analyst to supply a more descriptive one if desired.[9]

1 FILE-HAND TOOL-CIRCULAR MOTIONS							
2 GET COMPLEX		GB2)	10	GB6		GET COMPLEX	
3 GET COMPLEX		GB6)	33	*PC6		PUT W/ 2 CORR'NS	
4			270	PB6	18	PUT W/ 1 CORR'N	
5			252	A	18	APPLY PRESSURE	
6 REGRASP	9	R	54				
7 PUT EASY	9	PA6)	270	PB6	18	PUT W/ 1 CORR'N	
8			6	PA6		PUT EASY	

TOTAL TMU: 895

Figure 19-5 ADAM-2 output.

FAST APPLICATION OF STANDARD TIMES (FAST)

FAST comprises a series of software programs developed by Datawriter Corporation, 2793 Pheasant Road, Excelsior, MN 55331, that allows users to computerize a substantial number of work measurement functions (each is a separate software program). System Standard Data Manager (SDM) will integrate any or all of the packages into an integrated labor standards system. The functional software packages are as follows:

[9]*Ibid.*

Standard Data Manager (SDM)
MTM-1 Analyses (MT1)
MTM-2 Analyses (MT2)
Time Studies (TSX)
Work Sampling Studies (WSX)
Regression Analysis (RAX)
Modapts (MOD)
Master Standard Data (MSD)

The above items are available separately or in various systems called FAST-MTM, FAST-MOD, FAST-MSD, and FAST-TIME. They are also offered with or without guided training at the customer's facility.

The above described array of software packages differ significantly from those available in 1984–1985. Datawriter may continue to offer the old software, but in this writer's opinion (D.W.K.) only the new packages should be considered for current purchase.

FAST programs will run on either a floppy disk IBM-PC or a hard disk system. A demonstration package is available which contains the FAST Handbook describing all FAST modules and a demonstration disk of the various modules.

The FAST MTM system includes MT1 and MT2. MT1 software is designed to computerize much of the application of MTM-1, but it does not utilize the 4M coding scheme. However, it does produce some of the accuracy of 4M by essentially using the same distance increments. Moreover, the simultaneous motion table portion of the software contains some sophisticated options that are not currently found in 4M or ADAM. Also, MT1 does not have 4M's method improvement indexes. Datawriter states that any customer with at least one certified MTM applicator can use MT1 or MT2.

Some of the principal general procedures incorporated in the FAST-MTM system are the following:

Verification of quick standards via time study
Standards, allowances, and audits via work sampling
Formula generation using regression
Element where-used searches
Automatic updating
Can produce operator instructions
Rapid code searches
Keyword searches
Traceability by applicator and date
Table and formula capability
Supplementary datacard generator (facilitates creation of "macros" which Datawriter calls "macro-motions")

Can be accessed by customer's minicomputer or mainframe MRP system

FAST-MTM also includes SDM, which was previously mentioned. It helps customers create and maintain one or more data bases of coded standard elements called ELEMENTbooks. Standard elements in an ELEMENTbook can be in any of four formats: constant time, one-variable list, two-variable table, or time formula. Each ELEMENTbook can contain 4000 standard elements.

Labor standards can be generated from elements in an ELEMENTbook. SDM displays a blank labor standards form on the computer screen onto which the operator enters element numbers or element codes. SDM does all the necessary lookups and calculations, and constantly displays total time. If the user cannot remember an element, SDM can rapidly search the ELEMENTbook by code, partial code, or keyword.

Labor standards forms are saved to disk files called FORMSfiles. Each FORMSfile can hold 200 standards forms. Any labor standard or associated operator instructions (with or without element times) can be printed at any time. Standards forms can be recalled to the screen at any time for editing, using word-processor-type commands.

Any labor standard can optionally be added to an ELEMENTbook as a large macro element, and used just like any other standard element. Such macro elements allow the user to build multilevel standards, building complex labor standards from "subassemblies" of labor.

If an element time needs to be changed, SDM's where-used feature will tell the user which forms in a FORMSfile use the element. SDM then automatically edits just the labor standard forms the user wants to update. SDM can even move the updated standard to a new form, preserving the old standard form intact for reference, if so desired.

SDM can help the user find practically any labor standard form in a FORMSfile. It can search for all standards with a certain allowance value, or standards developed by a certain user, or created on a specified day, or in a given month or year, or that contain a keyword, or have a specified part number, or color, or size, etc.

It is claimed that the system is secure and traceable. ELEMENTbooks and FORMSfiles can only be accessed using passwords, which imprint a date and user ID on all transactions and on all printouts.

FAST-MTM also includes TSX Time Study and WSX Work Sampling. TSX permits taking time studies with a data collector and up loading to the computer. TSX will accommodate most available data collectors (Electro General Corp.'s Datamyte; Observational Systems' OS-3 recorder) and a new Datawriter DC-1 recorder (contains a 16K memory system). The user can also secure from Datawriter a software package that will convert the Radio Shack Model 100 handheld computer into a data collector that can be used with TSX.

WSX offers work sampling summarization of data collector data. Summaries can be developed from the active data, or subsets as fine as one subject for one activity for one hour.

The FAST-MTM system produces labor standards based on MTM-1 and MTM-2. Datawriter also offer FAST-MSD (based on MSD), FAST-MOD (based on MODAPTS), and FAST-TIME based on time study.

KARMA'S AUTOCAM—AUTOCOST

Automatic computer-aided-manufacturing (AutoCAM)/automatic cost optimization and simulation technique (AutoCOST) is a new and truly integrated software system for work and cost management that has just been introduced to the market. It is a very complex and complete software set, but it will not computerize the application of MTM, MOST, etc. However, it will accept work standards produced by these and other predetermined time systems as well as from time studies.

The system was developed by Karma, Inc. (Kern Automation Research and Management Associates) of Carson City, NV. It is written in COBOL and will run on a hard-disk-equipped IBM-PC if it is loaded with a Ryan McFarland COBOL Compiler. Karma can supply this third-party software if the user so desires.

This is truly a big system as can be seen by an examination of Figure 19-6. Another indication of its size and complexity is that aside from the AutoCAM/AutoCOST programs, several of which will be partially described, it utilizes important major third-party software—(1) AutoCAD (a popular computer-aided-design software product of Auto Desk, Inc. of Sausalito, CA) and (2) ENABLE (a somewhat new major integrated spreadsheet, data base management, graphics, word processing, and telecommunications management system, which is a product of The Software Group, Ballston Lake, NY). Again, Karma can supply copies of these *sets* of software programs.

A little serious study of Figure 19-6 will disclose that the Karma-designed and -supplied software systems plus the third-party sets of software is truly a very large and complex system. However, Karma claims to have embedded so many internal and well-designed help screens that learning to use the program is relatively easy in spite of its size and complexity—but it also is admitted that some of the available guided use instruction will be necessary for those who operate the system.

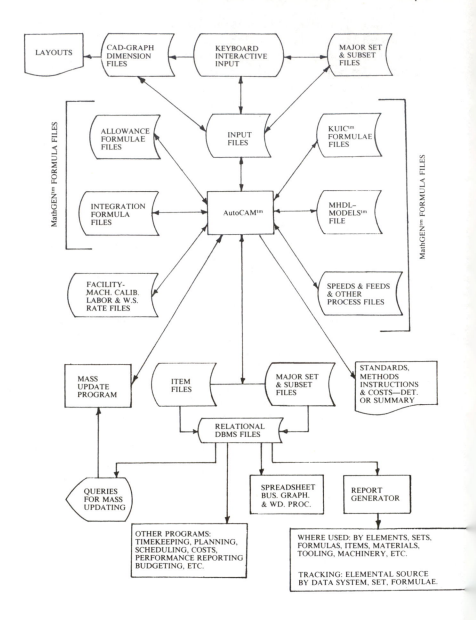

Fig. 19-6. A system/function diagram of the AutoCAM—AutoCOST system. (By W. L. Kern, KARMA, Inc., April 1, 1986.)

Speeds, Feeds, and Equipment Optimization Analysis

Like the subsequently described (in this chapter) Rath and Strong software set, Karma's AutoCAM/AutoCOST contains an internal file set of speeds, feeds, and other machining and process files. This set can be modified and/or expanded. Also, the user builds a file set of his manufacturing equipment and the operating characteristics of each machine. (Also associated with this file set are data on labor classifications and labor rates.)

Karma's software can determine, upon demand for a specified process set of instructions, the ideal and the actual process/machining time (including specification of feeds and speeds) in a manner somewhat similar to Rath and Strong's software. As to the relative size and quality of the two mentioned machining databases, this writer does not have enough data to render a judgment.

Generative Formulas

A simplified formula coding system called MathGEN is included with the AutoCAM System. All formulas, including hand handling (manual), machining, speeds and feeds, other formulas for simulation, process, allowances, scheduling, loading, costing, budgeting, as well as various performance efficiency formulas, are coded into the system through this routine. MathGEN includes codes for all mathematical or logical instructions that a user may want to use according to Karma.

The Elemental Standards System

The original work on the predetermined time system that is embedded in this set of programs dates back to 1964 when Mr. Willard Kern, CEO of Karma, organized and operated Management Science Inc. for the purpose of developing and marketing computer software for IE applications. His approach was to create and utilize a series of universal mathematical/statistical formulas to generate precise elemental standard times. Only numerical inputs were needed then and the same is true today for this system. Some of the readers will likely associate the above statements with UnivAtion Systems and UnivEl.

In the program supplied there are 72 hand-handling (manual work) formulas called the KUIC System (KARMA Universal Integration Concepts). These formulas will develop unique elemental times for the kind of macro functions associated with the higher-level predetermined time systems such as MTM-2 and MTM-3 (Get, Put, etc.)—but while similar to these categories, the system will not accept MTM-1, MTM-2, or MTM-3 motion coding as raw input.

They are quick to apply, yet very precise (Karma's statement), because the users level 2 or level 3 variables definition/determination are input to generate time. The hand-handling formulas will develop elemental time from a total of millions (Karma says quintillion) of combinations of second- or third-level variables.

Variables used by the KUIC System are linear distance, target complexity, weight or force of movement, hand and body motion combinations, walking, turning, and cranking. Elements generated are at a combined Get and/or Put level; as an example, if you walk, turn, bend, and get an item from varying targets and at varying distances, or if your walk, turn, bend, and place (or assemble) an item with varying targets and at various distances. Each of these examples may be input as a simple "Get" or "Put" element. Many other types of combined macro motion combinations are provided per Karma—comparable to level 2 or level 3 MTM elements.

Basic Functions of the System

1. Elemental time generation, maintenance, and tracking through formulas or other time data *sources* (MTM, MOST, Time Study, etc.).
2. Development of reports and information for various engineering or management uses.
3. Automatic database development of elemental standards, methods instructions, assembly sequences, workplace layouts, processes, planning data, scheduling data, and costing data.
4. Automatic updating and mass maintenance for all database information.
5. Formula coding and application for most mathematical and logical requirements for IE and Management needs.
6. The system can also generate time standards for materials handling and transportation activities.

Concluding Remarks

Willard Kern and Karma, even though they did not accommodate the usage of MTM-1, or any of the higher-level systems, did point the way most work measurement software providers should move by incorporating workplace analysis capability by using existing CAD software as well as by using an integrated software system like ENABLE to handle many other features of the system.

However, it is suggested that the reader very carefully study the data in Chapter 12 of *Advanced Work Measurement* on System Selection. Since this is a private system, this writer cannot be sure, but it seems obvious from the data available to the writer that the embedded predetermined time system is similar to MTM-2 and MTM-3 or some combination of these two systems. It seems to this writer that a majority would prefer a computer-aided MTM-1 system as the ultimate low-cost route to labor standards. Another alternative for Karma would be to offer their system in several versions: the current version, another to handle MTM-1, and still another for MOST, etc.

MOST

MOST was first developed as a manual technique by the Swedish Division of the H. B. Maynard Company, Inc., during 1967–1972, and was introduced in the United States in 1974. It is claimed to be MTM-1 and MTM-2 based. The MOST system is applicable to most any cycle length and degree of repetitiveness—as long as there are variations in motion pattern from one cycle to another. It is what is known as a proprietary system.

As to computerization, we quote from Chapter 13 of *Advanced Work Measurement*[10]:

> *Computer Configuration.* Through the use of a minicomputer, quite often located in the industrial engineering department, or a mainframe IBM in the data processing department, with on-line interactive access for the industrial engineer, MOST time standards can be calculated from workplace data and a method description. The computer thus will do all the work measurement, relieving the industrial engineer of practically all paperwork.
>
> The MOST computer systems, which are generic and second- and third-level, include all necessary steps and procedures to develop a complete time standard for an operation, such as:
>
> 1. Development of suboperation data based on MOST work measurement systems.
> 2. Calculation of time standards, including manual times, process times, and allowances.
> 3. Printing of method instructions and route sheets.
>
> The core of the program consists of comprehensive filing and editing procedures for retrieval, revision, and mass updating of time standards. The output formats can be tailor-made to customer conditions and requirements.
>
> *Process Time.* The computer can generate a series of optional modules to calculate process time for such operations as machining, welding, line balancing, and labor reporting. These times can be integrated at the proper point to give a complete standards calculation.
>
> *Simulation.* Finally, the computer gives the user simulation capabilities. Proposed work standard changes can be displayed to determine whether the new method or equipment will add or subtract time from the standard. Hypothetical conditions can be created for the method, workplace, and process.

MASTER STANDARD DATA (MSD)

MSD is a relatively popular high-level predetermined time system based on MTM. It was first described in the book *Master Standard Data* by R. M. Crossan and Harold W. Nance, copyrighted in 1962. The second edition of this text appeared in 1972 and the publisher is Robert E. Krieger Publish-

---

[10]For slightly more detail about the system see the referenced text, AWM, or for substantially more detail see *MOST Work Measurement Systems* by Kjell V. Zandin, Marcel Dekker, Inc., New York, 1980.

ing Co., Inc. The development work was done as a project of The Serge Birn Co. of which Mr. Nance is now President.

The Serge A. Birn Company has also developed a sophisticated software package known as MOD-II, which will handle work measurement systems that contain a specific code and time for each element. The software comes in formats suited to either manufacturing or office work measurement.

The software is designed for an IBM XT or AT, or compatible hardware. It can be used to create standard data, individual standards, and new standards from existing standards, and contains mass updating of all standards or revision of individual standards. Two kinds of printouts are available relating to standards—the standard detailed measurement or a description of how the operation is to be performed. The software manual is reported to be user friendly.

In addition to the work measurement software, the Birn Company also has a Personnel Productivity System that utilizes the standards created by MOD-II to prepare performance reports. Several kinds of these reports are available to the user.

RATH AND STRONG

Advanced Machining Logic (AML)

Rath and Strong, a management consulting firm, has developed AML to computerize the development of machining time standards that take into account such things as workpiece material, tool material, machining speeds and feeds, horsepower, tool geometry, setup rigidity, machine condition, coolant, and workpiece finish requirements.

The expert will understand that the user's material characteristics need to be considered. This is accomplished by relating the user's material to material characteristics in AML's data base which are modifiable by the user.

Machining operations included in AML are turning, milling, drilling, threading, grinding, gear hobbing/shaping, sawing, and single-point shaping.

The user loads into the data base his or her machine tools, including speeds, feeds, horsepower; materials; cutting tools; etc. AML maximizes and reports the results achieved with the designated machine tool(s) and also indicates whether the results are being limited to less than optimum by the user's equipment. It also provides a simulation capability that will answer "what if" questions, especially valuable in the evaluation of alternate machine tools, presently available or under consideration.

The system development took recognition of the fact that people seldom have the time to determine the one combination of speed and feed that will produce the required finish which making use of the available horsepower.

AML is a software system that can be used with conventional or numerically controlled machine tools. It also allows engineers, according to the company, to quickly and accurately establish time standards from their own machining data, or data supplied with the software. Included with AML is Computerized Standard Data (CSD), which maintains the data base. CSD is described in the next subsection. AML, according to Rath and Strong, can also be directly integrated into the 4M, Mod II system of the U.S./Canada MTM Association, at both the data base and functional levels.

CSD and CSD with AML are on-line, menu-driven, interactive software packages. They are written in COBOL and are available for the IBM mainframes, Data General, and Wang/VS minicomputers. There is a version that does not require a COBOL compiler for all IBM and IBM-compatible microcomputers with *at least* 5M of hard disk as a memory storage system.

Computerized Standard Data (CSD)

CSD is the other Rath and Strong software system. It computes, documents, and maintains methods and time standards. The software generates the standards using *your* data, regardless of the source, and interfaces with a firm's MRP-II system.

CSD is claimed, according to Rath and Strong, to meet Military Standard 1567A requirements for audit trail traceability and documentation of the makeup of standard costs—both 4M and CSD appear to have this capability.

It is claimed the CSD enables IE Departments to create and maintain standards substantially faster from the stored standard data elements than when using manual methods, either via a picklist of stored standard data or by modifying a quick copy of an existing standard. It creates detailed operator instructions, responds to methods and process changes, improves cost accountability, etc., per Rath and Strong.

CSD does not provide a routine for handling floor time studies or MTM studies—but it will handle standard data elements regardless as to how they were generated.

Software documentation is included, but whether it is user friendly this writer (D.W.K.) cannot say since he was not provided with a copy of the manual or the software system disks to check on some of the statements.

Features of the system include such things as a computerized rate sheet; multilevel where-used search and mass-update of standards by element, machine tool, material, and cutting tool; "what if" simulation; quick change for similar operations; identification of manual elemental times internal to process times; cost estimating of operations; automatic interfacing to user's MRP and MRP-II systems; and it can be used either as a real time system or run as a batch operation.

WOCOM

WOCOM is a registered trademark of a work measurement system developed by Science Management, Inc., the holding company for the Work-Factor Foundation. It is a proprietary system and detailed information can be found in *Work-Factor Time Standards.*[11]

The generic basic-level WOCOM system provides automated application of the two most recognized systems of predetermined times: MTM and Work-Factor. It is designed for simple operation and flexibility in meeting varying individual company needs and does not require previous experience in computer equipment or work measurement methods. Programs are available through the national time-sharing network or for operation on in-house computers. Information on the system is available from the Work-Factor Foundation.

The system consists of eight modular computer programs that can be used to analyze human and machine work, alternate work methods, and assembly line operations and maintain labor standards.[12]

GENERAL INDUSTRIAL ENGINEERING PROGRAMS

Learning Curve Programs

Learning curve programs are closely related to work measurement, hence, their location in this chapter. The program computerizes the learning curve equation found in Chapter 3 of *Advanced Work Measurement*[13] and covers both individual and organizational learning. Any industrial engineer can *easily* use the programs by entering only a few data items, and automatically (almost instantly), the usual answers desired—such as learning time, number of operations to be performed to reach specified learning level, etc.—appear. It is a 40,000 byte program written in BASIC and compiled to run on an IBM PC (or compatible) computer. The program was written by Beth Holtger under the direction of Dr. Walton Hancock, a coauthor of *Advanced Work Measurement* and Associate Dean and Professor at the University of Michigan, College of Engineering. At the A> prompt, one merely enters LC.BAS and the help screens will answer all

---

[11]Joseph H. Quick, James H. Duncan, and James A. Malcolm, Jr., *Work-Factor Time Standards,* McGraw-Hill Book Company, New York, 1962.
[12]*Advanced Work Measurement, op cit.*
[13]*Ibid.*

questions except the equations and their background—for this information consult Chapter 3 of *Advanced Work Measurement.*

Institute of Industrial Engineers (IIE)[14]

The Institute, to which most of us in this profession belong, is an excellent source of work measurement/industrial engineering microcomputer programs. We will first provide information on a series of programs that they regularly publish in *Industrial Engineering* (have been doing so for several years) in Dr. Gary Whitehouse's column "Mini-Micro Computers." These programs are of the kind that should be of interest to all practicing industrial engineers—they are virtually always written in BASIC. A representative lot of these will be described briefly.

Second, information will be provided on software programs that can be purchased from the institute—in this case they provide a program disk (available for the TRS80 Model I, III, and IV or the Apple II series or the IBM PCs) and a manual describing how to use each of the ordered program sets.

Mini-Micro Computer Programs

As mentioned above, these programs are generally written in BASIC. However, as the knowledgeable computer users know, some of the BASIC commands sometimes must be changed slightly when moving from one make of computer to another make.

Each month's issue contains some sort of IE-oriented program or information on how to adapt a particular program (like a Visicalc spreadsheet program) to IE needs. Each article contains the entire statement of the program so that the reader can key it into his/her PC.

A representative selection of these programs will be described. All are in a random order except the first one, which is strictly aimed at work measurement engineers.

Utility Program which Summarizes Stopwatch/Pace Data (In July 1983 issue of *Industrial Engineering*) by Mark Corser of Swift Textiles Inc., Columbus, GA. This program is written in BASIC and tested using CP/M (a disk operating system used in the Apple II series of computers). Handles 100 study elements and an unlimited number of time recordings and pace ratings.

In performing a time study, information is classed, paced, and recorded. Then comes the work of summarization. To assist the reader in judging the program's effectiveness the author provides an example of two similar time studies, one summarized in the traditional method with paper and pen and the other using the above described program and a PC.

[14]Institute of Industrial Engineers, 25 Technology Park/Atlanta, Norcross, GA 30092.

Description	Traditional	Using a PC
Length of Study	2½ hr	2½ hr
Number of Elements	84	84
Number of Time Recordings	400	400
Number of Pace Ratings	400	400
Time Required	6 hr	2 hr

The program is simple enough that it can be modified to provide other features without great difficulty, if you can program in BASIC.

General Software Adapted for I.E. Applications (In February 1984 issue of *Industrial Engineering*) by Ira H. Kleinfeld of the University of New Haven. The general purpose program examined is VISICALC, which is relatively low in cost and available for most all computers.

A simple example is used to illustrate the program's usefulness. For example, one can store employee names along the left side of the window identified as a matrix (these would be identified as "labels" in using the VISICALC program). Other labels as headings across the top would be Dates, Standard Hours of Production (produced by each employee), Actual Hours Worked (by each employee), Down Time, and Employee Efficiency. These are all entered (the hours) as "values" except for the Employee Efficiency, which is calculated automatically by a formula entered into the "cells" under the heading—it only needs to be entered into one cell and it can then be replicated in the other cells by using one of the special VISICALC commands.

The example described could be a daily chart in which a clerk enters the data into the computer. The computer could be programmed (using VISICALC) to produce a weekly chart automatically. There are many other possible uses for spreadsheet programs like VISICALC, Lotus 123, Symphony, Enable, Framework, etc. More information can be found in the article.

Break Even Analysis and Economic Order Quantity (In August 1982 issue of *Industrial Engineering*) by Steven M. Zimmerman of the University of South Alabama and Leo M. Conrad of Imagineering Concepts. Program is written in BASIC for the TRS80 Models I, III, and IV. This program is somewhat similar to one to be described later, which is available as a "package" from I IE (see next subsection). The actual article describing the above program is five pages in length.

Economic Analysis of Capital Investment Decisions (In February 1983 issue of *Industrial Engineering*) by Raymond D. Griggs of Howmet Aluminum Corporation. The program is written in BASIC and tested on the TRS80 Model III, 16K RAM computer.

Three methods of evaluation can be used when applying the program. The first method is to calculate the discounted rate of return, which provides the break-even interest rate that is required so that the present value of the returns generated by the project during its expected life equals the present value of the total cost of the investment.

The second approach provided by the program is to use the project payback period, which measures how quickly a capital outlay will be recovered by the investor. This measure does not indicate the duration of the returns or overall profitability or provide a rate-of-return measure.

The third method is the book return on investment which determines a percentage yield on investment based on actual projected cash flows. This method applies current dollars and does not take into consideration the time value of money. However, it does provide the projected effect of the capital investment on company financial statements.

The program has still another capability, it can serve equally as well as a model. Then by varying the program's tax rates, tax credits, depreciation schedules, project life, capital expenditures, projected net savings, etc., it is possible to determine and identify any parameter's singular or cumulative effect on resulting cash flows, paybacks, rates-of-returns, and overall profitability.

By now it should be obvious that the Mini-Micro Column is a source of many valuable and usable programs. All one needs to do is "key in" the programs. Back issues are available in many libraries.

IIE'S MICROSOFTWARE GROUPS OF PURCHASABLE PROGRAMS

Economic Analysis

Buy versus Lease Decision, takes into account the user's options.

Rate of Return Analysis determines the rate of achievable return from a cash flow which describes a particular project by a direct computer search.

Decision Tree Analysis describes a sequence of decisions as a decision tree, which includes the discounting of money and probabilistic futures.

After-Tax Cash Flow Analysis reports the after-tax evaluation of a project. It also determines the present value of the cash flow based upon the firm's annual rate of return.

Project Management Editor

Project Evaluation and Review Technique (PERT) provides a scheduled completion time for a project with as many as 300 activities and estimates the probability of meeting the specified schedule.

Critical Path Analysis determines the critical path and calculates free and total floats associated with a maximum of 500 activities.

Activity-on-Node Network Analysis calculates the critical path, the earliest and latest start times, and appropriate floats for a project management system described with the time-consuming elements represented by a maximum of 300 nodes.

Resource Allocation is a scheduling routine that considers both the activity time and the effect of limited resources. Allows 100 activities.

The above two described programs illustrate the adequacy of the programs to perform the described operations. Since space is limited the remainder of the list will not be as detailed as the above material.

Production Control
  Crew Size Analysis
  Assembly Line Balancing (uses COMSOL approach)
  Line of Balance Analysis
  Setup Time Sequencing (uses heuristic COMSOL approach)
Statistical Analysis
  The Mann–Whitney Test
  Statistical Data Analysis (calculates several standard descriptions plus it can
    develop a 500-observation histogram)
  Polynomial Curve Fit (uses least squares method)
  Distribution Fitting (uses Kolmogorov–Smirnov test)
Work Measurement
  Stopwatch Timestudy Analysis
  Learning Curve
  Work Sampling Analyzer
  Work Sampling Observation Generator
Forecasting
  Seasonal Forecasting Using Winters' Smoothing
  Exponential Smoothing (single, double, or triple)
  Multiple Regression
  Determination of Smoothing Constants for Forecasting Models
Operations Research
  Linear Programming (simplex method)
  Maximum Flow Analysis
  Inventory Models
  Queueing Models

HANDHELD COMPUTERS AND COMPUTERIZED DEVICES

First, we will examine several of the programmable handheld computers on the market and note some work measurement/industrial engineering available programs. The principal programmables are offered by Hewlett Packard and Texas Instruments.

The possibility of using such computers to produce time values (TMUs) through the application of the algebraic equations for the MTM-1 motion data was mentioned in a introductory chapter to the details of MTM-1 where the equations were stated—however, we should also state that these were developed by these authors.

The material which now follows is quoted directly from *Advanced Work Measurement*[15]:

The state of the art has advanced to such a degree in the area of handheld programmable calculators that it is possible to develop rather sophisticated work measurement programs for them. Although these programs do not supply method description in great detail, they are somewhat faster than manual sys-

[15]*Advanced Work Measurement, op. cit.*

tems and reduce the possibility of application error to an appreciable degree. Two MTM-based programs have been developed for these programmable calculators. They are known by the acronyms CUE and MATE. CUE will be described.

CUE

Cue was developed by Douglas M. Towne, General Analysis, Inc., Redondo Beach, California. It consists of data derived from MTM-1 by the process of averaging, elimination, and simplification. It is designed for the Texas Instruments TI-59 calculator and printer. The original version, CUE-I, does not provide for the analysis of simultaneous motions, but uses the predominant hand concept, in which the analyst determines which hand requires the most time to perform its motion.

The TI-59 has five function keys which will handle ten functions when used in conjunction with a shift key. The keys are labeled A, B, C, D, and E, and, when used with the shift key, A′, B′, C′, D′, and E′. The MTM motions and motion aggregates have been assigned to each of these function keys. A, for example, is the GET aggregate, while B is the Move motion. A 16-inch GET would be inputted by pressing 1-6-A, and 8-inch MOVE is 8-B, and APPLY PRESSURE (AP) 7-E.

The complete listing of motions for the CUE system are

A	GET	A′	TURN
B	MOVE	B′	Process Time
C	POSITION	C′	Finger Motions
D	Frequency Multiplier	D′	Horizontal Body Motions
E	Miscellaneous Motions	E′	Vertical Body Motions

The operation description is hand written but the output is printed on the printer to which the calculator is attached, as shown in Fig. 19-7.

OPERATION DESCRIPTION	CUE CODE	CUE	OUTPUT
Obtain & Pull Down Lever	12A	12.	GET
	6B	6.	MOVE
Get Next Casting to Machine	20A	20.	GET
(10 lbs.)	20.10B	20.1	MOVE
Wait For Machine Cycle	.05C′	0.0550	MIN
Loosen Holder	4A	4.	GET
	7E		AP1
	90A′	90.	TURN
Remove Finished Casting	4A	4.	GET
	A	0.	GET
To Bin	15.10B	15.1	MOVE
Place New Casting in Fixture	15A	15.	GET
	4B	4.	MOVE
	9C	9.	POSN
Turn & Tighten Holder	6A	6.	GET
	90A′	90.	TURN
	7E		AP1
		329.	TMU
		0.1973	MIN
		15	%PFD
		0.2269	TOT

Fig. 19-7. Annotated CUE output.

OTHER COMPUTERIZED DEVICES

The brief description of two data recorders available from the Datawriter Corporation of Excelsior, MN, was included in the earlier discussion of the FAST computer software systems and equipment.

The Electro General Corporation, 14960 Industrial Rd., Minnetonka, MN 55343, offers a data collector (handheld, about three times larger than the programmable handheld computers mentioned above) that for everyday time studies of up to 20 elements takes the user directly from data collection to immediate results. The unit called Datamyte 800 allows the user to assign up to 20 code numbers for various elements. Following this, one can begin keying in each element as it takes place. Time is recorded automatically to the nearest 0.01 min. Once data collection is complete, the results can be had immediately because the Datamyte does its own calculating. The results can all be recalled to a large LCD display by using the function keys or they can be printed.

Average time, frequency, min/max times, group time (foreigns), percentage of each element or group to total study time, and total real or rated time (depending on the model used) are all available. The raw data, however, cannot be printed or saved as this writer understands the system.

The same company also offers Datamyte 1000 PDE, which is a portable data entry terminal that takes the user directly from data collection to computer processing and reporting. In this case, the raw data are saved, offering full traceability.

Event Recorder

Observational Systems, Inc., 1103 Grand Avenue, Seattle, WA 98122, offers the OS-3 Event Recorder for use in time study, work sampling, and other work requiring event recording and generation of summary statistics and reports. This device prompts the user through setup and data collection and then automatically generates the summary statistics printout. The raw data can be saved to a cassette recorder.

Digital Time Study Boards

Several are available, all employing large ½ in. LCD displays. Two representative units are those offered by Meylan Corporation and Datawriter Corp.

BOOKS CONTAINING SOFTWARE PROGRAMS

There are a number of these on the market, but IIE (previously mentioned in this chapter) offer some of these which are authored and/or edited by Dr. Gary E. Whitehouse of the University of Central Florida. Three of them are (1) *Software for Engineers and Managers* (a collection), (2) Soft*cover Software: 28 Microcomputer Programs for I.E.s and Managers,* and (3) *Practical Partners: Microcomputers and the Industrial Engineer.*

SOME CONCLUDING REMARKS

Anytime one attempts to peer into the future, he or she usually has more chances for being wrong than right, but if we put our individual heads "into the sand" (like an ostrich), we tend to stop making progress. Also, at this particular point in this text, it is appropriate to try and look into the future because the general subject area not only has advanced faster then anyone imagined, but also its impact on our individual and collective lives has been great—and there are no signs that this impact is decreasing.

Earlier in this text, in the discussion of the 4M program, it was explained briefly that the usage of a computer using the 4M software was far more economical than applying MTM-1 and/or MTM-2 and/or MTM-3 manually (with paper and pen) for any user doing a reasonable amount of work measurement. Lately, microcomputers have become powerful enough (and low in cost) that 4M can be used without a minicomputer or mainframe computer. This should signal to a large degree that work on higher-level MTM-1 systems is no longer needed. In fact, the microcomputer programs for MTM-2, MTM-3, and other such variants, tend to have less and less value in the opinion of this writer (D.W.K). Also remember that still more powerful microcomputers will be available in the future; they will easily rival the current small minicomputers in power and speed, and at about the present cost (or less) than the current microcomputers at the upper end of the power range.

All this in turn should be taken into account when thinking about performing work measurement research. Since this text is mainly about MTM and MTM-1, it seems appropriate to suggest that further refining and expansion of the basic system would be appropriate. For example, it surely would make good sense to do further research on Position.

Some time ago work was performed on an area called Decide Action. It never was adopted into the MTM system because some additional research seemed to be needed. Work Factor, however, appears to have a method of handling mental decision times. The author (D.W.K.) again believes that such a research effort should be mounted. This kind of information is needed for setting standards in manufacturing, engineering, office, and service work.

Most of the resources of the U.S./Canada MTM association seem to be aimed at software development, and through the IMD at the development of specialized predetermined time systems based on MTM-1.

As to software, it might be appropriate to expand the 4M software into some of the areas covered by FAST and Auto-CAM–AutoCOST, as described briefly in this chapter. This would appear to be advantageous.

Whether any of the preceding gets done or not, work measurement will advance, and if one person or organization fails to seize an opportunity, someone else is almost sure to do so. One should continue to try and improve the basic "crown jewel" (MTM-1 in our case) and at the same time try to obsolete one or more of the MTM systems—maybe it is possible to even find and/or invent something better than MTM-1. A lot of seemingly wonderful things have disappeared because they were made obsolete by something that was far superior.

Now, let us look at other factors and try to develop some conclusions. First, in the U.S., Canada, and Western Europe high labor costs will further drive factories to automate and reduce direct labor—the only way most of these factories can survive. Automation and modernization of the factory is a must if a company is to be profitable. In a similar vein, how can companies fail to take advantage of incentive systems?

Elsewhere in this text the changing international competitive scene was discussed. Some U.S. firms seem to be taking "the easy out" and buying cheap foreign product and then putting their name on it and selling it here. This means that the firms are mere shells and no longer manufacture their own goods. If this trend continues, the U.S. will be in great trouble because we cannot continue to successfully exist without a substantial manufacturing segment. We must compete successfully and this means automation and spending capital dollars on manufacturing.

The only other application areas for predetermined time systems is the indirect factory work and clerical work that is manually oriented. If we are going to compete with Japan, Korea, Taiwan, Hong Kong, Singapore, etc., we *must* speed up the setting of labor standards for indirect factory labor and manually oriented clerical work.

Mr. Royal Dossett, President of Datawriter Corporation, said that in his opinion industrial engineers and managers were focusing on the wrong target in their work measurement efforts. I (D.W.K.) could not agree more with this thought. What is *really* needed is to set performance standards for the company as a whole and for its professional workers— and I do not think that this is impossible.

To set company or organization standards one needs to first engage in proper formal long- and short-range planning. For the company and its top executives, setting performance standards is then fairly simple.

For the remainder of the professional employees one can then use MBO techniques to set these standards. However, be warned that MBO without first performing correct long- and short-range planning is almost surely doomed to fail.

Those readers who are interested in more information on proper planning and the use of MBO should refer to Chapter 15, "Work Measurement Applicable to Professional Employees" (31 pages in length) of AWM.[16] Other references are provided in that chapter's material.

[16]*Ibid.*

Bibliography

To conserve space in this listing, calendar months are abbreviated and certain of the entries are coded. When the symbol ( )* appears, the asterisk footnotes the entry within the parentheses. The footnote directs the reader to the following codes:

Code	Entry Signified
IIE (Previously AIIE)	American Institute of Industrial Engineers, Inc.; 25 Technology Park/Atlanta; Norcross, Georgia 30071 (formerly located in Columbus, Ohio and New York, N.Y.). Also designates local chapter and conference locations which are identified in the listing.
JIE	*Journal of Industrial Engineering;* AIIE (see above). Final issue Vol. XIX, No. 12, Dec. 1968. Succeeded by *Industrial Engineering,* Vol. 1, No. 1, Jan. 1969.
ASME	American Society of Mechanical Engineers; United Engineering Center, 345 East 47th St., New York, N.Y. 10017.
MECH	*Mechanical Engineering;* ASME (see above).
MTMA	MTM Association for Standards and Research, 16–01 Broadway, Fairlawn, N.J. 07410 (formerly located in Pittsburgh, Pa. and Ann Arbor, Mich.). Also designates conference locations identified in the listing.
JMTM	*Journal of Methods-Time Measurement;* MTMA (see above). Final issue, Vol. XVIII, No. 4, 1973. Succeeded by *The MTM Journal* of International MTM Directorate at MTMA, Vol. I, No. 2, 1974.
FACT	*Factory.* McGraw-Hill Publishing Co., Inc.; New York, N.Y. 10036. Began Vol. 117, No. 1 in Jan 1959 as successor to *Factory Management and Maintenance* of final issue Vol. 116, No. 12 in Dec 1968.

Åberg, Ulf. "Frequency of Occurrence of Basic MTM Motions." *Research Committee Report No. 1,* Svenska MTM-Foreningen, Stockholm, Sweden, Aug 1962. Also published in (JMTM),* Vol. 8, No. 5, July–Aug 1963.

———. "Sequence of Motions in Manual Workshop Production." *Research Committee Report No. 2,* Svenska MTM-Foreningen, Stockholm, Sweden, July 1963. Also published in: (JMTM),* Vol. 9, No. 2, Nov–Dec 1963.

Åberg, Ulf and Hancock, Walton M. "Design Criteria of Predetermined Time Systems." International MTM Directorate, Solna, Sweden, 1968.

*See code at start of Bibliography.

Alford, Leon P. "Scientific Industrial Management." *Paper No. 341.*

Allan, W. G. "A New Dimension in Work Study." (JMTM)* Vol. 17, No. 4, Sept–Oct 1972.

———. "Decision Systems and Work Study." (JMTM)* Vol 17, No. 1, Jan–Feb 1972.

Allderige, John M. "Designing Models in Work Measurement." *Proc—13th Ann Nat (Gilbreth) Conf.* (AIIE)* Atlantic City, N.J., May 1962.

Allderige, John M. and Lomicka, Roy F. "Mathematical Man-Machine Analysis." (JIE)* Vol. 7, No. 3, May–June 1957.

Andersen, Donald S. and Edstrom, David P. "MTM Personnel Selection Tests Validation at Northwestern National Life Insurance Company," (JMTM)* Vol. 15, No. 3, May–June 1970.

Antis, William; Honeycutt, John M., Jr.; and Koch, Edward N. *The Basic Motions of MTM.* The Maynard Foundation, Naples, Fla. 1963, 4th ed., 1973.

Aquilano, Nicholas J. "A Physiological Evaluation of Time Standards for Strenuous Work as Set by Stopwatch Time Study and Two Predetermined Time Data Systems." (JIE)* Vol. 19, No. 9, Sept 1968.

———. "Setting Time Standards for Physically Tiring Jobs Through Physiological Measurements." *Proc—25th Ann Nat Conf* (AIIE)* New Orleans, La, May 1974.

———. "Why IE's Can't Measure Fatigue." (JIE)* Vol. 2, No. 3, Mar 1970.

(ASME)*. "Operation and Flow Process Charts." *ASME Standard 101,* 1947, Rev 1972.

Arnold, D. C. and Tinker, M. A. "The Fixation Pause of the Eyes." *Journal of Experimental Psychology* Vol. 25, 1939.

Ayoub, Mahmoud A. "The Problem of Occupational Safety." (JIE)* Vol. 7, No. 4, Apr 1975.

Ayoub, M. M. "Biomechanical Considerations in Industrial Work." *Paper No. 66-HUF-8.* (ASME)* Mar 1966.

Bailey, Gerald B. "Basic Motion Timestudy." *Proc—8th Ann Time Study and Methods Conf* (ASME)* and Society for Advancement of Management, Apr 1953.

———. "Comments on 'An Experimental Evaluation of the Validity of Predetermined Elemental Time Systems'." (JIE)* Vol. 12, No. 5, Sept–Oct 1961.

Bailey, Gerald B. and Presgrave, Ralph. *Basic Motion Timestudy,* New York: McGraw-Hill Book Company, 1958.

Barany, James W. "The Nature of Individual Differences in Bodily Forces Exerted During a Simple Motor Task." (JIE)* Vol. 14, No. 6, Nov–Dec 1963.

Barany, James W. and Greene, James H. "The Use of Self-Paced Tasks for the Evaluation of Work Methods." (JIE)* Vol. 11, No. 5, Sept–Oct 1960.

Barker, John. "Method of Establishing Standards Using Pre-Determined Times and Regression Analysis." *Proc—2nd Work Measurement and Methods Engineering Div Conf.* (AIIE)* Flint, Mich, Apr 1974.

Barnes, Donald W. "IE Lessons Come in Cans." (JIE)* Vol. 2, No. 2, Feb 1971.

Barnes, Ralph M. "Industrial Engineering Survey." *Industrial Engineering Report No. 101,* Univ. of Iowa, 1949.

———. "Industrial Engineering Survey—1967." (JIE)* Vol. 18, No. 12, Dec 1967.

———. *Motion and Time Study,* New York: John Wiley & Sons, 1937, 6th ed, 1968.

———. *Motion and Time Study Applications,* New York: John Wiley & Sons, 1942, 4th ed, 1961.

———. *Work Sampling,* Dubuque, Iowa: W. C. Brown and Company, 1956.

Barnes, Ralph M. and Andrews, Robert B. "Performance Sampling in Work Measurement." (JIE)* Vol. 6, No. 6, Nov–Dec 1955.

Barnes, Ralph M. and Amrine, Harold T. "The Effect of Practice on Various Elements Used in Screwdriver Work." *Journal of Applied Psychology* Vol. 26 No. 2, Apr 1942.

Barnes, Ralph M. and Mundel, Marvin E. "A Study of Hand Motions Used in Small Assembly Work," *Studies in Engineering Bulletin 16.* Univ of Iowa, 1939.

*See code at start of Bibliography.

——. "A Study of Hand Motions Used in Small Assembly Work." *Studies in Engineering Bulletin 17.* Univ of Iowa, 1939.

——. "Studies of Hand Motions and Rhythm Appearing in Factory Work." *Studies in Engineering Bulletin 12.* Univ of Iowa, 1938.

Barnes, Ralph M. and Perkins, J. S. "A Study of the Development of Skill During Performance of a Factory Operation." *Studies in Engineering Bulletin 22.* Univ of Iowa, 1940.

Barnes, Ralph M.; MacKensie, J. M., and Mundel, Marvin E. "Studies of One- and Two-Handed Work." *Studies in Engineering Bulletin 21.* Univ of Iowa, 1940.

Barnes, Ralph M.; Perkins, J. S.; and Juran, Joseph M. "A Study of the Effect of Practice on the Elements of a Factory Operation." *Studies in Engineering Bulletin 22.* Univ of Iowa, 1940.

Barta, Thomas A. "Using Multiple Linear Regression in the Analysis of Indirect Labor Work Measurement Data." *Proc—22nd Ann Nat Conf.* (AIIE)* Boston, Mass, May 1971.

Bartz, J. A. V. and Gianotti, C. R. "Computer Program to Generate Dimensional and Inertial Properties of the Human Body." *Paper No. 73-WA/Bio-3.* (ASME)* Nov 1973.

Bayha, Franklin H. "A Short Course of MTM Techniques." *Proc—1st Annual Management Conf.* (AIIE)* Fort Wayne, Ind, May 1955.

——. "Fundamental MTM Techniques." *Proc—2nd Ann Management Conf.* (AIEE)* Fort Wayne, Ind, Apr 1956.

——. "Progress Report on Development of New Simplified Data." *Proc—5th Ann MTM Conf.* (MTMA)* New York City, Oct 1956.

——. "Research Techniques—An Analysis." *Proc—6th Ann MTM Conf.* (MTMA)* New York City, Sept 1957.

——. "Your Standards Are As Good As Your Training!" *Proc—7th Ann Chapter Conf.* (AIIE)* Detroit, Mich, Feb 1965.

——. "Application of the New Supplementary Position Data." *Proc—13th Ann MTM Conf.* (MTMA)* New York City, Oct 1965.

——. "Proposal Report on Academic Training Services." (MTMA)* 1967.

——. "MTM-GPD Simultaneous Pattern (Sipa) Table." (MTMA)* Aug 1971.

——. "Decide Action: Exposition for Validation Purposes." (MTMA)* Jan 1975.

Bayha, Franklin H. and Foulke, James A. "Application Training Supplement No. 7—Position and Apply Pressure." (MTMA)* June 1964.

Bayha, Franklin H. and Hancock, Walton M. "A Comparison Between the Original and New Position Data." (JMTM)* Vol. 13, No. 1, Jan–Feb 1968.

——. "Application Guidelines on Decide Action Research." (MTMA)* Apr 1973.

——. "Application Training Supplement No. 12—Decide Action." (MTMA)* Aug 1972.

——. "Apply Pressure Technical Description." International MTM Directorate, Solna, Sweden, Feb 1972.

——. "Human Error Rates: An Approach to Practical Evaluation in Productive Processes." (MTMA)* Jan 1973.

——. "Position: Technical Description of Alternate Position Data." International MTM Directorate, Solna, Sweden, May 1972.

——. "Recommended Practice Guidelines—Apply Pressure." International MTM Directorate, Solna, Sweden, May 1968.

——. "Recommended Practice Guidelines—Position for Alternate Position Data." International MTM Directorate, Solna, Sweden, May 1968.

Bayha, Franklin H. and Karger, Delmar W. "Installing MTM—A New Approach." (JMTM)* Vol. I, No. 2, June 1954.

——. *Methods-Time Measurement Manual,* The Magnavox Company, Fort Wayne, Ind, 1953, Rev 1956.

*See code at start of Bibliography.

Bayha, Franklin H. and Mabry, John E. "Application Training Supplement No. 10—MTM-GPD General Purpose Data." (MTMA)* July 1966.

Bayha, Franklin H. and Stoll, Richard F. "Predetermined Time Standards." Chap. 5, Part 2 of *Foreman's Handbook,* New York: McGraw-Hill Book Company, 1967.

Bayha, Franklin H.; Eady, Karl E.; and Thompson, William C. "MTM Decision Diagrams." *MTM Training Aid.* (MTMA)* June 1970, 2nd ed, Jan 1975.

Bayha, Franklin H.; Foulke, James A.; and Hancock, Walton M. "Application Training Supplement No. 8—Position." (MTMA)* Jan 1965, Rev Jan 1973.

———. "Application Training Supplement No. 9—Apply Pressure." (MTMA)* Jan 1965.

———. "Supplementary Data on Position and Apply Pressure." (JMTM)* Vol. 10, No. 2, Mar–Apr 1965.

———. "Application Training Supplement No. 11—Apply Pressure." (MTMA)* Jan 1973.

Bayha, Franklin H.; Hancock, Walton M.; and Langolf, Gary D. "More Evaluation Parameters for MTM Systems." (JMTM)* Vol. 2, No. 1, 1975.

Bayha, Franklin H.; Erez, Shlomo; Hancock, Walton M.; and Lund, Knut. "A Comparison of the Swedish and U.S.–Canadian Data Cards." (JMTM)* Vol 15, No. 4, Sept–Oct 1970.

Berelson, Bernard and Steiner, Gary A. *Human Behavior,* New York: Harcourt, Brace, & World, 1964.

Berner, G. E. and Berner, D. E. "Relations of Ocular Dominance, Handedness, and the Controlling Eye in Binocular Vision." *Archives of Ophthalmology,* Vol. 50, 1953.

Betke, Richard L. "Application of Behavioral Sciences to the Practice of Industrial Engineering." (JIE)* Vol. 18, No. 5, May 1967.

Birdsong, Jackson H. "MTM and Rehabilitation: A Combination for Potential Profits." (JMTM)* Vol. 17, No. 4, Sept–Oct 1972.

Birdsong, Jackson H. and Chyatte, Samuel B. "Further Medical Applications of Methods-Time Measurement." (JMTM)* Vol. 15, No. 2, Mar–Apr 1970.

———. "The Assessment of Motor Performance in Brain Injury: A Bio-Engineering Method." *Proc—21st Ann Nat Conf.* (AIIE)* Cleveland, Ohio, May 1970.

Birn, Serge A. "Offices—A Challenge for Automated Work Measurement." *Proc—14th Ann Nat Conf.* (AIIE)*. Denver, Col, May 1963.

Birn, Serge A. and Hyde, William F. "The Consultant's Contribution to Industrial Engineering." (JIE)* Vol. 15, No. 6, Nov–Dec 1964.

Birn, Serge A.; Crossan, Richard M.; and Eastwood, Ralph W. *Measurement and Control of Office Costs.* New York: McGraw-Hill Book Company, 1961.

Block, Stanley M. "An Evaluation of Work Measurement Techniques." *Proc—7th Ann Chapter Conf.* (AIIE).* Detroit, Mich. Feb 1965.

———. "Effects of Visual Requirements Upon Certain Simultaneous Hand Motions." *Ph.D. dissertation,* Univ of Minn. July 1956.

———. "Recent Research on Blind Grasp" (JMTM)* Vol. 5, No. 4, Nov–Dec 1958.

Boggs, David H. and Simon, J. Richard. "Differential Effect of Noise on Tasks of Varying Complexity," *Journal of Applied Psychology* Vol. 52, No. 2, 1968.

Bonney, Maurice C. and Schofield, Norman A. "Computerized Work Study Using the SAMMIE/AUTOMAT System." (JMTM)* Vol. 17, No. 3, May–June 1972.

———. "Developments in the SAMMIE/AUTOMAT Computerized Work Study System." (JMTM)*, Vol. 18, No. 4, Sept–Oct 1973.

Boston Consulting Group, Inc. *Perspective on Experience,* Boston, Mass, 1968, 3rd ed., 1972.

Breen, W. W.; DeHaemer, M. J.; and Poock, Gary K. "Comparison of the Effect of Auditory versus Visual Stimulation on Information Capacity of Discrete Motor Responses." *Journal of Experimental Psychology* Vol. 82, No. 2, 1969.

*See code at start of Bibliography.

Brisley, Chester L. "Work Sampling: Past—Present—Future." *Proc—26th Ann Nat Conf.* (AIIE)* Washington, D.C., May 1975.

Bromley, D. B. *Psychology of Human Ageing,* New York: Penquin Books, 1974.

Brouha, Lucien. "Psychological Aspects of Work Measurement." *Proc—3rd Internat MTM Conf.* (MTMA)* New York City, Sept 1963.

———. "Physiological Assessment of Job Stress." *Proc—15th Ann Nat. Conf.* (AIIE).* Philadelphia, May 1964.

———. *Physiology in Industry,* New York: Pergamon Press, 1960.

Buffa, Elwood S. "The Additivity of Universal Standard Data Elements." (JIE)* Vol. 7, No. 5, Sept–Oct 1956.

———. "The Additivity of Universal Standard Data Elements, II." (JIE)* Vol. 8, No. 6, Nov–Dec 1957.

———. "The Electronic Time Recorder: A New Instrument for Work Measurement Research." (JIE)* Vol. 9, No. 2, Mar–Apr 1958.

———. "Toward a Unified Concept of Job Design." (JIE)* Vol. II, No. 4, July–Aug 1960.

Butcher, Philip J. "Predetermined Motion Times." *M. S. thesis,* Univ of Kansas, 1953.

Campbell, L. R. "The MTM of Learning." *Proc—11th Ann MTM Conf.* (MTMA)* New York City, Sept 1962.

Caplan, Stanley H. and Hancock, Walton M. "The Effect of Simultaneous Motions on Learning." (JMTM)* Vol. 9, No. 1, Sept–Oct 1963.

Carlson, Gary. "Predicting Clerical Error." *Datamation.* Feb 1963.

Carmichael, L. and Dearborn, W. F. *Reading and Visual Fatigue,* Boston: Houghton-Mifflin Company, 1947.

Carson, Gordon B.; Bolz, Harold A.; and Young, Hewitt H. eds *Production Handbook,* New York: Ronald Press Company, 1948, 3rd ed. 1972.

Chaffin, Don B. "An Historical Analysis of the Prediction of Physical Fatigue in Industrial Operations." (JMTM)* Vol. 12, No. 2, Mar–Apr 1967.

———. "Bio-Mechanics." *Proc—2nd Work Measurement and Methods Engineering Div Conf.* (AIIE)* Flint, Mich, Apr 1974.

———. "Electromyography—A Method of Measuring Local Muscle Fatigue." (JMTM)* Vol. 14, No. 2, Mar–Apr 1969.

———. "Industrial Man-Machine Research." *Proc—12th Ann Chapter Conf.* (AIIE)* Detroit, Mich, Oct 1970.

———. "Physical Fatigue: What it is—How it is Predicted." (JMTM)* Vol. 14, No. 3, May–June 1969.

———. "Prediction of Metabolic Energy Expenditure Rates." *Proc—Engineering Summer Conf No. 6913.* Univ of Mich, June 1969.

———. "The Development of a Prediction Model for the Metabolic Energy Expended During Arm Activities." *PhD dissertation,* Univ of Mich, 1967.

———. "The Prediction of Physical Fatigue During Manual Labor." (JMTM)* Vol. 11, No. 5, Nov–Dec 1966.

———. "What Basis Exists for Determining How Much We Can Safely Lift." *Proc—26th Ann Nat Conf.* (AIIE)* Washington, D.C., May 1975.

Chaffin, Don B. and Ayoub, M. M. "The Problem of Manual Materials Handling." (JIE)* Vol. 7, No. 7, July 1975.

Chaffin, Don B. and Baker, William H. "A Biomechanical Model for Analysis of Symmetric Sagittal Plane Lifting." *Transactions.* (AIIE)* Vol. 2, No. 1, Mar 1970.

Chaffin, Don B. and Garg, Arun. "A Biomechanical Computerized Simulation of Human Strength." *Proc—25th Ann Nat Conf.* (AIIE)* New Orleans, La, May 1974.

*See code at start of Bibliography.

Chaffin, Don B. and Hancock, Walton M. "Factors in Manual Skill Training." *Research Report 114.* (MTMA)* Aug 1966.

Chaffin, Don B.; Kilpatrick, Kerry E.; and Hancock, Walton M. "A Computer-Assisted Manual Work Design Model." *Proc—21st Ann Nat Conf.* (AIIE)* Cleveland, Ohio, May 1970. Also in *Transactions.* (AIIE)* Vol. 2, No. 4, Dec 1970.

Chapanis, Alphonse; Garner, W. R.; and Morgan, C. T. *Applied Experimental Psychology,* New York: John Wiley & Sons, 1949.

Christ, Charles F. "MRT—An Analytical Way to Measure Indirect Work." (JIE)* Vol. 1, No. 6, June 1969.

Christensen, Julian M. "The Evolution of the Systems Approach in Human Factors Engineering." *Human Factors.* Vol. 4, No. 1, Feb 1962.

Chyatte, Samuel B. and Birdsong, Jackson H. "Predetermined Motion Time Techniques as a Medical Measurement System." *Transactions.* (AIIE)* Vol 3, No. 3, Sept 1971.

Clark, Daniel O. "Industry Launching MTM Microscopic Research," (JMTM)* Vol. 14, No. 4, Sept–Oct 1969.

———. "Meet the MTM Family." (JMTM)* Vol. 17, No. 3, May–June 1972.

———. "The MTM Systems of the World." *Proc—23rd Ann Nat Conf.* (AIIE)* Anaheim, Calif, June 1972.

Clark, Thomas B. "Nonparametric Statistics." (JIE)* Vol. 7, No. 8, Aug 1975.

Clifford, Robert R. "Tests to Predict Worker Total Performance." *Proc—12th Ann MTM Conf.* (MTMA)* New York City, Sept 1964.

Clifford, Robert R. and Hancock, Walton M. "An Industrial Study of Learning." (JMTM)* Vol. 9, No. 3, Jan–Feb 1964.

Coan, James S. *Basic Basic, An Introduction to Computer Programming in Basic Language,* 2nd ed., Hayden Book Company Inc., Rochelle, N.J., 1978.

Cochran, Edward B. "Dynamic Labor Standards." *Proc—19th Ann Nat Conf.* (AIIE)* Tampa, Fla. May 1968.

———. "Learning: New Dimension in Labor Standards." (JIE)* Vol. 1, No. 1, Jan. 1969.

———. "New Concepts of the Learning Curve." (JIE)* Vol. 11, No. 4, July–Aug 1960.

Colbert, B. A. and Griffith, Gerald N. "MODAPTS—A Predetermined Time System That Can Be Memorized." *Proc—22nd Ann Nat Conf.* (AIIE)* Boston, Mass. May 1971.

Conrad, Robert B. "A Simulation Model for Time Measurement." *Transactions.* (AIIE)* Vol. 3, No. 2, June 1971.

———. "The Effect of Interruption on Standard Time." *Transactions.* (AIIE)* Vol. 2, No. 2, June 1970.

Conte, John A. "A Study of the Effects of Paced Audio-Rhythm on Repetitive Motion." (JIE)* Vol. 17, No. 3, Mar 1966.

Cornman, Guy L., Jr. "Fatigue Allowances—A Systematic Method." (JIE)* Vol. 2, No. 4, Apr 1970.

Correll, Donald S. and Barnes, Ralph M. "Industrial Application of the Ratio-Delay Method," *Advanced Management.* Vol. 15, No. 8 (Aug 1950) and Vol. 15, No. 9 (Sept 1950).

Crandall, Richard E. "Time Allocation—A Staff Department Problem." (JIE)* Vol. 14, No. 6, Nov–Dec 1963.

Cremer, Robert H.; Towne, Douglas M.; and Mason, Anthony K. "A Validation of Computer Developed Motion Patterns." (JMTM)* Vol. 12, No. 4, Sept–Oct 1967.

Crossan, Richard M. "A New Approach to Clerical Work Measurement." *Proc—7th Ann MTM Conf,* (MTMA)* New York City, Sept 1958.

———. "Let's Simplify MTM Application." *Proc—8th Ann MTM Conf,* (MTMA)* Detroit, Mich, Oct 1959.

*See code at start of Bibliography.

————. "Master Standard Data—A Step Toward Economic MTM Application." *Proc—10th Ann MTM Conf.* (MTMA)* New York City, Oct 1961.

Crossan, Richard M. and Nance, Harold W. *Master Standard Data,* New York: McGraw-Hill Book Company, 1962. Rev 1972

Crossman, E. R. F. W. "A Theory of the Acquisition of Speed-Skill." *Ergonomics.* Vol. 2, No. 2, Feb 1959.

————. "The Measurement of Perceptual Load in Manual Operations." *Ph.D. dissertation,* Univ of Birmingham, Eng, Oct 1956.

Dales, Harold E. *Work Measurement.* New York: Pitman Publishing Corp., 1972.

Davidson, Harold O. *Functions and Bases of Time Standards.* (AIIE)* 1952.

————. "On Balance—The Validity of Predetermined Time Systems." (JIE)* Vol. 13, No. 3, May–June 1962.

Davis, Harry L. and Miller, Charles I. "Work Physiology as an Aid to Job Design." *Proc—13th Ann Nat (Gilbreth) Conf,* (AIIE)* Atlantic City, N.J., May 1962.

Davis, Louis E. "A Study of Two-Handed Work with Variations in Weight Transported and in Transport Distances." *M.S. thesis.* State Univ of Iowa, 1942.

Davis, Louis E. and Schlegel, Charles. "The Effects of Auditory Visual Rhythmic Disturbances on Work Measurement Performance Rating." (JIE)* Vol. 11, No. 3, May–June 1960.

Davis, Roger T.; Wehrkamp, Robert F.; and Smith, Karl U. "Dimensional Analysis of Motion: I. Effects of Laterality and Movement Direction." *Journal of Applied Psychology* Vol. 35, No. 5, Oct 1951.

Davison, Hugh and Eggleton, Grace. *Principles of Human Physiology,* New York: Longman, Inc., 14th ed, 1968.

Deane, Richard H. "The IE's Role in Accommodating the Handicapped." (JIE)* Vol. 7, No. 7, July 1975.

deJong, J. R. "The Effects of Increasing Skill and Methods-Time Measurement," *Time and Motion Study.* Feb 1961.

————. "The Effects of Increasing Skill on Cycle-Time and Its Consequences for Time Standards." *Ergonomics.* Vol. 1, 1957.

Diehl, Peter and Howell, John R. "Engineering Teams: What Makes Them Go?" (MECH)* Vol. 96, No. 5, May 1974.

Dodge, R. and Cline, T. S. "The Angle Velocity of Eye Movements," *Psychology Review.* Vol. 8, 1901.

Doxie, Floyd T. and Ullom, K. John. "Human Factors in Designing Controlled Ambient Systems." (JIE)* Vol. 18, No. 11, Nov 1967.

Dyer, Dickey. "Mathematical Aspects of Work-Factor Allowances for Simultaneous Standard Elements of Work." (JIE),* Vol. 4, Feb 1953.

Eady, Karl E. "Translation of the Official MTM-2 and the MTM-3 Data Cards into the English System." (JMTM)* Vol. 17, No. 2, Mar–Apr 1972.

————. "What is the MTM Family Really Like?" (JMTM)* Vol. I, No. 2, 1974.

Edwards, Ward. "Optimal Strategies for Seeking Information: Models for Statistics, Choice Reaction Times, and Human Information Processing," *Journal of Mathematical Psychology* Vol. 2, No. 2, July 1965.

Edwards, Ward; Phillips, Lawrence D.; Hays, William L.; and Goodman, Barbara C. "Probabilistic Information Processing Systems: Design and Evaluation," *Transactions on System Science and Cybernetics.* New York: Institute of Electrical and Electronic Engineers, Vol. SSC-4, No. 3, Sept 1968.

Elmaghraby, Salah E. "The Role of Modeling in IE Design." (JIE)* Vol. 19, No. 6, June 1968.

Elrod, J. T.; Kubis, J. F.; and McLaughlin, E. J. "Time and Motion Study During Extended Space Missions." *Proc—20th Ann Nat Conf.* (AIIE)* Houston, Texas, May 1969.

*See code at start of Bibliography.

Evans, Frederick. "PMTS in Perspective." (JMTM)* Vol. 18, No. 3, 1973.
———. "Tape Data Analysis: An Extract from the Manual for MTM Instructors." (JMTM)*
    Vol. 15, No. 2, Mar–Apr 1970.
Evans, Frederick and Magnusson, Kjell-Eric. "MTM-2 Student Manual." United Kingdom
    MTM Assn, Warrington, Lancashire, Eng, May 1966.
Fairweather, Owen. "Management and Labor Look at MTM." *Proc—8th Ann MTM Conf.*
    (MTMA)* Detroit, Mich, Oct 1959.
———. "The Use of Predetermined Times in Labor Negotiations." (JMTM)* Vol. 2, No. 3,
    July–Aug 1955.
Fankhauser, Giovanni. "A New Point of View on the Leveling Procedure." (JMTM)* Vol. 8, No.
    4, May–June 1963.
———. "How Good Are MTM Standard Times?" (JMTM)* Vol. 12, No. 4, Sept–Oct 1967.
Faulkner, Terrance W. "Electromyography in Work Study." *Proc—19th Ann Nat Conf.* (AIIE)*
    Tampa, Fla, May 1968.
Fein, Mitchell. "A Rational Basis for Normal in Work Measurement." (JIE)* Vol. 18, No. 6,
    June 1967.
———. "Influencing the Motivation to Work." *Proc—26th Ann Nat Conf.* (AIIE)* Washington,
    D.C., May 1975.
———. "Rational Approaches to Raising Productivity." *Monograph, No. 5.* Work Measurement
    and Methods Engineering Div. (AIIE)* 1974.
———. "Short Interval Scheduling: A Labor Control Technique." (JIE)*, Vol. 4, No. 2, Feb.
    1972.
———. "What's Wrong With Incentives". *Proc—23rd Ann Nat Conf.* (AIIE)* Anaheim, Calif,
    June 1972.
———. "Work Measurement and Wage Incentives." (JIE)* Vol. 5, No. 9, Sept 1973.
———. "Work Measurement Today: Concepts of Normal Pace." (JIE)* Vol. 4, No. 9, Sept 1972.
———. "Work Measurement Today: Facing Up to the Problems." (JIE)* Vol. 4, No. 8, Aug
    1972.
Fenske, Russell W. "Extracting More Information from a Work Sampling Study." (JIE)* Vol. 18,
    No. 7, July 1967.
Figler, Richard B. "Regression and Correlation Analysis." (JIE)* Vol. 6, No. 10, Oct 1974.
Fitts, Paul M. "Functions of Man in Complex Systems." *Aerospace Engineering,* Vol. 21, No. 1,
    Jan 1962.
Fogel, Lawrence J. *Human Information Processing.* Englewood Cliffs, N.J.: Prentice-Hall, 1967.
Folsom, Richard G. "Technology and Humanism." (MECH)* Vol. 88, No. 1, Jan 1966.
Foulke, James A. "Advancements in Bioinstrumentation in Industrial Tasks." *Proc Engineering
    Summer Conf No. 6913,* Univ of Mich, June 1969.
———. "An Orientation to MTM Manual Learning Curve Research." (JMTM)* Vol. 11, No. 4,
    Sept–Oct 1966.
———. "Estimating Individual Operator Performance." (JMTM)* Vol. 15, No. 1, Jan–Feb
    1970.
Foulke, James A. and Hancock, Walton M. "Industrial Research on the MTM Element Apply
    Pressure." *Research Report 111* (MTMA)* Dec 1961.
Foulke, James A. and Turner, Roger N., Jr. "Computer-Aided MTM". *Proc—12th Ann Chapter
    Conf.* (AIIE)* Detroit, Mich, Oct 1970.
Franz, Donald and Nadler, Gerald E. "New Measurements to Determine the Effect of Task Fac-
    tors on Body Member Acceleration Patterns." (JIE)* Vol. 12, No. 5, Sept–Oct 1961.
Fuller, Frank H. "The Human Machine—Work & Environment." (MECH)* Vol. 85, No. 12,
    Dec 1963.

*See code at start of Bibliography.

Galbraith, Jay R. "Some Motivational Determinants of Job Performance." (JIE)* Vol. 18, No. 4, Apr 1967.

——. "The Use of Subordinate Participation in Decision-Making." (JIE)* Vol. 18, No. 9, Sept 1967.

Gavett, Robert. *Human Engineering and Motivation.* Englewood Cliffs, N.J.: Prentice-Hall, 1969.

Geller, E. Scott and Pitz, Gordon F. "Effects of Prediction, Probability, and Run Length on Choice Reaction Speed." *Journal of Experimental Psychology* Vol. 84, No. 2, 1970.

Geppinger, Helmut C. *Dimensional Motion Times.* New York: John Wiley & Sons, 1955.

Gershoni, Haim. "Stability of Workers' Micro-Methods." (JMTM)* Vol. 15, No. 4, Sept–Oct 1970.

——. "Use of the Tape Recorder for Teaching Workers Manual Tasks." (JMTM)* Vol. 16, No. 4, Sept–Oct 1971.

Gilbreth, Frank B. *Motion Study,* New York: D. Van Nostrand Company, 1911.

Gilbreth, Frank B. and Gilbreth, Lillian M. *Applied Motion Study.* New York: Sturgis and Walton Company, 1917.

——. "A Fourth Dimension for Measuring Skill for Obtaining the One Best Way to do Work." *Society of Industrial Engineering Bulletin,* Vol. 4, Nov 1923.

——. "Classify the Elements of Work." *Management and Administration* Vol. 8, 1924.

Gilbreth, Frederick M. "Crossing the Bridge—Some Pointers on Getting Action." (JMTM)* Vol. 7, No. 2. May–June 1959.

Gilbreth, Lillian M. "From Therbligs to Today." *Proc—1st Ann MTM Conf.* (MTMA)* New York City, Oct 1952.

——. "Scientific Control of Management." *Proc—5th Ann MTM Conf,* (MTMA)* New York City, Oct 1956.

——. "Frank as a Teacher." *Proc—13th Ann Nat (Gilbreth) Conf,* (AIIE)* Atlantic City. N.J., May 1962.

Glen, Thaddeus M. "The Prediction of Work Performance Capabilities of Mentally Handi-capped Young Adults." *PhD dissertation,* Univ of Mich, 1964.

Gockel, Thomas R. "Building Block Concept of Standard Data." *Proc—11th Ann MTM Conf.* (MTMA)* New York City, Sept 1962.

Godwin, G. A. "Basic Principles of the Work-Factor System of Time and Motion Study." *Machinery,* New York: Industrial Press, Vol. 85, Oct 1954.

Goel, Surender N. and Becknell, Robert H. "Learning Curves That Work." (JIE)* Vol. 4, No. 5, May 1972.

Goldman, Jay W. and Eisenberg, Hyman. "Can We Train More Effectively?" (JIE)* Vol. 14, No. 2, Mar–Apr 1963.

Goldman, Jay W. and Hart, L. W., Jr. "Information Theory and Industrial Learning." (JIE)* Vol. 16, No. 5, Sept–Oct 1965.

Gromberg, William. *A Trade Union Analysis of Time Study.* 2nd ed. Englewood Cliffs, N.J.: Prentice-Hall, 1955.

——. "Some Developments in the Relationship between Collective Bargaining and Industrial Engineering." (JIE)* Vol. 8, No. 5, Sept–Oct 1957.

——. "What Are the Specifications for an Effective System of Standard Data?" (JIE)* Vol. 5, No. 4, July 1954.

——. "Work Measurement Adapted To Its Environment." *Proc—24th Ann Nat (Silver) Conf.* (AIIE)* Chicago, May 1973.

Goode, Harry H. "Complexity, Research, and Industry." *Proc—3rd Ann MTM Conf,* (MTMA)* New York City, Oct 1954.

Goodman, Barbara Ettinger. "A Laboratory Study of Apply Pressure." (MTMA)* July 1959.

*See code at start of Bibliography.

Goodman, Barbara Ettinger and Foulke, James A. "Instrumentation Used In A Laboratory Study of Apply Pressure." (JMTM)* Vol. 5, No. 3, May–June 1958.

Gottlieb, Bertram. "A Critical Analysis of Predetermined Motion Time Systems." *Proc—16th Ann Nat Conf.* (AIIE)* Chicago, May 1965.

———. "A Fair Day's Work is Anything You Want It to Be." (JIE)* Vol. 19, No. 12, Dec 1968.

———. "Creative Incompetence or How to Limit the Peter Principle." (JIE)* Vol. 3, No. 4, Apr 1971.

———. "Sex, Equality and the Industrial Engineer." *Proc—26th Ann Nat Conf.* (AIIE)* Washington, D.C., May 1975.

———. "Work Measurement *Is* Professional Activity." *Proc—18th Ann Nat Conf.* (AIIE)* Toronto, Ontario, Can, May 1967.

Graham, Charles F. *Work Measurement and Cost Control.* Elmsford, N.Y.: Pergamon Press, 1965.

Greene, James H. and Morris, W. H. M. "The Design of a Force Platform for Work Measurement." (JIE)* Vol. 10, No. 4, July–Aug 1959.

———. "The Force Platform: An Industrial Engineering Tool." (JIE)* Vol. 9, No. 2, Mar–Apr 1958.

Greene, James H.; Morris, W. H. M.; and Wiebers, J. E. "A Method for Measuring Physiological Cost of Work." (JIE)* Vol. 10, No. 3, May–June 1959.

Griffith, Gerald N. "MODAPTS—A Predetermined Time System That Can Be Memorized." *Assembly Engineering.* Aug 1970.

Hale, David J. "The Relation of Correct and Error Responses in a Serial Choice Reaction Task." *Psychonomic Science.* Vol. 13(6), 1968.

———. "Repetition and Probability Effects in a Serial Choice Reaction Task." *Acta Psychologica,* No. 29, 1969.

Hamblin, Harry M. "Union Participation in Industrial Engineering." *Proc—7th Ann MTM Conf.* (MTMA)* New York City, Sept 1958.

Hancock, Walton M. "A Laboratory Study of Learning." *Proc—11th Ann MTM Conf.* (MTMA)* New York City, Sept 1962.

———. "New Research Techniques in Work Measurement." *Proc—12th Ann Nat Conf.* (AIIE)* Detroit, Mich, May 1961.

———. "Pinpointing Operator Learning Time." *Proc—3rd Internat MTM Conf.* (MTMA)* New York City, Sept 1963.

———. "Progress Report on Learning Curve Research." (JMTM)* Vol. 9, No. 2, Nov–Dec 1963.

———. "Results of Industrial Phase of Research on Learning." *Proc—12th Ann MTM Conf.* (MTMA)* New York City, Sept 1964.

———. "The Effects of Learning on Short-Cycle Operations." *Proc—10th Ann MTM Conf.* (MTMA)* New York City, Oct 1961.

———. "You Can Reduce Training Losses." A. T. Kearney & Company Seminar, Chicago, Nov 1963.

———. "The Learning Curve." *Industrial Engineering Handbook.* 3rd ed., New York: McGraw-Hill Book Company, 1971, pp 7-102–7-114.

———. "An Assessment of the Research Activities of the U.S./Canada MTM Association, 1948–1971." (JMTM)* Vol. 16, No. 5, Nov–Dec 1971.

———. "Human Factors and the Industrial Engineer." *Proc—Engineering Summer Conf No. 6620.* Univ of Mich, July 1966.

———. "Managerial Use of Computer Data." (JMTM)* Vol. 12, No. 4, Sept–Oct 1967.

———. "Selecting Predetermined Standard Time Systems: An Evaluation Methodology." *Proc—3rd Ann Systems Engineering Conf.* (AIIE)* Columbus, Ohio, Sept 1972.

*See code at start of Bibliography.

———. "The Cost of Training People." *Proc—Engineering Summer Conf No. 6913,* Univ of Mich, June 1969.

———. "The Design of Inspection Operations Using the Concepts of Information Theory." (JMTM)* Vol. 11, No. 4, Sept–Oct 1966.

———. "The Prediction of Learning Rates for Manual Operations." (JIE)* Vol. 18, No. 1, Jan 1967. Also in (JMTM)* Vol. 12, No. 2, Mar–Apr 1967.

———. "The System Precision of MTM-1." (JMTM)* Vol. 15, No. 3, May–June 1970.

Hancock, Walton M. and Foulke, James A. "A Description of the Electronic Data Collector and the Methods of its Application to Work Measurement." *Research Information Paper No. 1,* (MTMA)* 1961. Also published in: (JIE)* Vol. 13, No. 4, July–Aug 1962.

———. "Effects of Learning on Short-Cycle Operations." *Research Information Paper No. 2,* (MTMA)* 1961.

———. "Learning Curve Research on Short-Cycle Operations: Phase I. Laboratory Experiments." *Research Report 112.* (MTMA)* 1963.

———. "The Effects of Learning on Short-Cycle Operations." *Proc—10th Ann MTM Conf.* (MTMA)* New York City, Oct 1961.

———. "Computation of Learning Curves." (JMTM)* Vol. 11, No. 3, July–Aug 1966.

Hancock, Walton M. and Jelinek, Richard C. "The Staffing of Indirect Labor Situations Under Random Demands." (JMTM)* Vol. 14, No. 1, Jan–Feb 1969.

Hancock, Walton M. and Langolf, Gary D. "Productivity, Quality, and Complaint Considerations in the Selection of MTM Systems." (MTMA)* Aug 1973.

Hancock, Walton M. and Sathé, Prakash. "Learning Curve Research on Manual Operations: Phase II, Industrial Studies." *Research Report 113A* (113 Revised). (MTMA)* 1969.

Hancock, Walton M. and Tiffan, William R. "A Laboratory Study Concerning the Effect of Gloves on Performance Times." (JMTM)* Vol. 14, No. 2, Mar–Apr 1969.

Hancock, Walton M.; Foulke, James A.; and Miller, James M. "Accuracy Comparisons of MTM-1, GPD, MTM-2, MTM-3 and the AMAS Systems." (MTMA)* May 1973.

Hancock, Walton M.; Langolf, Gary D.; and Clark, Daniel O. "Development of a Standard Data System for Microscopic Work." *Proc—23rd Ann Nat Conf.* (AIIE)* Anaheim, Calif, June 1972.

Hancock, Walton M.; Clifford, Robert R.; Foulke, James A.; and Krystynak, Leonard F. "Learning Curve Research on Manual Operations: Phase II. Industrial Studies." *Research Report 113,* (MTMA)* July 1965.

Harris, Douglas H. and Chaney, Frederick B. *Human Factors in Quality Assurance,* New York: Wiley–Interscience Publishing, 1969.

Harris, Shelby J. and Smith, Karl U. "Dimensional Analysis of Motion: V. An Analytic Test of Psychomotor Ability." *Journal of Applied Psychology,* Vol. 37, No. 2, Apr 1953.

———. "Dimensional Analysis of Motion: VII. Extent and Direction of Manipulative Movements as Factors in Defining Motions." *Journal of Applied Psychology,* Vol. 38, No. 2, Apr 1954.

Hassan, M. Zia and Block, Stanley M. "A Study of Simultaneous Positioning." (JIE)* Vol. 18, No. 12, Dec 1967.

Hasselqvist, Olle. "Industrial Engineering and Productivity Services of Today and Tomorrow." (JMTM)* Vol. 2, No. 1, 1975.

———. "The International MTM Directorate—an International Cooperation Organization for MTM Matters." (JMTM)* Vol. 16, No. 1, Jan–Feb 1971.

———. "The MTM Method—With Special Reference to the Physically Handicapped." (JMTM)*, Vol. 16, No. 4, Sept–Oct 1971.

———. "The Swedish Program for Integrated Work Measurement Data." (JMTM)* Vol. 12, No. 4, Sept–Oct 1967.

*See code at start of Bibliography.

Hasselqvist, Olle: Söderström, Per; and Wiklund, Alf. *MTM:s Grundröreiser;* Utgivare, Svenska MTM-Gruppen, AB; Stockholm, Sweden; 1962.

Hayden, Spencer J. and Katzell, Raymond A. "Motivation Revisited." *Proc—13th Ann Nat (Gilbreth) Conf.* (AIIE)* Atlantic City, N.J., May 1962.

Hecker, Donald.;Green, Donovan; and Smith, Karl U. "Dimensional Analysis of Motion: X. Experimental Evaluations of a Time-Study Problem." *Journal of Applied Psychology.* Vol. 40, No. 4, Aug 1956.

Helmrich, Klaus. "Integrated Work Measurement with MTM and Work Classification." (JMTM)* Vol. 18, No. 4, Sept–Oct 1973.

Hildreth, G. "The Development and Training of Hand Dominance." *Journal of Genetic Psychology* Vol. 75, 1949.

Hitchcock, Frank J. "Do You Use Table X Correctly?" *Proc—8th Ann MTM Conf.* (MTMA)* Detroit, Mich, Oct 1959.

Hoag, La Verne L. "Prediction of Physiological Strain and Performance Under Conditions of High Physiological Stress." *PhD dissertation.* Univ of Mich, Ann Arbor, 1969.

———. "Stress, Strain, and Performance." *Proc—Engineering Summer Conf No. 6913,* Univ of Mich, June 1969.

Hodson, William K. "Accuracy of MTM-GPD." (JMTM)* Vol. 9, No. 1, Sept–Oct 1963.

———. "Industrial Engineering in an Era of Changing Concepts." *Proc—16th Ann Nat Conf.* (AIIE)* Chicago, May 1965.

———. "The 'Solid State' Industrial Engineer." *Proc—10th Ann MTM Conf.* (MTMA)* New York City, Oct 1961.

Hodson, William K. and Mattern, William J. "Universal Standard Data." *Industrial Engineering Handbook,* 2nd ed. New York: McGraw-Hill Book Company, 1963, pp. 3-194 to 3-216.

Hoffman, Thomas R. "Effect of Prior Experience on Learning Curve Parameters." (JIE)* Vol. 19, No. 8, Aug 1968.

Honeycutt, John M., Jr. "Comments on 'An Experimental Evaluation of the Validity of Predetermined Elemental Time Systems'." (JIE)* Vol. 13, No. 3, May–June 1962.

Hornbuch, Frederick W., Jr. "Why Management Needs Work Measurement." *Proc—21st Ann Nat Conf.* (AIIE)* Cleveland, Ohio, May 1970.

Hoxie, Robert Franklin. *Scientific Management and Labor,* New York and London: D. Appleton and Company, 1921.

Huiskamp, Janet; Smader, Robert C.; and Smith, Karl U. "Dimensional Analysis of Motion: IX. Comparison of Visual and Nonvisual Control of Component Movements." *Journal of Applied Psychology* Vol. 40, No. 3, June 1956.

Hummel, J. O. P. "Motion and Time Study in Leading American Industrial Establishments." *M. S. thesis.* Penn State Univ.

Hurni, Melvin L. "Management Science and the Manager." *Proc—Internat Conf.* (AIIE)* New York City, Sept 1963.

Hurst, Ronald. *Industrial Management Methods.* New York: International Publications Service, 1970.

I.I.E., ANSI (American National Standards Institute) Standard Z94.1–12, *Industrial Engineering Terminology,* 1972, 1982.

Ingenohl, Ingo. "Measuring Physical Effort." (JIE)* Vol. 10, No. 2, Mar–Apr 1959.

Ischinger, E. "Analysis of Some Differences Between One- and Two-Handed Industrial Work." *M.S. thesis,* Purdue Univ, 1950.

Jaffe, William J. "Dr. Lillian Moller Gilbreth." *Proc—13th Ann Nat (Gilbreth) Conf.* (AIIE)* Atlantic City, N.J., May 1962.

———. "Frank Bunker Gilbreth", *Proc—13th Ann Nat (Gilbreth) Conf.* (AIIE)* Atlantic City, N.J., May 1962.

*See code at start of Bibliography.

Jennings, Paul. "The Conflict Between Industrial Engineering and Industrial Relations." (JIE)* Vol. 4, No. 9, Sept 1972.

Jinich, Carlos and Niebel, Benjamin W. "Synthetic Leveling—How Valid?" (JIE)* Vol. 2, No. 5, May 1970.

Johnson, Richard A.; Kast, Fremont E.; and Rosenzweig, James E. *The Theory and Management of Systems,* New York: McGraw-Hill Book Company, 1963.

Johnson, Wendell. *People in Quandaries.* New York: Harper and Brothers, 1946.

Jones, W. Dale. "New Time Study Tool." *Advanced Management,* Vol. 18, Apr. 1953.

Kane, Don. *Planned Multiple Machine Assignments,* Argus Cameras, Ann Arbor, Mich, 1956.

Karger, Delmar W. "Background of Predetermined Time Systems." (JMTM)* Vol. 3, No. 2, May–June 1956.

———. "Developing Your MTM Training Program." *Proc—3rd Ann MTM Conf.* (MTMA)* New York City, Oct.1954.

———. "Dividends from Research," *Proc—3rd Internat MTM Conf.* (MTMA)* New York City, Sept 1963.

———. "Electronic Development," *Proc—6th Ann MTM Conf.* (MTMA)* New York City, Sept 1957.

———. *The New Product.* New York: Industrial Press Inc., 1960.

———. "Work Measurement Application and MTM." (JMTM),* Vol. 10, No. 1, Nov–Dec 1964.

Karger, Delmar W. and Albrecht, Donald D. "The Measurement of Professional Work." (JIE)* Vol. 3, No. 8, Aug 1971.

Karger, Delmar W. and Bayha, Franklin H. *Engineered Work Measurement,* New York: Industrial Press Inc., 1957, 2nd ed, 1966, 1977.

———. *La Mesure Rationelle du Travail MTM et Systèmes du Temps Prédeterminés,* trans Georges R. A. Lapoirie, Gauthier-Villars, Paris, France, 1962.

Karger, Delmar W. and Hancock, Walton M. *Advanced Work Measurement,* Industrial Press Inc., New York, 1982.

Karger, Delmar W. and Murdick, Robert G. *Managing Engineering and Research,* New York: Industrial Press Inc., 1963. 2nd ed. and 3rd ed., 1969 and 1980.

———. *New Product Venture Management,* New York: Gordon & Breach Science Publishers, 1972.

Karlson, K. G. "Human Engineering and The International MTM Directorate." (JMTM)* Vol. I, No. 4, 1974.

Karpovich, Peter V. and Sinning, Wayne E. *Physiology of Muscular Activity,* Philadelphia: W. B. Saunders Company, 7th ed. 1971.

Keachie, E. C. "Cost and the Learning Curve." *American Association of Cost Engineers Bulletin.* Vol. 3, No. 2, June 1961.

Keliner, Arthur D. "Creativity: How to Manage It." (MECH)* Vol. 92, No. 2, Feb 1970.

Kerkhoven, C. L. M. "Ergonomics: Study of Fatigue of Workers in Industry." (JIE)* Vol. 17, No. 11, Nov 1966.

———. "Time and Motion Study III." *Proc—Internat Conf.* (AIIE)* New York City, Sept 1963.

Khalil, Tarek M. "Design Tools and Machines to Fit the Man." (JIE)* Vol. 4, No. 1, Jan 1972.

Kilpatrick, Kerry E. "A Model for the Design of Manual Work Stations." *PhD dissertation,* Univ of Mich, 1971.

———. "Computer-Aided Workplace Design." (JMTM)* Vol. 14, No. 4, Sept–Oct 1969.

Kinack, Ronald J. "Activity Evaluation Technique—A Quick and Easy Procedure for Developing Time Standards." (JIE)* Vol. 18, No. 5, May 1967.

Kinch, Robert E. "Human Factors in the Measurement of Total System Effectiveness." (JIE)* Vol. 14, No. 6, Nov–Dec 1963.

*See code at start of Bibliography.

——. "The Systems Concept in Management." (JIE)* Vol. 18, No. 5, May 1967.

Kirwin, Ralph S. "Maximum Performance in a Daywork Plant." *Proc—3rd Ann Management Conf.* (AIIE)* Fort Wayne, Ind, May 1957.

Knight, A. A. and Dagnall, P.R. "Precision in Movements." (JMTM)* Vol. 14, No. 1, Jan–Feb 1969.

Koch, William C. "Correlation of Purdue Pegboard Dexterity Test Scores with Short-Cycle Operations." *(Unpublished) Report,* Univ of Mich, 1964.

Koepke, C. A. and Whitson, L. S. "Power and Velocity Developed in Manual Work." (MECH)* Vol. 62, 1940.

Konz, Stephan A. "Fitting the Job to the Man." (JIE)* Vol. 3, No. 1, Jan 1971.

——. "Design of Work Stations." (JIE)* Vol. 18, No. 7, July 1967.

Konz, Stephan A. and Dickey, George L. "Manufacturing Assembly Instructions: Part II. Combinations and Forms of Media." (JIE)* Vol. 18, No. 1, Jan 1967.

Konz, Stephan A. and Goel, Satish C. "The Shape of the Normal Work Area in the Horizontal Plane." *Transactions.* (AIIE)* Vol. 1, No. 1, Mar 1969.

Konz, Stephan A. and Redding, Stephen "The Effect of Social Pressure on Decision Making." (JIE)* Vol. 16, No. 6, Nov–Dec 1965.

Konz, Stephan A.; Jeans, Carl E.; and Rathore, Ranveer S. "Arm Motions in the Horizontal Plane." *Transactions.* (AIIE)* Vol. I, No. 4, Dec 1969.

Konz, Stephan A.; Braun, E.; Jachindra, K.; and Wichlan, D. "Human Transmission of Numbers and Letters." (JIE)* Vol. 19, No. 5, May 1968.

Konz, Stephan A.; Dickey, George L.; McCutchan, Carl M.; and Daniels, Roger W. "Manufacturing Assembly Instructions." (JIE)* Vol. 17, No. 5, May 1966.

Konz, Stephan A.; Dickey, George L.; McCutchan, Carl M.; and Koe, Bruce. "Manufacturing Assembly Instructions: Part III. Abstraction, Complexity, and Information Theory." (JIE)* Vol. 18, No. 11, Nov 1967.

Koontz, Harold and O'Donnell, Cyril. *Essentials of Management.* New York: McGraw-Hill Book Company, 1974.

Krick, Edward V. "An Investigation of Simultaneous Motions as Treated in the Methods-Time Measurement System." *M.S. thesis,* Cornell Univ, Ithaca, N.Y., 1955.

——. "Hyperenthusiastic and Hypercritical Writings on Predetermined Motion Times." (JIE)* Vol. 9, No. 3, May–June 1958.

——. *Methods Engineering,* New York: John Wiley & Sons, 1962.

Kuttan, Appu and Nadler, Gerald E. "Operator Performance Studies III: Three-Dimensional Equations for the Hand Motion Path." *Transactions.* (AIIE)* Vol. 1, No. 3, Sept 1969.

Kyser, Robert C., Jr. "Applying IE and Behavioral Science." (JIE)* Vol. 7, No. 4, Apr 1975.

Lambert, P. "Practice Effect of Non-Dominant vs. Dominant Musculature in Acquiring Two-Handed Skill." *Research Quarterly* Vol. 22, Mar 1951.

Lang, Andrews M. "An MTM Analysis of Performance Rating Systems." *Research Report 104* (MTMA)* Apr 1952.

——. "A Research Report on Standards for Reading Operations." *Research Report 102* (MTMA)* July 1951.

——. "Latest MTM Research Developments." *Proc—1st Ann MTM Conf.* (MTMA)* New York City, Oct 1952.

——. "MTM Newsletter." (MTMA)* Nos. 1 through 5, circa 1950–51.

——. "Methods-Time Measurement Bulletin." (MTMA) Oct 1951 and Feb. 1953.

Lang, Andrews M. and Biel-Nielsen. H. Erik. "Preliminary Research Report on Disengage." *Research Report 101* (MTMA)* Feb 1951.

Langolf, Gary D. "Human Motor Performance in Precise Microscopic Work—Development of Standard Data for Microscopic Assembly Work", *PhD dissertation,* Univ of Mich, 1973.

*See code at start of Bibliography.

Lanier, Brian O. "Randomized Time Study." (JIE)* Vol. 6, No. 6, June 1974.

Lansburgh, Richard H. and Spriegel, William R. *Industrial Management.* New York: John Wiley & Sons, 1940.

Lave, Roy E., Jr. "Timekeeping for Simulation." (JIE)* Vol. 18, No. 7, July 1967.

Lavigueur, Lawrence L. "Standard Data for Eye Time." (JMTM)* Vol. 2, No. 1, Jan–Feb 1955.

Lazarus, Irwin P. "A System of Predetermined Human Work Times." *Ph.D. dissertation,* Purdue Univ.. 1952.

Lee, Robert C. and Moore, James M. "CORELAP—COmputerized RElationship LAyout Planning." (JIE)* Vol. 18, No. 3, Mar 1967.

Leerburger, B. A., Jr. "Scientific Management: Story of a Revolution." (FACT)* Oct 1960.

Lehman, Melvin. "Make This Learning-Curve Calculator." (JIE)* Vol. 2, No. 3, Mar 1970.

———. "What's Going On in Product Assembly?" (JIE)* Vol. 1, No. 4, Apr 1969.

Levin, Robert, "Using MTM for Operator Training." *Proc—2nd Ann MTM Conf.* (MTMA),* New York City, Oct 1953.

Likert, Rensis. *New Patterns of Management,* New York: McGraw-Hill Book Company, 1961.

Likert, Rensis and Seashore, Stanley E. "Making Cost Control Work." (JMTM)* Vol. 10, No. 3, June–July 1965.

Lipton, Paul R. "An Application of Factorial Experimentation to the Work Measurement Process." (JIE)* Vol. 18, No. 8, Aug 1967.

Lipton, Paul R. and Duffy, Daniel J. "An Application of Statistical Control to a Work Measurement Problem." (JIE)* Vol. 18, No. 9, Sept 1967.

Louden, J. Keith. "The Manager—Today and Tomorrow." *Proc—6th Ann MTM Conf.* (MTMA)* New York City, Sept 1957.

———. *Wage Incentives.* New York: John Wiley & Sons, 1944.

Lowe, Cecil W. *Industrial Statistics,* (2 volumes). Boston: Cahners Publishing Co., 1970.

Lowry, Stewart M.; Maynard, Harold B.; and Stegemerten, Gustave J. *Time and Motion Study and Formulas for Wage Incentives.* 3rd ed, New York: McGraw-Hill Book Company, 1940.

Luckiesh, M. and Moss, F. R. "The Extent of the Perceptual Span in Reading." *Journal of General Psychology* Vol. 25, 1941.

Lynch, H. "Basic Motion Timestudy." (JIE)* Vol. 4, Aug 1953.

Mabry, John E. "MTM-GPD General Purpose Data Original Development and Future Expansion." (JMTM)* Vol. 9, No. 1, Sept–Oct 1963.

———. "MTM-2: International Combined MTM Data." (JMTM)* Vol. 11, No. 1, Mar–Apr 1966.

Magnusson, Kjell-Eric. "The Development of MTM-2, MTM-V, and MTM-3." (JMTM)* Vol. 17, No. 1, Jan–Feb 1972.

Magnusson, Kjell-Eric and Engström, Kåre. "MTM for Machine Tools—MTM-V." (JMTM)* Vol. 15, No. 5, Nov–Dec 1970.

Maher, James J. "The Simplicity of MTM-2 Application." (JMTM)* Vol. 15, No. 1, Jan–Feb 1970.

Malik, Zafar A. "Formal Long Range Planning and Organizational Performance." *PhD dissertation,* Rennselaer Polytechnic Inst, Troy, N.Y., 1974.

Mansoor, E. M. "An Investigation into Certain Aspects of Rating Practice." (JIE)* Vol. 18, No. 2, Feb 1967.

Mariotti, John J. "Proxemics—How Close is Your Neighbor?" *Proc—1st Work Measurement and Methods Engineering Conf.* (AIIE)* East Lansing, Mich, Nov 1972.

Martin, Donald D. "The 'Space' Effect." (JIE)*, Vol. 7, No. 8, Aug 1975.

Martin, James B. and Chaffin, Don B. "Biomechanical Computerized Simulation of Human Strength in Sagittal-Plane Activities," *Transactions,* (AIIE)*, Vol. 4, No. 1, Mar 1972.

Martin, John C. "A Better Performance Rating System", (JIE)*, Vol. 2, No. 8, Aug. 1970.

———. "4-M Data Pre-Determined Time Systems", *Proc—2nd Work Measurement and Methods Engineering Conf,* (AIIE)*, Flint, Mich, Apr 1974.

———. "New Developments in Predetermined Time Training and Application", (JMTM)*, Vol. 10, No. 1, Nov–Dec 1964.

———. "Practical Advantages Through Applications of 4M Data", (JMTM)*, Vol. I, No. 4, 1974.

———. "The 4M Data System", (JIE)*, Vol. 6, No. 3, Mar 1974.

———. "Standards and Wage System Audits." *Industrial Engineering Handbook,* 3rd ed. New York: McGraw-Hill Book Company, 1971, pp. 6-158 to 6-170.

Mason, Anthony K. and Britto, Ney O. "Errors in Estimated Productivity as a Function of the Quality of the Labor Standards and the Measure of Productivity." *Proc—21st Ann Nat Conf.* (AIIE)* Cleveland, Ohio, May 1970.

Mason, Anthony K. and Towne, Douglas M. "Toward Synthetic Methods Analysis." (JIE)* Vol. 18, No. 1, Jan 1967.

Mattern, William J.; Knott, Kenneth; and McDonald, Russell W. "MTM-GPD and Other Second Generation Data." *Industrial Engineering Handbook,* 3rd ed., New York: McGraw-Hill Book Company, 1971, pp. 5-102 to 5-121.

Maynard, Harold B. "Design for Leadership." *Proc—3rd Internat MTM Conf.* (MTMA)* New York City, Sept 1963.

———. "IE Today Sketched by Survey." (JIE)* Vol. 3, No. 11, Nov 1971.

Maynard, Harold B. (ed-in-ch). *Handbook of Business Administration.* New York: McGraw-Hill Book Company, 1967.

———. (ed-in-ch). *Handbook of Modern Manufacturing Management.* New York: McGraw-Hill Book Company, 1970.

———. (ed-in-ch). *Industrial Engineering Handbook.* New York: McGraw-Hill Book Company, 1956, 3rd ed. 1971.

———. "Predetermined Time Systems." *Proc—Internat Conf.* (AIIE)* New York City, Sept 1963.

———. "Industrial Engineering—Training for Management." (JIE)* Vol. 10, No. 1, Jan–Feb 1959.

———. "The Ever-Growing Value of MTM." (JMTM)* Vol. 7, No. 4, Nov–Dec 1960.

———. "The Future of MTM." *Proc—1st Ann MTM Conf.* (MTMA)* New York City, Oct 1952.

———. *Top Management Handbook,* New York: McGraw-Hill Book Company, 1960.

———. "When Top Management Supports MTM." *Proc—9th Ann MTM Conf.* (MTMA)* Toronto, Can, June 1960.

Maynard, Harold B. and Hodson, William K. "Standard Data Concepts." *Industrial Engineering Handbook,* 3rd ed., New York: McGraw-Hill Book Company, 1971.

Maynard, Harold B. and Stegemerten, Gustave J. *Guide to Methods Improvement.* New York: McGraw-Hill Book Company, 1944.

———. *Operation Analysis,* New York: McGraw-Hill Book Company, 1939.

Maynard, Harold B.; Aiken, William M.; and Lewis, John F. *Practical Control of Office Costs,* Greenwich, Conn: Management Publishing Corp, 1960.

Maynard, Harold B.; Stegemerten, Gustave J.; and Schwab, John L. *Methods-Time Measurement.* New York: McGraw-Hill Book Company, 1948.

McCormick, Ernest J. *Human Factors Engineering.* New York: McGraw-Hill Book Company, 1957, 3rd ed. 1970.

McCormick, Ernest J. and Tiffin, Joseph. *Industrial Psychology.* 6th ed., Englewood Cliffs, N.J.: Prentice-Hall, 1975.

*See code at start of Bibliography.

McCugh, R. M. "Time and Energy Relationship of One- and Two-Handed Activity." *M.S. thesis,* Washington Univ, 1932.

McKechnie, D. F. "Short Run, Long Cycle Operations Can Be Measured." *Mill and Factory,* Oct 1959.

McKinley, James W. "Method of Analyzing and Controlling Indirect Labor." *Proc—2nd Work Measurement and Methods Engineering Div Conf.* (AIIE)* Flint, Mich, Apr 1974.

Meister, David. "Methods of Predicting Human Reliability in Man-Machine Systems." *Human Factors.* Dec 1964.

Melton, Arthur W. "Implications of Short-Term Memory for a General Theory of Memory." *Journal of Verbal Learning and Verbal Behavior* 2 (1963).

Meredith, G. P. "Theory of the Therblig." *Occupational Psychology* Vol. 27, London, England, July 1953.

Miller, George A. "The Magical Number Seven, Plus or Minus Two: Some Limits on Our Capacity for Processing Information." *The Psychological Review* Vol. 63, No. 2, Mar 1956.

Miller, James W. "Toward the Objective Measurement of the Man-Machine Interface." *Proc—13th Ann Chapter Conf.* (AIIE)* Detroit, Mich, Oct 1971.

Minter, A. L. "The Estimation of Energy Expenditure." (JMTM)* Vol. 15, No. 5, Nov–Dec 1970.

Mirer, Steven J. "Clerical Work Measurement." *M.S. thesis,* Rensselaer Polytechnic Inst. Troy, New York, 1963.

Mogensen, Allen H. "Work Simplification: Past—Present—Future." *Proc.—26th Ann Nat Conf.* (AIIE)* Washington, D.C., May 1975.

Morrow, Robert A. and Langolf, Gary D. "MTM-Magnification Time System." (JMTM)* Vol. 18, No. 2, Mar–Apr 1973.

Morton, Thomas J. "Clerical Work Measurement and People." (JMTM)* Vol. 2, No. 1, 1975.

Motycka, Joseph. "Learning Curves—Pricing and Controlling." *Proc—16th Ann Nat Conf.* (AIIE)* Chicago, May 1965.

———. "New Advances in Time Study—Electronic Data Processing." *Proc—12th Ann Nat (Gantt) Conf.* (AIIE)* Detroit, Mich, May 1961.

(MTMA)*. "Application Training Course." (MTM-ATC Manual), 1964 (Current Rev).

———. "Application Training Supplement No. 6—Position." Oct 1960.

———. "MTM Basic Specifications." (JMTM)* Vol. 5, No. 4, Nov–Dec 1958.

———. *Proc—1st Ann MTM Conf.* New York City, Oct 1952.

———. *Proc—2nd Ann MTM Conf,* New York City, Oct 1953.

———. *Proc—3rd Ann MTM Conf,* New York City, Oct 1954.

———. *Proc—4th Ann MTM Conf,* Chicago, Oct 1955.

———. *Proc—5th Ann MTM Conf,* New York City, Oct 1956.

———. *Proc—6th Ann MTM Conf,* New York City, Sept 1957.

———. *Proc—7th Ann MTM Conf,* New York City, Sept 1958.

———. *Proc—8th Ann MTM Conf,* Detroit, Mich, Oct 1959.

———. *Proc—9th Ann MTM Conf,* Toronto, Can, June 1960.

———. *Proc—10th Ann MTM Conf,* New York City, Oct 1961.

———. *Proc—11th Ann MTM Conf,* New York City, Sept 1962.

———. *Proc—3rd Internat MTM Conf,* New York City, Sept 1963.

———. *Proc—12th Ann MTM Conf,* New York City, Sept 1964.

———. *Proc—13th Ann MTM Conf,* New York City, Oct 1965.

———. *Proc—14th Ann MTM Conf,* New York City, Oct 1966.

———. *Proc—15th Ann MTM Conf,* New York City, Oct 1967.

———. *Proc—16th Ann MTM Conf,* New York City, Oct 1968.

———. *Proc—17th Ann MTM Conf,* New York City, Oct 1969.

*See code at start of Bibliography.

——. *Proc—18th Ann MTM Conf,* New York City, Oct 1970.

——. *Proc—19th Ann MTM Conf,* New York City, Oct 1971.

——. *Proc—20th Ann MTM Conf,* Atlanta, Georgia, Oct 1972.

——. *Proc—21st Ann MTM Conf,* New Orleans, Louisiana, Oct 1973.

——. *Proc—22nd Ann MTM Conf,* Reston, Va, Oct 1974.

——. *Proc—23rd Ann MTM Conf,* New York City, Oct 1975.

——. *Proc—MTM Seminar,* Philadelphia, 13 May 1964.

——. *Proc—MTM Seminar,* Detroit, Mich, 5 June 1964.

——. *Proc—MTM Seminar,* Los Angeles, 26 Mar 1965.

——. *Proc—MTM Seminar,* Washington, D.C., Apr 1967.

——. *Proc—MTM Seminar,* Los Angeles, 28 Mar 1968.

——. *Proc—MTM Seminar,* New York City, 7 June 1968.

——. *Proc—MTM Seminar,* Los Angeles, 27 Feb 1969.

——. *Proc—MTM Western Conf,* Anaheim, Calif, Apr 1970.

——. *Proc—MTM Western Conf,* Anaheim, Calif, Apr 1971.

——. *Proc—MTM Western Conf,* San Diego, Calif, Apr 1972.

——. *Proc—MTM Western Conf,* Los Angeles, Apr 1973.

——. *Proc—MTM Western Conf,* Los Angeles, Apr 1974.

——. *Proc—MTM Western Conf,* San Francisco, Apr 1975.

——. "MTM-2 Application Training Course." 1972 (Current Rev).

——. "MTM-3 Application Training Course." 1972 (Current Rev).

——. "MTM-GPD General Purpose Data Instruction Manual: Vol. I Basic Data." 1962 (Current Rev).

——. "MTM-GPD General Purpose Data Instruction Manual: Vol. II Multi-Purpose Data." 1964 (Current Rev).

——. "MTM-GPD General Purpose Data Instructor's Manual." 1970 (Current Rev).

——. "MTM Magnification Res Proj." *Final Report,* Magnification Res Consortium, Dec 1972.

——. "MTM Magnification Res Proj." *Interim Report,* Magnification Res Consortium, Oct 1970.

(MTM Consortium). "A Magnified View of Work Measurement." (JIE)* Vol. 2, No. 1, Jan 1970.

Muckstadt, John A. "Independence of MTM Elements." *Proc—12th Ann MTM Conf.* (MTMA)* New York City, Sept 1964.

Mundel, Marvin E. "Allowing in Time Standards for Weight Handled." (JIE)* Vol. 8, No. 4, July–Aug 1957.

——. "Joint Audits of Standards with the Union." *Proc—26th Ann Nat Conf.* (AIIE)* Washington, D.C., May 1975.

——. *"Motion and Time Study."* Englewood Cliffs, N.J.: Prentice-Hall, 1950, 4th ed, 1960.

Mundel, Marvin E. and Lazarus, Irwin P. "Predetermined Time Standards in the Army Ordnance Corps." (JIE)* Vol. 5, No. 6, Nov 1954.

Murphy, Wilton W.; Bird, Robert G.; and Garcia, Daniel H. "Biomechanics Data for a Man-Amplifier." (MECH)* Vol. 89, No. 8, Aug 1967.

——. "Measurements of Body Motions." *Paper No. 66-WA/BHF-2,* (ASME)* Nov 1966.

Murrell, K. F. H. *Human Performance in Industry,* New York: Reinhold Publishing Corp., 1965.

Nadler, Gerald E. *Motion and Time Study,* New York: McGraw-Hill Book Company, 1955.

——. "Time Study Rating Evaluation." (JIE)* Vol. 4, No. 4, Nov 1953.

——. "Work Design: A Philosophy for Applying Work Principles." (JIE)* Vol. 10, No. 3, May–June 1959.

——. "A Strategy for Designing Management Systems." *Proc—16th Ann Nat Conf.* (AIIE)* Chicago, May 1965.

*See code at start of Bibliography.

————. "A Systems Concept: Definition, Strategy, and Productivity Program for Any Organization." *Proc—21st Ann Nat Conf.* (AIIE)* Cleveland, Ohio, May 1970.

————. "Systems Engineering and Concern for People: Compatible or Contradictory?" *Proc—20th Ann Nat Conf.* (AIIE)* Houston, Texas, May 1969.

————. "The Myopia of Measurement." *Proc—24th Ann Nat (Silver) Conf.* (AIIE)* Chicago, May 1973.

————. *Work Design.* Homewood, Illinois: Richard D. Irwin, 1963.

————. *Work Simplification.* New York: McGraw-Hill Book Company, 1957.

————. *Work Systems Design: The IDEALS Concept.* Homewood, Illinois: Richard D. Irwin, 1967.

Nadler, Gerald E. and Denholm, Donald H. "Therblig Relationships: I. Added Cycle Work and Context Therblig Effects." (JIE)* Vol. 6, No. 2, Mar–Apr 1955.

Nadler, Gerald E. and Goldman, Jay W. "Operator Performance Studies: II. Learning Analysis from Three-Plane Motions." (JIE)* Vol. 14, No. 5, Sept–Oct 1963.

Nance, Harold W. "There Is a Need for Guided Application." *Proc—9th Ann MTM Conf.* (MTMA)* Toronto, Can, June 1960.

————. "Four Myths of Office Measurement." (JMTM)* Vol. 11, No. 2, May–June 1966.

Nance, Harold W. and Nolan, Robert E. *Office Work Measurement,* New York: McGraw-Hill Book Company, 1971.

Nanda, Ravinder. "Developing Variability Measures for Predetermined Time Systems." (JIE)* Vol. 18, No. 1, Jan 1967.

————. "The Additivity of Elemental Times." (JIE)* Vol. 19, No. 5, May 1968.

Narayen, Badri. "Decision Trees for MTM Application." *Technical Report,* Office of Research Admin. Univ of Mich, Ann Arbor, Mich, Apr 1968.

Nash, A. E. "Ratio-Delay—A Prize Tool." *Proc—1st Ann Management Conf.* (AIIE)* Fort Wayne, Ind, May 1955.

Neale, Fred J. *Primary Standard Data.* London, England: McGraw-Hill Publishing Company, Ltd., 1967.

————. "Simplified MTM—Its Uses and Developments in the United Kingdom." (JMTM)* Vol. 9, No. 2, Nov–Dec 1963.

Niebel, Benjamin W. *Motion and Time Study.* Homewood, Illinois: Richard D. Irwin, 1955, 5th ed, 1972.

————. *Laboratory Manual for Motion and Time Study* 3rd ed. Homewood, Illinois: Richard D. Irwin, 1962.

Niebel, Benjamin W.; Kassab, Samuel J.; and Noll, J. Brian. "The Influence of Distance Between Dual Fixtures on Time in Performing Simultaneous Positions." (JMTM)* Vol. 17, No. 5, Nov–Dec 1972.

Nordgren, Kurt. "Swedish Unions and MTM." (JIE)* Vol. 18, No. 5, May 1967.

Novak, E. Rogers. "Management Strategy for Motivation." (JMTM)* Vol. 2, No. 2, 1975.

Oberg, Erik; Jones, Franklin D.; Horton, Holbrook L.; and Schubert, Paul B. eds. *Machinery's Handbook,* New York: Industrial Press Inc., 1914, 20th ed, 1975.

Officer of Manpower Economics. "Measured Daywork." Great Britain Dept of Employment. London: Her Majesty's Printing Office, 1973.

Offner, David H. "Bionics: A Creative Aid to Engineering Design." (MECH)* Vol. 96, No. 7, July 1974.

Otis, Irvin and Mead, Kermit K. "Industrial Work Standards." *Proc—26th Ann Nat Conf.* (AIIE)* Washington, D.C., May 1975.

Pachella, Robert G. and Pew, Richard W. "Speed-Accuracy Tradeoff in Reaction Time: Effect of Discrete Criterion Times." *Journal of Experimental Psychology* Vol. 76, No. 1, 1968.

*See code at start of Bibliography.

Paterson, D. G. and Tinker, M. A. "Eye Movements in Reading Type Sizes in Optimal Line Widths." *Journal of Educational Psychology* Vol. 34, 1943.

———. "Influence of Line Width on Eye Movements." *Journal of Experimental Psychology* Vol. 27, 1940.

———. "Influence of Size of Type on Eye Movements." *Journal of Applied Psychology* Vol. 26, 1942.

Peer, S. and Kennedy, W. B. "Video Tape Analysis." (JIE)* Vol. 5, No. 2, Feb 1973.

Pegels, C. Carl. "On Startup or Learning Curves: An Expanded View." *Transactions.* (AIIE)* Vol. 1, No. 3, Sept 1969.

Pishkin, Vladimir. "Availability of Feedback-Corrected Error Instances in Concept Learning." *Journal of Experimental Psychology* Vol. 73, No. 2, 1967.

Pomerantz, James R. "Eye Movements Affect the Perception of Apparent (Beta) Movement." *Psychonomic Science,* Vol. 19(4), 1970.

Pomeroy, Richard W. "Adapting Methods-Management Techniques to Extreme Fluctuations in Workload: A Case Study." (JIE)* Vol. 18, No. 7, July 1967.

Poock, Gary K. "Prediction of Elemental Motion Performance Using Personnel Selection Tests." *Research Report 115,* (MTMA)* Jan 1968.

Poock, Gary K. and Wiener, Earl L. "Music and Other Auditory Backgrounds During Visual Monitoring." (JIE)* Vol. 17, No. 6, June 1966.

Presgrave, Ralph. *Dynamics of Time Study.* New York: McGraw-Hill Book Company, 1945.

Punton, John W. "An Analysis of Research Into Position." (JMTM)* Vol. 4, No. 2, May–June 1957.

Quick, Joseph H.; Duncan, James H.; and Malcolm, James A., Jr. *Work-Factor Time Standards,* New York: McGraw-Hill Book Company, 1962.

Ramsey, Jerry D. and Ayoub, M. M. "Occupational Heat Stress." (JIE)* Vol. 7, No. 9, Sept 1975.

Ramsey, Jerry D. and Purswell, J. L. "The Correlation of Mechanical and Physiological Measures of Work for Movement of the Upper Limb." *Transactions.* (AIIE)* Vol. 3, No. 2, June 1971.

Ramsing, Kenneth and Downing, Ronald. "Assembly Line Balancing with Variable Element Times." (JIE)* Vol. 2, No. 1, Jan 1970.

Rand, Christian. "A Hyper-Audio Pulse Triggered Input Device for Work Measurements, Standard Data Sampling, and Biological Response Studies." (JIE)* Vol. 7, No. 2, Mar–Apr 1956.

Raouf, Abdul. "Informational Load and Angle of Moves in a Combined Manual and Decision Task." *Proc—26th Ann Nat Conf.* (AIIE)* Washington, D.C., May 1975.

———. "Information Processing Rate and Prediction of Decision Times: an Experimental Investigation." (JMTM)* Vol. 18, No. 1, Jan–Feb 1973.

———. "Study of Visual Performance Times for Inspection Tasks." *PhD dissertation.* Univ of Windsor, Ont, Can. 1970.

Raphael, David L. "A Basic Introduction to Methods-Time Measurement." *Proc—1st Ann Management Conf.* (AIIE)* Fort Wayne, Ind. May 1955.

———. "An Analysis of Short Reaches and Moves." *Research Report 106* (MTMA)* Sept 1953.

———. "A Research Methods Manual," *Research Report 107,* (MTMA)* Apr 1954.

———. "A Study of Arm Movements Involving Weight." *Research Report 108* (MTMA)* Mar 1955.

———. "A Study of Positioning Movements: I. The General Characteristics: II. Special Studies Supplement." *Research Report 109* (MTMA)* 1957.

———. "A Study of Positioning Movements: III. Application to Industrial Work Measurement." *Research Report 110* (MTMA),* Nov 1957.

———. "MTM Research Activities," (JMTM)* Vol. 3, No. 5, Jan–Feb 1957.

*See code at start of Bibliography.

————. "Position Application." *Proc—6th Ann MTM Conf.* (MTMA)* New York City, Sept 1957.

————. "Position Research." *Proc—5th Ann MTM Conf.* (MTMA)* New York City, Oct 1956.

————. "Recent MTM Research Developments." *Proc—2nd Ann MTM Conf.* (MTMA)* New York City, Oct 1953.

————. "Research and MTM," (JMTM)* Vol. 2, No. 5, Jan–Feb 1956.

————. "Research—An MTM Application." *Proc—4th Ann MTM Conf.* (MTMA)* Chicago, Oct 1955.

————. "Research Developments in Move with Weights." *Proc—3rd Ann MTM Conf.* (MTMA)* New York City, Oct 1954.

————. "Short Reaches and Moves—A Summary of the Complete Report." (JMTM)* Vol I, No. 1, Apr 1954.

Raphael, David L. and Clapper, Grant C. "A Study of Simultaneous Motions." *Research Report 105* (MTMA)* Sept 1952.

Raphael, David L. and Stoll, Richard F. "Application Training Supplements Nos. 1–5: Reading, Short Reaches and Moves, Grasp, Release, and Moves with Weight." (MTMA)* Mar 1955.

Ratner, R. A. "Effect of Variations in Weight Upon Move Times." *M.S. thesis,* Iowa State College, 1951.

Restle, Frank, "Training of Short-Term Memory." *Journal of Experimental Psychology* Vol. 83, No. 2, 1970.

Reynolds, Nancy Z. "Dr. Lillian Moller Gilbreth 1878–1972." (JIE)* Vol. 4, No. 2, Feb 1972.

Richardson, James A. and Riland, Lane H. "The Relevance of Behavioral Science Research to Industrial Engineering." (JIE)* Vol. 17, No. 11, Nov 1966.

Richardson, Wyman R. "RAWS Multiple Regression Analysis." *Institute of Science and Technology Report 5432-4-R,* Univ of Mich, Ann Arbor, Mich, Aug 1963.

Robens, Alfred. *Human Engineering,* Mystic, Conn: Verry Lawrence Inc., 1971.

Robinson, Gordon H. "Visual Search by Operators in Man-Machine Systems." *Proc—23rd Ann Nat Conf.* (AIIE)* Anaheim, Calif, June 1972.

Roozbabar, A. "A Physiological Evaluation of Lifting Activities." *Proc—25th Ann Nat Conf.* (AIIE)* New Orleans, La, May 1974.

Rubin, Gerald; von Trebra, Patricia Ann; and Smith, Karl U. "Dimensional Analysis of Motion: III. Complexity of Movement Pattern." *Journal of Applied Psychology* Vol. 36, No. 4, Aug 1952.

Sadosky, Thomas L. "Prediction of Cycle Time for Combined Manual and Decision Tasks." *Research Report 116* (MTMA)* 1969.

Salem, M. D., Jr. "Multiple Linear Regression Analysis for Work Measurement of Indirect Labor." (JIE)* Vol. 18, No. 5, May 1967.

Salvendy, Gavriel. "Handedness and Psychomotor Performance." *Transactions.* (AIIE)* Vol. 2, No. 3, Sept 1970. Also in *Proc—21st Ann Nat Conf.* (AIIE)* Cleveland, Ohio, May 1970.

————. "Hand Size and Assembly Performance." *Transactions.* (AIIE)* Vol. 3, No. 1, Mar 1971.

————. "Learning Fundamental Skills—A Promise for the Future." *Transactions.* (AIIE)* Vol. 1, No. 4, Dec 1969. Also in *Proc—20th Ann Nat Conf.* (AIIE)* Houston, Texas, May 1969.

Salvendy, Gavriel, Ed., *Handbook of Industrial Engineering,* John Wiley & Sons, N.Y., N.Y. 1982.

Salvendy, Gavriel and Pilitsis, John. "Psychophysiological Techniques: A Tool for the Establishment of Work Standards." *Proc—22nd Ann Nat Conf.* (AIIE)* Boston, Mass, May 1971.

Schmeck, Ronald R. "Error-Produced Frustration as a Factor Influencing the Probability of Occurrence of Further Errors." *Journal of Experimental Psychology* Vol. 82, No. 2, 1970.

Schmid, Merle D. "Work Measurement Sampling." *Proc—7th Ann Chapter Conf.* (AIIE)* Detroit, Mich, Feb 1965.

*See code at start of Bibliography.

———. "Simplified Standards for the Smaller Plant—Work Measurement Sampling." *Proc—14th Ann Nat Conf.* (AIIE)* Denver, Colo, May 1963.

———. "Work Measurement Sampling." *Proc—2nd Work Measurement and Methods Engineering Div Conf.* (AIIE)* Flint, Mich, Apr 1974.

Schmid, Richard O. "An Overview of MTM-2 and MTM-3." *Proc—19th Ann MTM Conf.* (MTMA)* New York City, Oct 1971.

———. "Improving Design Performance Through Visualization With MTM-3." *Proc—25th Ann Nat Conf.* (AIIE)* New Orleans, La, May 1974.

———. "Simplified MTM Simplifies Method Improvement." (JIE)* Vol. 4, No. 7, July 1972.

Schmidtke, Heinz and Stier, Fritz "An Experimental Evaluation of the Validity of Predetermined Elemental Time Systems." (JIE)* Vol. 7, No. 3, May–June 1961.

———. "Response to the Comments on 'An Experimental Evaluation of the Validity of Predetermined Elemental Time Systems'." (JIE)* Vol. 14, No. 3, May–June 1963.

Schnitzler, William F. "Frank B. Gilbreth—A Friend of the Labor Movement." (JIE)* Vol. 19, No. 7, July 1968.

Schoeller, V. Donald. "Learn to Manage Yourself First." *Proc—13th Ann Nat (Gilbreth) Conf.* (AIIE)* Atlantic City, N.J., May 1962.

———. "The Industrial Engineer and Management Performance Standards." *Proc—16th Ann Nat Conf.* (AIIE)* Chicago, May 1965.

———. "The Management of Time." (JIE)* Vol. 19, No. 4, Apr 1968.

Schor, Robert M. "Information to the Union: The Industrial Engineer's Duty to Disclose." *Proc—21st Ann Nat Conf.* (AIIE)* Cleveland, Ohio, May 1970.

Schutt, William H. *Process Engineering,* New York: McGraw-Hill Book Company, 1945.

———. *Time Study Engineering,* New York: McGraw-Hill Book Company, 1943.

Schutz, Rodney K. and Chaffin, Don B. "Cyclic Work-Rest Exercise's Effect on Local Muscle Fatigue Rates." *Proc—23rd Ann Nat Conf.* (AIIE)* Anaheim, Calif, June 1972.

Schwab, John L. "The Measurement of Human Effort—Part II." *Proc—15th Ann Nat Conf.* (AIIE)* Philadelphia, May 1964.

Schwan, Harry T. "Maintenance Costs Can Be Reduced." *Proc—3rd Ann Management Conf.* (AIIE)* Fort Wayne, Ind, May 1957.

Schwartz, Shula. "The Learning Curve and Its Application." *Library Bibliography No. 20,* Chance Vought Aircraft Co, Dallas, Texas.

Scoutten, E. F. "MTM Versus Union Goals." *Proc—11th Ann MTM Conf.* (MTMA)* New York City, Sept 1962.

Secor, Harry W. "The Use of Regression Analysis Techniques to Increase Standards Application Efficiency." (JIE)* Vol. 17, No. 1, Jan 1966.

Segur, A. B. "Labor Costs at the Lowest Figure." *Manufacturing Industries,* Vol. 13, 1927.

Sellie, Clifford. "Comments on 'An Experimental Evaluation of the Validity of Predetermined Elemental Time Systems'," (JIE)* Vol. 12, No. 5, Sept–Oct 1961.

Shapero, A. "Studies in Multiple Grasping and Positioning of Small Parts." *Proc—3rd Ann Industrial Engineering Inst,* Univ of Calif, Berkeley, 1951.

Shaw, Anne G. "Dr. Lillian Gilbreth." *Proc—13th Ann Nat (Gilbreth) Conf.* (AIIE)* Atlantic City, N.J., May 1962.

Siegler, N. "A Study of Some Factors Affecting Element Time in Manual Work." *Ph.D. dissertation,* Univ of Minn, 1954.

Simerson, Floyd W. "Work Simplification in a Changing World." *Proc—16th Ann Nat Conf.* (AIIE)* Chicago, May 1965.

Simon, Fred. "Management and Labor Look at MTM." *Proc—8th Ann MTM Conf.* (MTMA)* Detroit, Mich, Oct 1959.

Simon, J. Richard and Smader, Robert C. "Dimensional Analysis of Motion: VIII. The Role of

Visual Discrimination in Motion Cycles." *Journal of Applied Psychology* Vol. 39, No. 1, Feb 1955.

Sirota, David. "The Conflict Between IE and Behavioral Science." (JIE)* Vol. 4, No. 6, June 1972.

Skargard, Karl. "Integrated Work Measurement with MTM." (JMTM)* Vol. 11, No. 3, July–Aug 1966.

Smader, Robert C. and Smith, Karl U. "Dimensional Analysis of Motion: VI. The Component Movements of Assembly Motions." *Journal of Applied Psychology* Vol. 37, No. 4, Aug 1953.

Smalley, Harold E. "Comments on 'Work Measurement in Perspective-Universal Time Data.'" (JIE)* Vol. 15, No. 1, Jan–Feb 1964.

———. "Another Look at Work Measurement." (JIE)* Vol. 18, No. 3, Mar 1967.

Smalley, Harold E. and Freeman, John R. *Hospital Industrial Engineering.* New York: Reinhold Publishing Corp., 1966.

Smith, B. J. "Validation of Certain Tests of Ocular Dominance." *Ph.D. dissertation,* Univ of Minn, 1954.

Smith, Karl U. and Rubin, Gerald "Principles of Design of the Sequential Interval Recorder," *Perceptual and Motor Skills* Vol. 4, No. 1, Mar 1952.

Smith, Karl U. and Wehrkamp, R. F. "A Universal Motion Analyzer Applied to Psychomotor Performance." *Science,* Vol. 113, 1951.

Smith, Leo A. and Barany, James W. "An Elementary Model of Human Performance on Paced Visual Inspection Tasks." *Transactions.* (AIIE)* Vol. 2, No. 4, Dec 1970.

Smith, Patricia Ann and Smith, Karl U. "Effects of Sustained Performance on Human Motion." *Perceptual and Motor Skills,* Vol. 5, 1955.

Smith, William A., Jr. "Accuracy of Manual Entries in Data-Collection Devices." *Journal of Applied Psychology* Vol. 51, No. 4, 1967.

———. "Data Collection Systems: Part I. Characteristics of Errors." (JIE)* Vol. 18, No. 12, Dec 1967.

———. "Data Collections Systems: Part II. Environmental Effects on Accuracy." (JIE)* Vol. 19, No. 1, Jan 1968.

Snakenberg, Ted R. "The Use of Multi-Variable Charts to Simplify MTM Standard Data." (JMTM)* Vol I, No. 5, Dec 1954.

Society for Advancement of Management. "How to Use Performance Rating Films." New York City, 1952.

Solberg, R. B. "Time Allowances for the Handling of Weights for Use in Stopwatch Time Studies." *M.S. thesis,* Purdue Univ, 1946.

Staros, Anthony and Peizer, Edward "The Clinical Engineer." (MECH)* Vol. 96, No. 6, June 1974.

Stender, John L. "The IE's Role in OSHA." *Proc—26th Ann Nat Conf.* (AIIE)* Washington, D.C., May 1975.

Stewart, John M. "Productivity and Its Many Different Meanings." *Proc—24th Ann Nat (Silver) Conf.* (AIIE)* Chicago, May 1973.

Stohlmann, Donald G. "Development and Charting of Predetermined Time Data." (JIE)* Vol. 4, May 1953.

———. "How to Reach Your Cost Reduction Goal." *Proc—1st Ann Management Conf.* (AIIE)* Fort Wayne, Ind, May 1955.

Stoll, Richard F. "Future of MTM Research or More Specifically Basic Motion Research." *Proc—6th Ann MTM Conf.* (MTMA)* New York City, Sept 1957.

———. "General Purpose Data (MTM-GPD)—The Standard Data System the MTM Association Developed and Sponsors." *Proc—15th Ann Nat Conf.* (AIIE)* Philadelphia, May 1964.

———. "The MTM Association and Its Work." *Proc—1st Ann Management Conf.* (AIIE)* Fort Wayne, Ind, May 1955.

Stone, Mervyn. "Models for Choice-Reaction Time." *Psychometrika,* Vol. 25, No. 3, Sept 1960.

Storment, John. "Industrial Application of Biomechanics." *Proc—22nd Ann Nat Conf,* (AIIE)* Boston, Mass, May 1971.

Stukey, Arthur E. "Work Measurement in Perspective—Universal Time Data." (JIE)* Vol. 15, No. 1, Jan–Feb 1964.

Swanson, James M. and Briggs, George E. "Information Processing as a Function of Speed versus Accuracy." *Journal of Experimental Psychology* Vol. 81, No. 2, 1969.

Swensson, Richard G. "The Elusive Tradeoff: Speed versus Accuracy in Visual Discrimination Tasks." *Perception & Psychophysics,* Vol. 12(1A), 1972.

Tabernacle, J. B. and Wharton, F. "Control Charts Aid Work Study Training." (JIE)* Vol. 2, No. 2, Feb 1970.

Taggert, John B. "Comments on 'An Experimental Evaluation of the Validity of Predetermined Elemental Time Systems'." (JIE)* Vol. 12, No. 6, Nov–Dec 1961.

Taylor, Frederick Winslow. *Shop Management,* New York: Harper & Brothers, 1919.

———. "The Present State of the Art of Industrial Management." *Transactions.* (ASME)* Vol. 34, 1912.

Thomas, Marlin U. "Some Probabilistic Aspects of Performance Times for a Combined Manual and Decision Task." *PhD dissertation,* Univ of Mich, 1971.

Thomopoulos, Nick T. and Lehman, Melvin "The Mixed Model Learning Curve." *Transactions.* (AIIE)* Vol. 1, No. 2, June 1969.

Thompson, David A. "Information Theory in Mental Work Measurement." *AIIE Western Regional Conference,* San Francisco, California, Oct 1963.

———. "The Development of a Quantified Difficulty Measure for Precision Work," (JMTM),* Vol. 8, No. 5, July–Aug. 1963.

Thompson, David A. and Acker, Martin. "A Detailed Physiological Difficulty Analysis of Manual Work." (JMTM)* Vol. 7, No. 3, Sept–Oct 1960.

Thompson, David A. and Applewhite, Phillip B. "Objective Effort Level Estimates in Manual Work." (JIE)* Vol. 19, No. 2, Feb 1968.

Thompson, William C. "MTM in Micro-Assembly." (JMTM)* Vol. 15, No. 3, May–June 1970.

Thorson, P. J. "Work Sampling—Sample Size." *Industrial Engineering News,* Dept. of Mechanical Engineering, Mich State Univ, No. 29, Summer 1956.

Tichauer, Erwin R. "Anatomy of Predetermined Time Systems." *Proc—20th Ann Nat Conf.* (AIIE)* Houston, Texas, May 1969.

———. "Ergonomic Aspects of Biomechanics." *The Industrial Environment—Its Evaluation and Control,* U.S. Dept. of Health, Education, and Welfare, Washington, D.C., 1973, chap. 32.

———. "Gilbreth Revisited." *Paper No. 66-WA/BHF-7.* (ASME)* Nov 1966.

———. "Human Capacity, A Limiting Factor in Design." *Institution of Mechanical Engineers,* London, England, July 1963.

———. "Industrial Engineering in the Rehabilitation of the Handicapped." (JIE)* Vol. 19, No. 2, Feb. 1968.

———. "The Biomechanics of the Arm-Back Aggregate Under Industrial Working Conditions." *Paper No. 65-WA/HUF-1.* (ASME)* Nov 1965.

———. "The Role of Biomechanics in the Utilization of the Human Resources of Industry." *Proc—26th Ann Nat Conf.* (AIIE)* Washington, D.C., May 1975.

Tichauer, Erwin R.; Gage, Howard; and Harrison, Lynne B. "Electromyographic

*See code at start of Bibliography.

Kinesiology—A New Tool for Work Measurement." *Proc—22nd Ann Nat Conf.* (AIIE)* Boston, Mass, May 1971.

———. "Use of Biomechanical Profiles in Work Measurement." (JIE)* Vol. 4, No. 5, May 1972.

Tinker, Miles A. "Time Taken by Eye Movements in Reading." *Journal of Genetic Psychology* Vol. 48, 1936.

Tinker, Miles A. and Paterson, D. G. "Eye Movements in Reading Black Print on White Background and Red Print on Dark Green Background." *American Journal of Psychology* Vol. 57, 1944.

Tomlinson, R. W. "Ergonomics Methods in Work Study." (JMTM)* Vol. 18, No. 1, Jan–Feb 1973.

Towne, Douglas M. "Recent Advances in Computer-Aided Work Measurement." *Proc—22nd Ann Nat Conf.* (AIIE)* Boston, Mass, May 1971.

Townsley, Claude, J. "MTM Usage—Procedure and Examples." (JMTM)* Vol. 2, No. 3, July–Aug 1955.

Travis, K. C. "The Latency and Velocity of the Eye in Saccadic Movements." *Psychology Monograms,* Vol. 47, 1936.

Turban, Efraim. "Incentives During Learning—An Application of the Learning Curve Theory and a Survey of Other Methods." (JIE)* Vol. 19, No. 12, Dec 1968.

Turner, Roger N., Jr. "AMAS—A Methods Analysis System Based on Analyst-Oriented Job Descriptions." *PhD dissertation.* Univ of Mich, 1971.

Tyler, Leona E., et al. *The Psychology of Human Differences,* New York: Appleton–Century–Crofts, 1947; Englewood Cliffs, N.J.: Prentice–Hall, 3rd ed, 1965.

Urwick, Lyndall. "The Value of MTM to Top Management." *Proc—1st Ann MTM Conf.* (MTMA)* New York City, Oct 1952.

Vaill, Peter B. "Industrial Engineering and Socio-Technical Systems." (JIE)* Vol. 18, No. 9, Sept 1967.

Vance, Stanley. *Industrial Administration,* New York: McGraw-Hill Book Company, 1959.

Varkvisser, J. "The Relation Between Motionpattern and Performance Time of Therbligs." (JMTM)* Vol. 5, No. 6, Mar–Apr 1959.

von Trebra, Patricia Ann and Smith, Karl U. "Dimensional Analysis of Motion: IV. Transfer Effects and Direction of Movement." *Journal of Applied Psychology* Vol. 36, No. 5, Oct 1952.

Waddell, Harry Lee, ed. "Work Sampling." (FACT)* July 1952.

Walker, R. Y. "The Eye Movements of Good Readers." *Psychology Monograms,* Vol. 44, 1933.

Webb, Paul. "Dissociation of Heat Production and Heat Loss in Working Men." *Paper No. 66-WA/HT-45.* (ASME)* Nov 1966.

Wehrkamp, Robert F. and Smith, Karl U. "Dimensional Analysis of Motion: II. Travel-Distance Effects." *Journal of Applied Psychology* Vol. 36, No. 3, June 1952.

Weindling, Joachim I. "Some Aspects of the Design of Task Oriented Social Structures." (JIE)* Vol. 17, No. 6, June 1966.

Welford, A. T. *Ageing and Human Skill,* Cambridge, Eng.: Oxford University Press, 1958.

———. *Fundamentals of Skill,* London, Eng: Methuen & Co., Ltd., 1968.

Wendelborg, Owe. "MTM for the Construction Industry—MTM-Bygg." (JMTM)* Vol. 17, No. 5, Nov–Dec 1972.

Werkmeister, W. H. *An Introduction to Critical Thinking.* Lincoln, Neb: Johnsen Publishing Company, 1948.

White, Kendall C. "MTM-A Keystone for Standards." *Proc—11th Ann MTM Conf.* (MTMA)* New York City, Sept 1962.

———. "Predetermined Elemental Motion Times." *Paper No. 50-A-88* (ASME)* Nov 1950.

Whitmore, S. *Measurement and Control of Indirect Labour,* New York: American Elsevier Publishing Co., 1971.

*See code at start of Bibliography.

Wilkes, J. W. "The Effect of Job Difficulty Factors on Time, Energy, and Acceleration of Motion Patterns." *Sc.D. dissertation,* Washington Univ, 1954.

Wilkinson, John J. "Measuring and Controlling Maintenance Operations." (JMTM)* Vol. 11, No. 1, Mar–Apr 1966.

———. "Measurement of Low Quantity Work." *Industrial Engineering Handbook.* 3rd ed, New York: McGraw-Hill Book Company, 1971, pp. 4–15 to 4–30.

Williams, Harold. "Developing a Table of Relaxation Allowances." (JIE)* Vol. 5, No. 12, Dec 1973.

Wirta, R. W. "EMG Control of External Power." *Paper No. 65-WA/HUF-3.* (ASME)* Nov 1965.

Wolfberg, Stanley T. "Standard Data—Catalysis or Catastrophe," *Proc—10th Ann MTM Conf.* (MTMA)* New York City, Oct 1961.

———. "Topsy-Turvy Control Concept." (JIE)* Vol. 15, No. 2, Mar–Apr 1964.

Woodson, W. E. *Human Engineering Guide,* Los Angeles: University of California Press, 1956.

Work-Factor Company. *Information About the Work-Factor System,* New York City, 1950.

Yard, Roger L., Sr. "GPD—Latest Development of the MTM Association." (JMTM)* Vol. 8, No. 5, July–Aug 1963.

———. "New Methods of Developing Standard Data." (JMTM)* Vol. 11, No. 3, July–Aug 1966.

Yost, Edna. *Frank and Lillian Gilbreth—Partners for Life.* (ASME)* New York City, 1949.

Young, Hewitt H. "The Relationship Between Heart Rate and the Intensity of Work for Selected Tasks." (JIE)* Vol. 7, No. 6, Nov–Dec 1956.

Young, Samuel L. "Misapplications of the Learning Curve Concept." (JIE)* Vol. 17, No. 8, Aug 1966.

Zalusky, John. "Union IE Charges the Profession is Shallow." (JIE)* Vol. 2, No. 12, Dec 1970.

Zandin, Kjell V. *MOST Work Measurement Systems,* Marcel Decker, Inc., N.Y., N.Y., 1980.

Zeyher, Lewis R. *Production Manager's Handbook of Formulas and Tables,* Englewood Cliffs, N.J.: Prentice-Hall, 1972.

*See code at start of Bibliography.

Index